WASTEWATER TREATMENT PROCESS MODELING

Prepared by the **Wastewater Treatment Process Modeling** Task Force
of the **Water Environment Federation**

Andrew R. Shaw, P.E., *Chair*

Bruce R. Johnson, P.E., BCEE, *Vice-Chair*

Rasheed Ahmad, Ph.D., P.E.

Vinod Barot, Ph.D.

Evangelia Belia, Ph.D., ing.

Lorenzo Benedetti, Ph.D.

José R. Bicudo, Ph.D., P.E.

Katya Bilyk, P.E.

Joshua P. Boltz, Ph.D., P.E., BCEE, F.IWA

Lucas Botero, P.E., BCEE

Akram Botrous, Ph.D., P.E.

Ken Brischke, P.E.

Marie S. Burbano, Ph.D., P.E., BCEE

S. Rao Chitikela, Ph.D., P.E., BCEE

Rhodes R. Copithorn

John B. Copp, Ph.D.

Timur Deniz, Ph.D., P.E.

Leon S. Downing, Ph.D., P.E.

Eric Evans

Richard Finger

M. Kim Fries

M. Truett Garrett, Jr., Sc.D., P.E.

Sylvie Gillot, Ph.D.

Matthew J. Gray, P.E.

Russell D. Grubbs

Samuel S. Jeyanayagam, Ph.D., P.E., BCEE

Dimitri Katehis, Ph.D., P.E.

Frank Kulick

Thomas J. Kutcher, P.E.

Ron J. Latimer

Carlos Lopez

Ting Lu, Ph.D.

Robert Nerenberg, Ph.D., P.E.

Daniel A. Nolasco, M.Eng., M.Sc., P.Eng.

Ana J. Peña-Tijerina, Ph.D., P.E., BCEE

Heather M. Phillips, P.E.

William Randall, P.E.

Gary Revoir

Leiv Rieger, Ph.D., P.Eng.

Joel C. Rife, P.E.

Murat Sarioglu

Peter Schauer

Oliver Schraa, M.Eng.

Art Umble

K.C. "Kumar" Upendrakumar, P.E., BCEE

Peter A. Vanrolleghem, Ph.D., ing.

Ifetayo Venner, P.E.

Milind V. Wable, Ph.D., P.E., BCEE

Paul Wood, P.E.

Under the Direction of the **Municipal Design Subcommittee** of the
Technical Practice Committee

2013

Water Environment Federation
601 Wythe Street
Alexandria, VA 22314–1994 USA
http://www.wef.org

WASTEWATER TREATMENT PROCESS MODELING

WEF Manual of Practice No. 31
Second Edition

Prepared by the Wastewater Treatment Process Modeling Task Force of the Water Environment Federation

WEF Press

Water Environment Federation   Alexandria, Virginia

New York   Chicago   San Francisco   Athens   London
Madrid   Mexico City   Milan   New Delhi
Singapore   Sydney   Toronto

Cataloging-in-Publication Data is on file with the Library of Congress.

McGraw-Hill Education books are available at special quantity discounts to use as premiums and sales promotions, or for use in corporate training programs. To contact a representative please visit the Contact Us page at www.mhprofessional.com

Wastewater Treatment Process Modeling, MOP 31, Second Edition

1 2 3 4 5 6 7 8 9 0   DOC/DOC   1 2 0 9 8 7 6 5 4 3

ISBN 978-0-07-179842-6
MHID 0-07-179842-0

Water Environment Research, WEF, and WEFTEC are registered trademarks of the Water Environment Federation.

This book is printed on acid-free paper.

About WEF

Founded in 1928, the Water Environment Federation (WEF) is a not-for-profit technical and educational organization of 36,000 individual members and 75 affiliated Member Associations representing water quality professionals around the world. WEF members, Member Associations, and staff proudly work to achieve our mission to provide bold leadership, champion innovation, connect water professionals, and leverage knowledge to support clean and safe water worldwide. To learn more, visit www.wef.org.

For information on membership, publications, and conferences, contact

Water Environment Federation
601 Wythe Street
Alexandria, VA 22314-1994 USA
(703) 684-2400
http://www.wef.org

Manuals of Practice of the Water Environment Federation®

The WEF Technical Practice Committee (formerly the Committee on Sewage and Industrial Wastes Practice of the Federation of Sewage and Industrial Wastes Associations) was created by the Federation Board of Control on October 11, 1941. The primary function of the Committee is to originate and produce, through appropriate subcommittees, special publications dealing with technical aspects of the broad interests of the Federation. These publications are intended to provide background information through a review of technical practices and detailed procedures that research and experience have shown to be functional and practical.

Water Environment Federation Technical Practice Committee Control Group

Jeanette Brown, P.E., BCEE, D. WRE, *Chair*
Eric Rothstein, *Vice-Chair, Publications*
Stacy J. Passaro, P.E., BCEE, *Vice-Chair, Distance Learning*
R. Fernandez, *Past Chair*

J. Bannen
R. Copithorn
S. V. Dailey, P.E.
J. Davis
E. M. Harold, P.E.
M. Hines
R. Horres
D. Medina
D. Morgan
J. Newton
P. Norman
Christine A. Pomeroy, Ph.D., P.E.
R. Pope
K. Schnaars
C. Stacklin, P.E.
Andrew R. Shaw, P.E.
J. Swift
J.E. Welp
N. Wheatley

Contents

Chapter 1   History of Process Modeling

Chapter 2   Modeling Fundamentals

Chapter 3   Unit Process Model Descriptions

Chapter 4   Process Modeling Tools

Chapter 5   Dedicated Experiments and Tools

Chapter 6 Overview of Available Modeling and Simulation Protocols

Chapter 7   Project Definition

Chapter 8   Building a Facility Model

Chapter 9 Using Models for Design, Optimization, and Control

List of Figures

List of Tables

Preface

Over the past 20 years, mathematical modeling of wastewater treatment processes has become the default tool for process design in many engineering firms throughout the world and is beginning to be used at operating facilities to help make day-to-day operating decisions. Increased computer processing power and user-friendly simulation software make it possible to model many of the complexities of a water resource recovery facility (WRRF) using personal computers. These simulators can be used to develop mass-balance models of the plant, linking several unit processes together and modeling their interactions. In addition, they can be used to carry out dynamic simulations to investigate diurnal and other transient behavior of a WRRF, such as the effect of wet weather.

With an increased use of process models through user-friendly simulators, there has been widespread acknowledgment in the industry that good training and expert guidance is needed to ensure that these models are developed, used, and documented correctly. This manual provides a broad range of information to help process engineers, operators, regulators, and owners understand general modeling concepts, terminology unique to computer modeling, and practical guidance and ideas on how to use process models for design and operation of small, medium, and large WRRFs. The modeling approach presented in this manual is consistent with the unified protocol proposed by the International Water Association task group on good modeling practice.

This publication was produced under the direction of Andrew R. Shaw, P.E., *Chair*, and Bruce R. Johnson, P.E., BCEE, *Vice-Chair*.

The principal authors of this publication are as follows:

Chapter 1	M. Truett Garrett, Jr., Sc.D., P.E.
	Ting Lu, Ph.D.
Chapter 2	Carlos Lopez
	Evangelia Belia, Ph.D., ing.
Chapter 3	Daniel A. Nolasco, M.Eng., M.Sc., P.Eng.
	Joshua P. Boltz, Ph.D., P.E., BCEE, F.IWA

Chapter 4	John B. Copp, Ph.D.
	Katya Bilyk, P.E.
Chapter 5	Peter A. Vanrolleghem, Ph.D., ing.
	Leiv Rieger, Ph.D., P.Eng.
Chapter 6	Leiv Rieger, Ph.D., P.Eng.
	Sylvie Gillot, Ph.D.
	Peter A. Vanrolleghem, Ph.D., ing.
Chapter 7	Marie S. Burbano, Ph.D., P.E., BCEE
	Ana J. Peña-Tijerina, Ph.D., P.E., BCEE
Chapter 8	Heather M. Phillips, P.E.
	Matthew J. Gray, P.E.
Chapter 9	Peter Schauer
	Ken Brischke, P.E.
	Ron J. Latimer
	Oliver Schraa, M.Eng.
Appendix A	Daniel A. Nolasco, M.Eng., M.Sc., P.Eng.
	Joshua P. Boltz, Ph.D., P.E., BCEE, F.IWA
Appendix B	Daniel A. Nolasco, M.Eng., M.Sc., P.Eng.
	Joshua P. Boltz, Ph.D., P.E., BCEE, F.IWA
Appendix C	Daniel A. Nolasco, M.Eng., M.Sc., P.Eng.
	Joshua P. Boltz, Ph.D., P.E., BCEE, F.IWA
Appendix D	Daniel A. Nolasco, M.Eng., M.Sc., P.Eng.
	Joshua P. Boltz, Ph.D., P.E., BCEE, F.IWA
Appendix E	Marie S. Burbano, Ph.D., P.E., BCEE
	Ana J. Peña-Tijerina, Ph.D., P.E., BCEE
Appendix F	Bruce R. Johnson, P.E., BCEE

Authors' and reviewers' efforts were supported by the following organizations:

BioChem Technology, King of Prussia, Pennsylvania

Black & Veatch, Kansas City, Missouri

Brentwood Industries, Reading, Pennsylvania

Carollo Engineers, Phoenix, Arizona

CDM Smith, Cambridge, Massachusetts

Cemagref–HBAN, France

CH2MHill, Inc., Englewood, Colorado

City of Nacogdoches, Texas

City of Olathe, Kansas

Clean Water Services, Tigard, Oregon

Department of Natural Resources, Des Moines, Iowa

Department of Watershed Management, City of Atlanta, Georgia

Donohue & Associates, Sheboygan, Wisconsin

EnviroSim Associates Ltd., Hamilton, Ontario, Canada

GHD, Bowie, Maryland

Greeley and Hansen, Chicago, Illinois

Hazen & Sawyer, Raleigh, North Carolina

Hydromantis Environmental Software Solutions, Inc., Hamilton, Ontario, Canada

Johnson Controls, Inc., Westerville, Ohio

Lockwood, Andrews & Newnam, Inc., Houston, Texas

Malcolm Pirnie, the Water Division of ARCADIS, Tampa, Florida

Metropolitan Sewer District of Greater Cincinnati, Ohio

modelEAU-Université Laval, Quebec City, Québec, Canada

MWH Americas, Inc., Denver, Colorado, and Dubai, United Arab Emirates

NJS Consultants, Tokyo, Japan

NOLASCO y Asociados S.A., Buenos Aires, Argentina

O'Brien & Gere, Richmond, Virginia

Primodal Inc., Hamilton, Ontario, Canada, and Kalamazoo, Michigan

Region of Waterloo Water Services, Kitchener, Ontario, Canada

Stantec, Rocklin, California

University of Notre Dame, South Bend, Indiana

Veolia Water North America, Indianapolis, Indiana

Village Creek Water Reclamation Facility, Fort Worth, Texas

WATERWAYS srl, Impruneta, Italy

Chapter 1

History of Process Modeling

1.0 INTRODUCTION

Modeling is important to the engineering design of modern water resource recovery facilities (WRRFs) that are experiencing increasing demands on effluent quality. A wastewater process model consists of a number of mathematical equations that describe reactions and reaction rates of biological, chemical, and physical phenomena

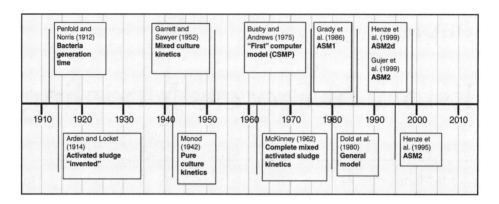

FIGURE 1.1   Activated sludge modeling timeline.

of various unit processes. Modern computers and software packages can solve the equations, either directly or by iterations (Olsson and Newell, 1999).

Activated sludge process models have developed over the years to the point where they are now used to develop simulators that are widely used for process design, facility evaluations, troubleshooting, and training. Simulators imitate operation of a real world process using models of the key characteristics of the process. There are general-purpose simulators for the final packaging of "in-house" model development and application-specific simulators for total facility simulation, including models for other unit processes (e.g., biofilm processes). These simulators are discussed in Chapter 4.

A history of modeling is provided here to give design engineers an appreciation of the developmental milestones of these models and to provide references to further investigate the fundamental steps and assumptions used to develop them. Figure 1.1 presents a timeline of some of the milestones described in the following sections.

2.0   MATHEMATICAL EXPRESSIONS OF PROCESSES

2.1   Bacterial Growth Rate and Substrate Concentration

In 1912, generation time was the measure used to express the speed of bacterial growth. Generation time is related to the net specific growth rate, as follows:

$$\mu_{net} = \ln(2)/g \qquad (1.1)$$

where

μ_{net} = the net specific growth rate, h^{-1}

g = the generation time, h

Specific growth rate means dX/XdT because dX/dT is the growth rate. The net specific growth rate is the observed growth rate. Because a decay rate is observed in the absence of food, $b = -dX/XdT$, it is presumed that there is a true specific growth rate and that decay occurs simultaneously during the observed growth. The true specific growth rate is the net specific growth rate plus the decay rate. As such, in modeling, where the decay of solids is used, it is necessary to use the true specific growth rate (net specific growth rate plus decay rate). Penfold and Norris (1912) found that the generation time of *Bacillus typhosus* at 37 °C varied inversely with an increase in food concentration (peptone). This implies that the net specific growth rate reaction is first order with respect to substrate concentration up to a maximum value and zero order at higher substrate concentrations.

Monod (1942) reported the results of studies on the growth of pure cultures of bacteria in dilute solutions of various sugars (200 mg/L or less). He measured the concentration of substrate and the concentration of cells during the growth of a bacterial culture and determined the rate of growth as divisions per hour. Figure 1.2 presents a graph constructed from the combined results of five experiments on the growth of *Bacillus coli* on glucose at 37 °C (Monod, 1942).

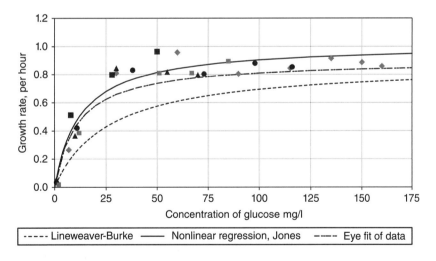

FIGURE 1.2 The net specific growth rate of *B. coli* on glucose at 37 °C from Monod (1942) with Monod curves determined by Lineweaver–Burke (1934), eye fit of data, and nonlinear regression (Jones, 1992).

As Figure 1.2 shows, the relationship is similar to that reported by Penfold and Norris (1912). However, Monod (1942) introduced the use of the Michaelis–Menten equation.

Menten and Michaelis (1913) developed their equation to relate the reaction rate of an enzyme with the concentration of substrate. The equation, as applied to bacterial growth, is now known as the *Monod equation* to relate the net specific growth rate to the substrate concentration, as follows:

$$\mu_{net} = \hat{\mu}_{net} \times S/(K_s + S) \tag{1.2}$$

where

μ_{net} = the net specific growth rate, d^{-1}
$\hat{\mu}_{net}$ = the maximum net specific growth rate (at the extant temperature), d^{-1}
S = the substrate concentration, kg/m^3
K_s = the substrate concentration at which the growth rate is half the maximum growth rate, kg/m^3

The constants K_s and $\hat{\mu}_{net}$ may be determined by linearization methods such as Lineweaver–Burke (1934), by fitting the constants by eye, and by nonlinear regression as described by Jones (1992).

In Figure 1.2, the Lineweaver–Burke curve (using all data) misses the data by a factor of 2 or more. This is because the trendline is disproportionally affected by the reciprocals of erroneous values of low substrate concentration and growth rate. The nonlinear regression method overcomes this problem. However, the value of $\hat{\mu}_{net}$ by linearization or regression will exceed the values of the data. For fast-growing heterotrophs, this may not be significant; however, for slow-growing autotrophs and anaerobes, it can be important. As such, selecting constants by eye may give the desired result. This indicates that it is prudent to inquire about the method used to produce K_s and $\hat{\mu}_{net}$ when using published data. However, the Monod equation is considered good enough and has been presented as fact in many scholarly publications. Moreover, the equation has been the basis for advancement in modeling of anaerobic and aerobic bacterial growth. The Monod equation readily fits into programming languages. In addition, the equation serves as a switching function for other substances that limit the rate of growth such as oxygen, phosphorus, nitrogen, and so on. Figure 1.2 illustrates the notions that biological relationships are complex and that mathematical relationships to simulate them are approximate. Table 1.1 lists the constants of the Monod equation by the different methods that are used for data in Figure 1.2.

Table 1.1   Constants of the Monod equation by different methods (for data in Figure 1.2).

Method	μ_{net}, h^{-1}	K_s, g, m^{-3}
Lineweaver–Burke	0.88	26
Eye fit of data	0.9	11
Regression	1.015	12.3

2.2   Inhibition of Bacterial Growth

The Monod equation (Eq. 1.2) is used to relate the growth rate to the substrate concentration when there is no inhibition of growth. However, inhibition may significantly affect growth rates. For instance, in anaerobic digesters, high concentrations of un-ionized volatile acids may cause failure even with a sufficiently long solids retention time (SRT), according to the model of Andrews and Greaf (1971). The various types of inhibition of enzymes and bacterial growth may be represented by modifications of the Michaelis–Menten equation (http://www.uvm.edu./'mcase/courses/chem205/lecture13.pdf).

In competitive inhibition, the inhibitor and substrate have an affinity for the same active site on an enzyme and compete for access to the site. The type of inhibitor affects K_s, but not $\hat{\mu}$, as follows:

$$\mu = \hat{\mu} \times S/(a \cdot K_s + S) \tag{1.3}$$

where a is a function of inhibitor concentration.

In noncompetitive inhibition, binding of the inhibitor to the enzyme reduces its activity, but does not affect binding of the substrate. The maximum net growth rate, $\hat{\mu}$, increases, but K_s remains the same, as follows:

$$\mu = a\,\hat{\mu} \times S/(K_s + S) \tag{1.4}$$

In uncompetitive inhibition, the inhibitor binds to the enzyme-substrate complex and causes both $\hat{\mu}$ and K_s to decrease, as follows:

$$\mu = a \cdot \hat{\mu} \times S/(a \cdot K_s + S) \tag{1.5}$$

In mixed noncompetitive inhibition, the binding of the inhibitor to the enzyme influences the binding of substrate. Both $\hat{\mu}$ and K_s are changed, as follows:

$$\mu = a \cdot \hat{\mu} \times S/(a \cdot K_s + a' \cdot S) \tag{1.6}$$

2.3   Temperature Effect on Bacterial Growth

The following van't Hoff–Arrhenius equation (http://chemwiki.ucdavis.edu/ Physical_Chemistry/Kinetics/Reaction_Rates/Temperature_Dependence_of_ Reaction_Rates/Arrhenius_Equation) is often used to describe the temperature dependence of reaction rates:

$$k = A \exp \frac{-E}{[RT]} \tag{1.7}$$

where
  k = the reaction rate, s^{-1}, for first order reactions
  A = a factor, s^{-1}
  E = the activation energy, kJ/mol
  R = the universal gas constant, J/mol/K
  T = temperature, K

According to Metcalf and Eddy (2003), the equation may be used to derive the following equation to calculate the effect of temperature on the rate of a reaction:

$$k_T = k_{20} \cdot \theta^{(T-20)} \tag{1.8}$$

where
  k_T = the rate at the desired temperature, T
  k_{20} = the rate at 20 °C

Although θ is assumed to be constant, it is often cited in the literature as 1.047 for the biochemical oxygen demand (BOD) reaction. For temperatures from 4° to 20 °C, it is 1.135 (Metcalf and Eddy, 2003). The ratio of the reaction rates for a 10 °C increase in temperature is called "Q10" and is often cited as 2. For this value of Q10, the value of θ is 1.072.

2.4   Decay of Bacterial Cells

Decay and endogenous respiration exist in cultures of bacterial cells and in other biological processes. Heukelekian et al. (1951) quantified this as follows as they found the pounds of volatile solids produced per pound of BOD decreased linearly with the pounds of volatile suspended solids (VSS) in the facility:

$$\text{lb VSS produced/d} = 0.5 \cdot \text{lb BOD fed/d} - 0.055 \cdot \text{lb VSS} \tag{1.9}$$

Dold et al. (1980) modeled endogenous decay as "death and regeneration", that is, a portion of the dead cells are hydrolyzed and available to the active bacterial mass. This was included in the first models produced by the International Water Association (IWA) task group, Activated Sludge Model No. 1 (ASM1) (Henze et al., 1987) and

Activated Sludge Model No. 2 (ASM2) (Henze et al., 1995). Van Loosdrecht and Henze (1999) reviewed the literature on bacterial maintenance, endogenous respiration, lysis, decay, and predation and concluded that the death regeneration concept is most likely not related to the dying of bacteria. Rather, processes such as storage polymer metabolism and predation make up a significant part of endogenous respiration.

3.0 WASTEWATER TREATMENT PROCESSES

3.1 Aerobic Processes

3.1.1 *Aerobic Suspended Growth Processes*

Some of the fundamental mathematics of biological processes that became the activated sludge process were developed around the time of the Arden and Lockett (1914) publication, which first described the activated sludge process. Penfold and Norris (1912) had shown that the generation time of bacteria was related to the reciprocal of the concentration of the substrate down to a minimum generation time that remained constant at higher substrate concentrations. Monod (1942) found a similar relationship, but proposed the use of the Michaelis and Menten (1913) equation to describe the relationship between concentration of substrate and the rate of bacteria growth.

Garrett and Sawyer (1952) demonstrated that the kinetics of bacterial growth for pure cultures (Monod, 1942; Penfold and Norris, 1912) is applicable to the mixed culture environment of the activated sludge process. They presented the results obtained in experiments with fill-and-draw and continuous-flow systems using seeding from wastewater from Boston, Massachusetts. Values of maximum net specific growth rate were reported as follows: 2 d^{-1} at 10 °C, 5 d^{-1} at 20 °C, and 8 d^{-1} at 30 °C.

At low values of substrate BOD, the relationship between substrate concentration and growth rate was linear and the growth of solids was linear with the substrate removed. This results in the following activated sludge loading relationship:

$$S_e = k \cdot S_0 \cdot Q/(X_a \cdot V_a) \qquad (1.10)$$

where

S_e = effluent BOD, kg/m^3

k = a constant

S_0 = the influent BOD, kg/m^3

Q = the influent flow, m^3/d

X_a = mixed liquor suspended solids, kg/m^3

V_a = the aeration tank volume, m^3

This disproved the intuitive relationship proposed by Ridenour (1946) and supported by the National Research Council Report (1946), which was expressed as

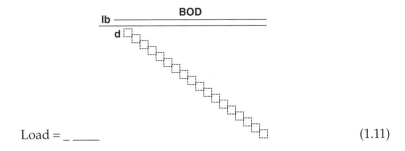

$$\text{Load} = \_\,\underline{} \tag{1.11}$$

1000 lb aeration solids·hours aeration

which may be expressed in terms of flow, concentration, mixed liquor solids, and aeration tank volume, as follows:

$$\text{Load} = k\, S_0 \cdot Q^2/(Xa \cdot Va^2) \tag{1.12}$$

where the variables are the same as in Eq. 1.10.

McKinney (1962) advanced the mathematics of completely mixed activated sludge to include the following fractions of mixed liquor volatile suspended solids: active organisms, inert endogenous residue, and particulate unbiodegradable organics from the influent.

During the 1950s and 1960s, facility operating data were not helpful in verifying these early models because facility discharge permits specified reporting the uninhibited BOD test with samples taken before chlorination. Many of the BOD values obtained in these tests were erroneously high because of the oxygen demand of ammonia in the sample and in the dilution water caused by nitrification in the BOD test bottle. This was corrected in the late 1970s when the U.S. Environmental Protection Agency required reporting of carbonaceous BOD (CBOD) in effluent because CBOD was the test that had been used in the stream models, which were the basis of waste-load allocations. The chemical oxygen demand (COD) test was often used in educational institutions and by some consultants because it was not subject to the nitrification problem. Servizi and Bogan (1963) considered COD the best test because it allowed a mass balance to be calculated over the activated sludge system.

In the activated sludge process, where solids are recycled to the aeration tank, the mixed liquor SRT may be directly controlled by removing (wasting) solids from the aeration tank as mixed liquor. Thus, the mass of solids in the mixed liquor aeration tank(s) divided by the mass of solids wasted per day is equal to the mixed liquor aeration tank volume divided by the daily flow of the waste mixed liquor because the concentrations are identical. Garrett (1958) described this procedure as *hydraulic growth rate control*. The significance of SRT is that it is the reciprocal of the average net growth rate. Ekama (2010) stressed the importance of hydraulic control of sludge age in the design and operation of activated sludge facilities required to remove ammonia or nitrogen and phosphorus. Pitman (1999) stated that the waste sludge should not be taken from the return sludge in biological nutrient removal (BNR) facilities, but rather from the mixed liquor at the end of the aeration tank(s). At some BNR facilities in Johannesburg, South Africa, the sludge age is automatically controlled on a year-round basis without operator intervention by controlling the flow of waste mixed liquor (adjusted by online temperature measurements).

3.1.2   Aerobic Biofilm Processes

The trickling filter, which has been used for nearly 100 years, is one of the main biofilm processes in the municipal and industrial wastewater field. It is a nonsubmerged biofilm biological reactor using rock or manufactured packing over which wastewater is distributed continuously (Metcalf and Eddy, 2003). In contrast to a suspended growth system in which diffusion is lumped in the half saturation value, the biofilm is subject to mass transfer limitations. Substrate removal (electron donor or receiver) that occurs within the depth of the attached growth biofilm is governed by the same kinetics as suspended growth. However, the rate of substrate removal by diffusion is as governed by Fick's law. The concentration of the various substrates will vary with depth in the biofilm and the driving force for diffusion is provided by the concentration gradient. Thus, the rate of substrate removal in biofilm is governed by normal Monod kinetics plus diffusion, which can actually be the limiting factor. Biofilm models typically allow the biofilm to grow until a steady state is reached between growth of the biofilm and its loss because of sloughing and shearing (Lu et al., 2007).

Suspended growth and biofilm systems are both aggregates of suspended solids containing different groups of microorganisms. Modeling has covered many aspects of the activated sludge process, but only limited research has been conducted to describe the diversity of multiple biomass systems (Lou and de los Reyes, 2005; Lu et al., 2007; Saikaly and Oerther, 2004). Lu et al. (2007) studied variations in the microbial population in biofilms using a biofilm model. The results showed that competition for growth-limiting nutrients generated oscillations in the abundance of planktonic and sessile microbial populations, which allowed for a greater diversity in the population. Imhoff and Fair (1940) described contact aerators as contact beds submerged in the wastewater they treated. The contact media included stone, laths, corrugated aluminum sheets, and so on. The bed was aerated by diffusers beneath the bed. Clyde C. Hays (Classen, 1946) developed a contact aeration WRRF design that used corrugated aluminum sheets. The first facility was installed at Elgin, Texas, in 1939. The National Research Council Report (1946) evaluated Hays facilities built at military installations.

Submerged aerobic biofilm systems are currently being promoted as integrated fixed film/activated sludge systems. Often, they are used to enhance the nitrifying capacity of activated sludge facilities by placing the media in existing aeration tanks. The media require primary sedimentation or fine screening of the influent to prevent clogging. Systems are also used for industrial waste and wastewater treatment aboard ship. The development of new plastic media and a better understand of fixed film processes has made the process attractive again. Takacs et al. (2007) presented a biofilm model that applies to nonsubmerged and submerged biofilm processes. Chapter 3 contains further discussion of current biofilm models.

3.2 Anaerobic Processes

3.2.1 Anaerobic Suspended Growth Processes

After the cesspool and pit privy, the septic tank was developed in France by John Mouras and patented in 1881. In 1911, Karl Imhoff invented the two-story tank that bears his name. An improvement over the septic tank, the new tank separated the flow of wastewater from solids digesting in the lower compartment. Separate sludge digestion tanks followed for large facilities because of cheaper construction and more practical heating of the contents, gas could be collected from either type. However, there were problems with process stability.

Lawrence and McCarty (1970) and Andrews (1969) describe application of the Monod equation to models of anaerobic digestion. Andrews and Greaf (1971) modeled the toxicity of non-ionized acetic acid by using a Haldane (1930) inhibition function instead of the Monod equation. Following the success of activated sludge models, IWA formed a task group to develop a model for anaerobic digestion; the product was known as *Anaerobic Digestion Model No. 1* (Batstone et al., 2002).

3.2.2 Anaerobic Biofilm Processes

McCarty (1968) described anaerobic treatment of soluble wastewater using a packed bed biofilm process. Young and McCarty (1969) further discussed the use of the anaerobic filter for wastewater treatment. Cakir (2001) discussed anaerobic treatment of low-strength wastewater by anaerobic processes that increase SRT such as upflow anaerobic filters and upflow anaerobic sludge blanket reactors.

4.0 SETTLING AND THICKENING OF ACTIVATED SLUDGE

Coe and Clevenger (1916) published a procedure for designing thickeners used in the mining industry. Their plot of settling velocity of a sample against the dilution (ratio of water to solids) for a series of sample dilutions is illustrated in Figure 1.3. To determine the limiting dilution, draw a line from the desired discharge dilution, D, on the abscissa and tangent to the curve of settling velocity. The point of tangency is the limiting dilution, L, and the intersection of the extension of the line at the Y-axis is the underflow rate, U, in feet/hour. Using the selected discharge dilution, Coe and

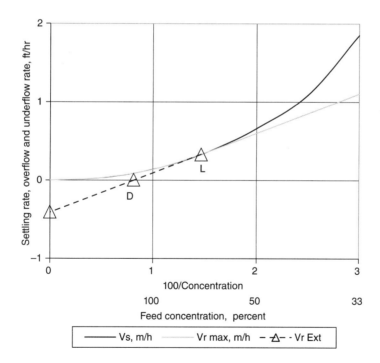

FIGURE 1.3   Coe and Clevenger (1916) plot of settling velocity vs 1/concentration, after Eq. 1.9 ("DL" is a line from the desired underflow concentration tangent to the underflow rate "Vs U", which is a negative overflow rate).

Clevenger then determined the capacity (pounds per hour per square foot, presently called *solids flux*) for the series of dilutions according to the following formula:

$$C_s = 62.35 \times R/(D_s - D) \qquad (1.13)$$

where

C_s = the capacity for dilution D_s (x value of point L)

R = the rate of settling in feet per hour of the sample dilution D_s (y value of point L)

D = the discharge dilution

The curve shows a minimum capacity at the same dilution as the limiting dilution, L, in Figure 1.3. This was a total flux plot without showing the flux of the discharge rate that was used. Concha and Burger (2003) stated that the Coe and Clevenger procedure "with certain corrections, continues to be the most reliable method of thickener design to date".

Kynch (1952) published a kinematical theory of sedimentation. The basic assumption of Kynch's theory is that, at any point in a thickener, the settling velocity is a function of the local solids concentration only. Yoshioka et al. (1957) found, in batch settling tests, "that above a certain concentration the relationship between settling rate and concentration in the case when there is an effect of raking action will be approximately linear on log-log paper". This permitted the plotting of the flux vs settling velocity over a wide range of underflow values as shown in Figure 1.4. This has become the standard curve for design. The Yoshioka procedure is essentially

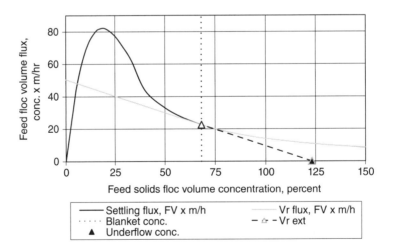

FIGURE 1.4   Floc volume flux plot (after procedures by Yoshioka et al. [1957] and Scott [1968]).

the same as Coe and Clevenger's, with the improvement that the flux curve does not have to be redrawn for different values of the underflow rate as required with Coe and Clevenger's total flux plot. According to Yoshioka et al. (1957), the detention time of the sludge does not affect the underflow concentration in contrast to the conclusion of Coe and Clevenger (1916); however, Coe and Clevenger recommend designing for a sufficient depth of sludge to provide for steady operation.

Michaels and Bolger (1962) found that sedimentation of flocculated kaolin suspensions depends on the volume of the flocs, not the volume of the individual particles as in the Kynch analysis. Previously, Richardson and Zaki (1954) found that the settling velocity varied as a power of the void fraction, as follows:

$$V_s = V_{so} \times (1 - k \times c)^{4.65} \qquad (1.14)$$

where
  k = a constant characteristic of the suspension with the dimensions, mL/g
  c = the solids concentration, g/mL
  V_s = the settling velocity at concentration c
  V_{so} = the Stokes settling velocity of the average floc

Note that the dimensions of k are the same as the dimensions of SVI. Scott (1968) found the Richardson and Zaki (1954) relationship applied for values of kc less than 0.4 and used the log-log relationship for higher values of kc.

Manning et al. (1999) reported on facility-scale tests at overflow rates that exceeded the settling rate of the mixed liquor solids by as much as 70%. They found that the mixed liquor does not split at the bottom of the feed well as assumed in the popular multilayer model, but flows downward to the floor or a sludge blanket of higher solids concentration even when the overflow rate exceeds the settling rate of the solids. Some solids settle into the sludge blanket and the rest form a layer at mixed liquor concentration. This action was shown in the model by Busby and Andrews (1975), typically by computational fluid dynamic models. A modification of the model of Garrett et al. (1984) was formulated to include the thin blanket formation at overflow rates exceeding the settling rate of the mixed liquor solids. The new model allows design for a maximum safe sludge blanket depth at higher overflow rates or higher mixed liquor DSV30 than if the clarifier influent flux were limited to the settling flux curve.

The equations of Manning et al. (1999) are

$$V_s = a[(100 - DSV30)/100]^n, \text{ for DSV30 less than 40\%} \qquad (1.15)$$

and

$$V_s = a'(\text{DSV30})^{-n'}, \text{ for DSV30 greater than } 40\% \tag{1.16}$$

Takacs et al. (1991) found that dilute mixed liquor solids do not settle as fast as predicted by the Vesilind equation and proposed the following two-exponent equation to better fit the data:

$$V_s = V_o'[e^{-rh(X-X\min)} - e^{-rp(X-X\min)}] \tag{1.17}$$

This equation is applicable for determining effluent suspended solids of primary and tertiary settling tanks and secondary settling tanks.

5.0   MODERN APPLICATIONS OF MODELS

5.1   Steady-State Conditions

The personal computer brought us the spreadsheet, a tool that completes calculations with each "enter" keystroke or on command at the completion of all entries. Steady-state spreadsheet models are often prepared by individual designers to determine tank volumes based on a set of input values and system requirements. Most models cover the dynamic condition and the simulator may run the model for a steady-state condition such as average flow.

5.2   Dynamic Conditions

The development of models of dynamic conditions was made possible by the advent of the digital computer and software to solve simulations involving simultaneous differential equations. Storer and Gaudy (1969) discussed computational analysis of transient responses to shock loading of heterogeneous populations of bacteria. Smith and Eilers (1970), at the Federal Water Pollution Control Administration office in Cincinnati, Ohio, prepared one of the first reports on the simulation of the activated sludge process. Andrews (1971) presented a discussion of kinetic models for biological waste treatment processes, and Jones (1971) reported on a mathematical model of the contact-stabilization process.

Graduate theses at Clemson University (Clemson, South Carolina) by Bryant (1972), Busby (1973), Stenstrom (1975), and Tracy (1973) led to publications by Busby and Andrews (1975) of a dynamic model using the Continuous System Modeling Program involving stored mass, active mass, and a settler model, where feed was introduced at the top of the sludge blanket. Tracy and Keinath (1974) presented a dynamic model of thickening activated sludge. Early models were also presented by Goodman and Englande (1974), Grady and Roper (1974), and Lawrence and McCarty (1970). This work was synthesized by Dold et al. (1980), who presented a general model for the aerobic activated sludge process. Modeling the nitrification process was not difficult: the model based on the Monod kinetics of Downing et al. (1964) could be directly incorporated to the general aerobic model. Following practical developments in biological denitrification using wastewater and endogenously generated organics as an electron donor by Ludzack and Ettinger (1962), Christensen-Henze and Harremoes (1977) presented a significant review of biological denitrification. This provided valuable information to Van Haandel et al. (1981) to extend the aerobic activated sludge model to include biological denitrification.

In 1982, upon the recommendation of IWA (then the International Association on Water Pollution Research and Control), a task group was formed to develop practical mathematical models for the design and operation of WRRFs. Members of the task group, Grady et al. (1986), presented details of the model, which became known as *Activated Sludge Model No. 1*, and was later published by the same authors as the International Association on Water Quality (London, U.K.) Scientific and Technical Report No. 1 (Henze et al., 1987). This was followed by Activated Sludge Model No. 2 (Henze et al., 1995) and Activated Sludge Model No. 2d (Henze et al., 1999). In separate efforts, Barker and Dold (1997) and Wentzel et al. (1989) developed biological phosphorus models that are used in some of the more popular simulator packages.

An additional task group was formed to prepare a comprehensive model of anaerobic digestion. The task group considered the breakdown of the various organics in sludge (proteins, fatty acids, and carbohydrates) and the bacteria attacking them leading to methanogenesis. The report is titled *Anaerobic Digestion Model No. 1 (ADM1)* (Batstone et al., 2002).

5.3 Development of Commercial Simulators

With the development of more powerful personal computers, commercial simulators such as GPS-X, a product of Hydromantis, Inc. (Patry and Takacs, 1990), and Biowin produced by Envirosim Associates Ltd. (Barker and Dold, 1997), which provide a platform for full-scale facility simulation, software is commonly used by designers and operators in the wastewater field. The steps to build a model are now straightforward: a user first produces a process diagram of the modeled WRRF, establishes physical parameters and operating parameters, runs the model, and analyzes the results. With such a friendly user phase, operators can optimize the wastewater process on their own personal computers by modeling different scenarios. Understanding the theory behind the model is essential to running the model and interpreting the model results (Shaw et al., 2007).

6.0 REFERENCES

Andrews, J. F. (1969) Dynamic Model of the Anaerobic Digestion Process. *J. Sanit. Eng. Div., Proc. Am. Soc. Civil Eng.,* **95** (1), 95–116.

Andrews, J. F. (1971) Kinetic Models of Biological Waste Treatment Processes. *Biotechnol. Bioeng. Symp. 2*; Wiley & Sons: New York.

Andrews, J. F.; Greaf, S. P. (1971) Dynamic Modeling and Simulation of the Anaerobic Digestion Process. In *Anaerobic Biological Treatment Processes. Advances in Chemistry Series 105*; American Chemical Society: Washington, D.C.

Arden, E.; Lockett, W. T. (1914) Experiments on the Oxidation of Sewage Without the Aid of Filters. *J. Soc. Chem. Ind.* (G.B.), **33**, 523–539.

Barker, P. S.; Dold, P. L. (1997) General Model for Biological Nutrient Removal Activated Sludge Systems: Model Presentation. *Water Environ. Res.,* **69** (5), 969–984.

Batstone, D. J.; Keller, J.; Angelidaki, L.; Kalyuzhnvi, S.; Pavlostathis, S. G.; Rozzi, A.; Sanders, W.; Siegrist, H.; Vavilin, V. (2002) *Anaerobic Digestion Model No. 1 (ADM1)*; IWA Publishing: London, U.K.

Bryant, J. O. (1972) Continuous Time Simulation of the Conventional Activated Sludge Wastewater Renovation System. Ph.D. Thesis, Clemson University, Clemson, South Carolina.

Busby, J. B. (1973) Dynamic Modeling and Control Strategies for the Activated Sludge Process. Ph.D. Thesis, Clemson University, Clemson, South Carolina.

Busby, J. B.; Andrews, J. F. (1975) Dynamic Modeling and Control Strategies for the Activated Sludge Process. *J.—Water Pollut. Control Fed.,* **47**, 1055–1080.

Cakir, F. Y. (2001) Anaerobic Treatment of Low Strength Wastewater. Power Point presentation. www.seas.ucla.edu/stenstro/s/s1 (accessed January 2013).

Christensen-Henze, M.; Harremoes, P. (1977) Biological Denitrification of Sewage: A Literature Review. *Prog. Water Technol.,* **8**, 509–555.

Classen, A. G. (1946) Other Treatment Methods and Disposal of Sewage. In *Manual for Sewage Plant Operators*; L. C. Billings, Ed.; Texas Water Works and Sewerage Short School: Austin, Texas; pp. 258–271.

Coe, H. S.; Clevenger, G. H. (1916) Methods for Determining the Capacities of Slime Settling Tanks. *Trans. Am. Inst. Min. Eng.*, **55**, 356–383.

Concha, F.; Burger, R. (2003) Thickening in the 20th Century: A Historical Perspective. *Minerals Metallurgical Process.*, **20**, 57–67.

Dold, P. L.; Ekama, G. A.; Marais, G. V. R. (1980) A General Model for the Activated Sludge Process. *Prog. Water Technol.*, **12**, 44–77.

Downing, A. L.; Painter, H. A.; Knowles, G. (1964) Nitrification in the Activated Sludge Process. *J. Proc., Inst. Sew. Purif.*, **63**, 130–153.

Ekama, G. A. (2010) The Role and Control of Sludge Age in Biological Nutrient Removal Activated Sludge Systems. *Water Sci. Technol.*, **61**, 1645–1652.

Garrett, M. T., Jr. (1958) Hydraulic Control of Activated Sludge Growth Rate. *Sew. Ind. Wastes*, **30**, 253.

Garrett, M. T., Jr.; Ma, J.; Yang, W.; Hyare, G.; Norman, T.; Ahmad, Z. (1984) Improving the Performance of Houston's Southwest Wastewater Treatment Plant, U.S.A. *Water Sci. Technol.*, **16**, 317.

Garrett, M. T., Jr.; Sawyer, C. N. (1952) Kinetics of Removal of Soluble Substrates by Activated Sludge. *Proceedings of the 7th Indiana Waste Conference*; Lafayette, Indiana, May 7–9; Purdue University: Lafayette, Indiana; p 51.

Goodman, B. L.; Englande, A. J., Jr. (1974) A Unified Model of the Activated Sludge Process. *J.—Water Pollut. Control Fed.*, **46**, 312–332.

Grady, C. P. L., Jr.; Roper, R. E. (1974) A Model for the Bio-Oxidation Process which Incorporates the Viability Concept. *Water Res.*, **8** (7), 471–483.

Grady, C. P. L., Jr.; Gujer, W.; Henze, M.; Marais, G. V. R.; Matsuo, T. (1986) A Model for Single Sludge Wastewater Treatment Systems. *Water Sci. Technol.*, **18**, 47–61.

Haldane, J. B. S. (1930) *Enzymes*; Longmans, Green, and Co.: London, U.K.

Henze, M.; Grady, C. P. L.; Gujer, W.; Marais, G. V. R.; Matsuo, T. (1987) *Activated Sludge Model No. 1*; IAWPRC Scientific and Technical Report No. 1; International Association on Water Pollution Research and Control: London, U.K.

Henze, M.; Gujer, W.; Mino, T.; Matsuo, T.; Wentzel, M. C.; Marais, G. V. R. (1995) *Activated Sludge Model No. 2;* IAWQ Scientific and Technical Report No. 3; International Association on Water Quality: London, U.K.

Henze, M.; Gujer, W.; Mino, T.; Matsuo, T.; Wentzel, M. C.; Marais, G. V. R.; van Loosdrecht, M. C. M. (1999) Activated Sludge Model No. 2d, ASM2d. *Water Sci. Technol.,* **39** (1), 165–182.

Heukelekian, H.; Orford, H. E.; Manganeli, R. (1951) Factors Affecting the Quantity of Sludge Production in the Activated Sludge Process. *Sew. Ind. Wastes,* **23** (8), 945–958.

Imhoff, K; Fair, G. M. (1940) *Sewage Treatment;* Wiley & Sons: New York.

Jones, M. E. (1992) Analysis of Algebraic Weighted Least-Squares Estimators for Enzyme Parameters. *Biochem. J.* (G.B.), **288**, 533–538.

Jones, P. H. (1971) A Mathematical Model for Contact Stabilization-Modification of the Activated Sludge Process. In *Advances in Water Pollution Research. Proceedings of the 5th International Conference on Water Pollution Research;* Pergamon Press: New York.

Kynch, G. J. (1952) A Theory of Sedimentation. *Trans. Faraday Soc.* (G.B.), **48**, 166–176.

Lawrence, A. W.; McCarty, P. L. (1970) Unified Basis for Biological Treatment Design and Operation. *Am. Soc. Civ. Eng., J. Sanit. Eng. Div.,* **96**, 757–778.

Lineweaver, H.; Burke, D. (1934) The Determination of Enzyme Dissociation Constants. *J. Am. Chem. Soc.,* **56** (3), 658–666.

Lou, I. C.; de los Reyes, F. L. (2005) Integrating Decay, Storage, Kinetic Selection, and Filamentous Backbone Factors in a Bacterial Competition Model. *Water Environ. Res.,* **77** (3), 287–296.

Lu, T.; Saikaly, P. E.; Oerther, D. B. (2007) Modeling the Competition of Planktonic and Sessile Aerobic Heterotrophs for Complementary Nutrients in Biofilm Reactor. *Water Sci. Technol.,* **55** (8–9), 227–235.

Ludzack, F. J.; Ettinger, M. B. (1962) Controlling Operation to Minimize Activated Sludge Effluent Nitrogen. *J.—Water Pollut. Control Fed.,* **34**, 920–931.

Manning, W. T., Jr.; Garrett, M. T., Jr.; Malina, J. F., Jr. (1999) Sludge Blanket Response to Storm Surge in an Activated-Sludge Plant. *Water Environ. Res.,* **71** (4), 432–442.

McCarty, P. L. (1968) Anaerobic Treatment of Soluble Wastes. In *Advances in Water Quality Improvement*; Gloyna, E. F., Eckenfelder, W. W., Jr., Eds.; University of Texas Press: Austin, Texas; pp. 336–352.

McKinney, R. E. (1962) Mathematics of Complete Mixing Activated Sludge. *Am. Soc. Civ. Eng., J. Sanit. Eng. Div.*, **88** (SA3), 87–113.

Menten, L.; Michaelis, M. I. (1913) Die Kinetik der Invertinwirkung. *Biochem Z*, **49**, 333–369.

Metcalf and Eddy, Inc. (2003) *Wastewater Engineering: Treatment and Reuse*; McGraw-Hill: New York.

Michaels, A. S.; Bolger, J. C. (1962) Settling Rates and Sediment Volumes of Flocculated Kaolin Suspensions. *Ind. Eng. Chem. Fundam.*, **1**, 24–33.

Monod, J. (1942) *Recherches sur la Croissance des Cultures Bactériennes*; Herman et Cie: Paris, France.

National Research Council Report (1946) Sewage Treatment at Military Installations. *Sew. Works J.*, **18**, 997–1016.

Olsson, G.; Newell, B. (1999) *Wastewater Treatment Systems, Modeling, Diagnosis, and Control*; IWA Publishing: London, U.K.

Patry, G. G.; Takacs, I. (1990) Simulator-Based Modeling of Wastewater Treatment Plants. *Proceedings Annual Conference and 1st Biennial Environmental Specialty Conference, CSCE*, **1**, 491–505.

Penfold, W. J.; Norris, D. (1912) The Relation of Concentration of Food Supply to the Generation Time of Bacteria. *J. Hyg.*, **12**, 527–531.

Pitman, A. R. (1999) Management of Biological Nutrient Removal Plant Sludges— Change the Paradigm? *Water Res.*, **33** (5), 1141–1146.

Richardson, J. F.; Zaki, W. N. (1954) Sedimentation and Fluidization, Part I. *Trans. Inst. Chem. Eng. (G.B.)*, **32,** 35.

Ridenour, G. M. (1946) Some Fundamental Concepts in Expression of Loadings and Efficiencies. *Sew. Works J.*, **18**, 641.

Saikaly, P. E.; Oerther, D. B. (2004) Bacterial Competition in Activated Sludge: Theoretical Analysis of Varying Solids Retention Times on Diversity. *Microbial Ecol.*, **48** (2), 74–284.

Scott, K. J. (1968) Thickening of Calcium Carbonate Slurries. *Ind. Eng. Chem. Fundam.*, **7** (3), 484.

Servizi, J. A.; Bogan, R. H. (1963) Free Energy as a Parameter in Biological Treatment. *Am. Soc. Civ. Eng., J. Sanit. Eng. Div.*, **89** (SA3), 17–40.

Shaw, A.; Phillips, H. M.; Sabherwal, B.; deBarbadillo, C. (2007) Succeeding at Simulation. *Water Environ. Technol.*, April, 54–58.

Smith, R.; Eilers, R. G. (1970) *Simulation of the Time-Dependent Performance of the Activated Sludge Process Using Digital Computer*; U.S. Department of the Interior, Federal Water Pollution Control Administration: Cincinnati, Ohio.

Stenstrom, M. K. (1975) A Dynamic Model and Computer Compatible Control Strategies for Wastewater Treatment Plants. Ph.D. Thesis, Clemson University, Clemson, South Carolina.

Storer, F. F.; Gaudy, A. F., Jr. (1969) Computational Analysis of Transient Response to Quantitative Shock Loadings of Heterogeneous Populations in Continuous Culture. *Environ. Sci. Technol.*, **3**, 143–149.

Takacs, I.; Bye, C. M.; Chapman, K.; Dold, P. L.; Fairlamb, P. M.; Jones, R. M. (2007) Biofilm Model for Engineering Design. *Water Sci. Technol.*, **55** (8-9), 329–336.

Takacs, I.; Patry, G. G.; Nolasco, D. (1991) A Dynamic Model of the Clarification-Thickening Process. *Water Res.*, **25** (10), 1263–1271.

Tracy, K. D. (1973) Mathematical Modeling of Unsteady State Thickening of Compressible Slurries. Ph.D. Thesis, Clemson University, Clemson, South Carolina.

Tracy, K. D.; Keinath, T. M. (1974) Dynamic Model for Thickening of Activated Sludge. *AIChE Symp. Ser.*, **136** (70), 291–308.

Van Haandel, A. C.; Ekama, G. A.; Marais, G. V. R. (1981) A General Model for the Activated Sludge Process Part 3: Single Sludge Denitrification. *Water Res.*, **15**, 1135–1152.

Van Loosdrecht, M. C. M.; Henze, M. (1999) Maintenance, Endogenous Respiration, Lysis, Decay, and Predation. *Water Sci. Technol.*, **39** (1) 107–117.

Wells, W. N.; Garrett, M. T., Jr., (1971) Getting the Most from an Activated Sludge Plant. *Public Works Magazine*, **102** (5), 63.

Wentzel, M. C.; Dold, P. L.; Ekama, G. A.; Marais, G. V. R. (1989) Enhanced Polyphosphate Organism Cultures in Activated Sludge Systems. Part III: Kinetic Model. *Water SA*, **15** (2), 89–102.

Yoshioka, N.; Hotta, Y.; Tanaka, S.; Naito, S.; Tsugami, S. (1957) Continuous Thickening of Homogeneous Flocculated Slurries. *Chem. Eng. J.*, **21**, 66–75.

Young, J. C.; McCarty, P. L. (1969) The Anaerobic Filter for Waste Treatment. *J.—Water Pollut. Control Fed.*, **41** (5), 160–173.

7.0 SUGGESTED READINGS

Dick, R. I.; Young, K. W. (1972) Analysis of Thickening Performance of Final Settling Tanks. *Proceedings of the 27th Indiana Waste Conference*; Purdue University: West Lafayette, Indiana; p. 34.

Eadie, G. S. (1942) The Inhibition of Cholinesterase by Physostigmine and Prostigmine. *J. Biol. Chem.*, **146**, 85–93.

Vesilind, A. P. (1968) Design of Thickeners from Batch Settling Tests: Practical Considerations. *Water Sew. Works*, **115** (7), 302–307.

Chapter 2

Modeling Fundamentals

1.0 INTRODUCTION

The primary purpose of process modeling in engineering practice is to use mathematical models that represent treatment processes to perform system optimization, troubleshooting, and design. Therefore, it is critical that model users have a good understanding of all the assumptions and limitations of the models to interpret and use modeling results correctly and in accordance with the objectives of the modeling project.

23

Process modeling uses many ideas and terms that may be unfamiliar to design engineers. This chapter contains an overview of the fundamental concepts and terminology that are commonly used in process modeling. The purpose of the overview is to give process engineers a broad understanding of the basics of process modeling and an appreciation of the general principles involved. Readers who want to understand these fundamentals in more depth are directed to the following references: for activated sludge models, International Water Association (IWA) Scientific and Technical Reports by Henze et al. (2000); for anaerobic digestion modeling, *Anaerobic Digestion Model No. 1 (ADM1)* by Batstone et al. (2002); and, for biofilm modeling, *Mathematical Modeling of Biofilms* by Wanner et al. (2006). Chapter 1 also provides a discussion of the genesis of many of the expressions used in process modeling.

2.0 MODELS VERSUS REALITY

Computer models, process or otherwise, attempt to represent the real world in the "virtual" world of the computer. Any model produced on a computer is, necessarily, an approximation of reality and is not the real world. Assumptions must be made in setting up the model, and the person or persons developing the model must decide which parameters and features to include. A model cannot be used to design or optimize a feature if it is not included in the model (i.e., a model of a fixed-volume, completely mixed reactor cannot be used to properly describe a system in which the reactor volume varies). Therefore, it is important to realize that models vary in complexity according to their intended use; it is also important that process engineers, as model users, realize the assumptions and limitations of the models they use to make good use of models for design or troubleshooting of wastewater treatment systems.

Figure 2.1 presents a diagram of the information flow between the process model and the real world. Different types of water resource recovery facility (WRRF) process data need to be entered into the model or compared to the outputs from the model. Data from the real world include influent data, physical sizes of facilities, operating data, and effluent data. Within the model, these must be translated into numbers that can be used by the models through influent fractions and WRRF configuration interfaces. Similarly, the numbers produced by the models have to be converted into outputs that can be compared to real world data, such as facility effluent concentrations. An important use of model outputs is the comparison of effluent concentrations predicted by the model with effluent limits specified in facility permits to evaluate environmental permit and regulatory compliance.

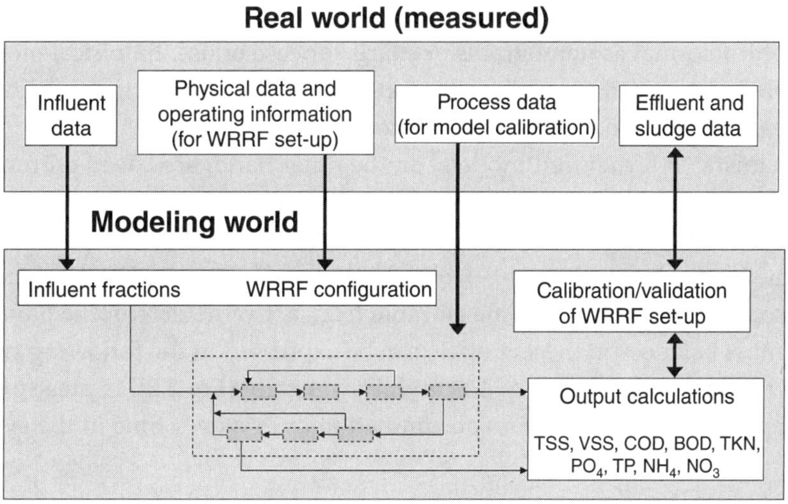

Real world (measured)

FIGURE 2.1   Diagram showing distinction between the "real world" and the "modeling world" and the interactions between them (courtesy of IWATGGMP, 2008). (WRRF = water resource recovery facility; TKN = total Kjeldahl nitrogen; PO_4 = orthophosphate; NH_4 = ammonium; NO_3 = nitrate; and TP = total phosphorus).

Most models used for design are termed *mechanistic models*, meaning that they are made up of a series of equations that describe the behavior of the process based on an understanding of the behavior of the system's components. These models are also termed *physics-based* because they use equations that describe the physical behavior, or "physics", of the processes in the model. Another approach is the *statistical model*, which does not use a fixed set of equations, but instead develops a set of statistical rules based on historical data to determine how facility inputs affect facility outputs. These models are also known as *black-box* or *empirical models* because they typically are developed without the user considering the fundamental processes occurring in the system.

Empirical (experience-based) models express relationships between key measured variables such as solids residence time (SRT), sludge production, effluent quality, oxygen demand, and so on. Because of the complex nature of underlying processes and the different, often incompatible units of variables that can be readily measured by standard analytical methods (volatile suspended solids [VSS], biochemical oxygen demand [BOD], total suspended solids [TSS], etc.), empirical models typically are not mass-balance based. Rather, they are often expressed in correlations or in graphical format as nomograms or charts for ease of use. Empirical models and the knowledge they contain play an important role in engineering work, even with the detailed mechanistic models available today (WEF, 2010).

Mechanistic or structured models, on the other hand, are based on fundamental laws and principles that govern a process or transformations occurring as part of the process. Mechanistic models use mass balances and kinetics performed on a consistent set of basic components (called *state variables* in mathematical modeling) that are sometimes not directly measurable (e.g., active heterotrophic biomass). In general, mass balances in these models can be expressed in the following two ways (WEF, 2010): in steady state (Eq. 2.1) and dynamically (Eq. 2.2). In steady state for each component, there is no mass accumulation or loss over time in the system, as follows:

$$0 = \text{Input (kg/d)} - \text{Output (kg/d)} \pm \text{Reaction rate (kg/d)} \qquad (2.1)$$

where the reaction rate is the production or loss of the component considered because of chemical or biological reactions. Steady-state mechanistic models can only be used to calculate stable conditions, for example monthly averages.

Dynamic models, on the other hand, express the mass change in time, and thus can be used to calculate dynamic events such as diurnal variation, storms, peak oxygen demand, or similar events, as follows:

$$\text{Mass change (kg/d)} = \text{Input (kg/d)} - \text{Output (kg/d)} \pm \text{Reaction rate (kg/d)} \qquad (2.2)$$

The focus of this chapter is mechanistic, structured models that are used for calculation of facility-wide degradable organic carbon and nutrient transformation/removal performance and mass balances, typically implemented in simulation software.

3.0 MODEL REPRESENTATION OF WASTEWATER TREATMENT PROCESSES

This section describes the main processes that occur in wastewater treatment and shows how these processes are represented in mathematical models. Processes are transformations of wastewater components (e.g., nutrients and organic carbon) as a result of biological, chemical, or physical reactions. A description of how these models are implemented is included in Chapter 4.

3.1 Processes

There are several fundamental processes that occur in all biochemical systems. In mathematical models used for the simulation of wastewater treatment processes, modelers have selected to quantitatively describe processes that, based on current understanding, are the most important for the engineering application (nitrification, biological phosphorus removal, etc.). This section gives an overview of the processes included in the most commonly used simulators.

3.1.1 Biological Reactions

Wastewater treatment is centered on biological transformations of carbon, nitrogen, and phosphorus compounds, carried out by bacteria living in biological facilities. Mathematical representation of bacterial metabolism is a key element of modeling wastewater treatment processes, and it is an area of continuous development within the modeling community. While different models may use slightly different approaches to mathematically describe biological reactions in wastewater treatment, the main biological processes of interest for modeling purposes are discussed herein.

3.1.1.1 Biomass Growth, Substrate Utilization, and Yield

This process describes the growth of microorganisms using substances found in wastewater as energy and/or carbon sources. *Biomass growth* and *substrate utilization* are, in most cases, coupled; as a result, the removal of one unit of substrate results in production of Y units of biomass. The "Y" represents the true *growth yield* of the microbial community. True growth accounts for substrate utilization for cell maintenance purposes.

The description of growth in the models includes expressions of the four elements required for growth: carbon source, inorganic nutrients, energy source (electron donor), and a source of reducing power (electron acceptor). The mathematical expressions of the models (Section 3.2) describe the relationships of these four components for each process (Grady et al., 1999). For example, heterotrophic bacterial growths use biodegradable chemical oxygen demand (COD) as carbon and energy sources, oxygen as an electron acceptor for reducing power, and nitrogen and phosphorus as limiting nutrients. Another example is autotrophic growth, where the energy sources are ammonia and nitrite, the carbon source is inorganic carbon from alkalinity, and nitrogen and phosphorus are the limiting nutrients.

3.1.1.2 Maintenance, Endogenous Metabolism, Decay, Lysis, and Death

Maintenance describes the energy required by microorganisms for functions other than growth. Such functions include motility, maintenance of ionic gradients, and molecular transport. The source of this energy can be external or internal. In the presence of sufficient external energy sources, the microorganism will use a portion of the available energy for maintenance and the rest for growth. With the availability of external energy diminishing, growth will continuously decrease. In the event that no external energy source is available for maintenance, internal energy sources will be used and maintenance needs will be covered by *endogenous metabolism*.

Cell lysis results from the loss of the ability of a microorganism to maintain the physical integrity of its cellular structure. As a result, the cell wall is ruptured and the microorganism dies. Cell lysis is another process that describes reduced biomass growth as compared to the theoretical cell yield (Y).

Death describes the state of a microorganism that has lost the ability to metabolize substrates. There is debate among researchers regarding the amount of microorganisms that are really dead in a facility. There is evidence to suggest that dead cells do not remain intact long, but break up into substrates and biomass debris. The inactive organic part of the mixed liquor in wastewater can be attributed to the accumulation of biomass debris rather than dead cells (Grady et al., 1999).

Mathematical models use different approaches to account for maintenance and decay mechanisms. Some models (e.g., Activated Sludge Model No. 1 [ASM1]) use a death-regeneration approach, where bacterial decay produces biodegradable and nonbiodegradable material that is consumed by active bacteria or accumulated in the system. Other models (e.g., ASM3) use an endogenous respiration approach in which decay mechanisms for bacteria and their storage products account for decay and maintenance.

3.1.1.3 Storage

Bacteria often synthesize internal storage products that serve as energy or carbon sources. One of the more common storage inclusions involves poly-β-hydroxyalkanoate (PHA), which, in bacteria, functions as a carbon and energy storage product. Glycogen is another common carbon and energy storage product. Storage requires energy, which is obtained from respiration (Henze et. al., 2000).

Given the right environmental conditions, some organisms can accumulate granules containing long chains of phosphate. Similar to PHA storage, polyphosphate storage requires energy. Poly-β-hydroxyalkanoate and polyphosphate storage and release are key processes in enhanced biological phosphorus removal.

Sulfur globules are also found in a variety of bacteria capable of oxidizing reduced sulfur compounds, such as hydrogen sulfide and thiosulfate. Oxidation of these compounds is linked to either energy metabolism or photosynthesis.

3.1.1.4 Hydrolysis

Hydrolysis plays two key roles in biochemical systems. First, it releases cellular components following cell lysis. Second, through hydrolysis, complex organic materials are broken down by extracellular enzymes to produce low-molecular-weight compounds that can be transported inside the cell and metabolized. Current activated sludge and biofilm models include several hydrolysis reactions that cover the hydrolysis of complex organic molecules and entrapped organic nutrients, such as nitrogen and phosphorus. In anaerobic digestion models, particulate carbohydrates and proteins are hydrolyzed to produce simple sugars and amino acids; hydrolysis of lipids generates long chain fatty acids.

3.1.1.5 Fermentation, Acidogenesis, Acetogenesis, and Methanogenesis

Fermentation and *acidogenesis* describe processes by which, under strict anaerobic conditions, bacteria generate simple acids. During acidogenesis, amino acids and soluble sugars are degraded by fermentative reactions primarily to acetate, propionate, and butyrate products (Batstone et al., 2002). *Acetogenesis* describes the degradation of higher organic acids to acetate. During aceticlastic *methanogenesis*, acetate is used by bacteria to form methane and carbon dioxide (Batstone et al., 2002).

3.1.1.6 Ammonification

Ammonification describes the conversion of organic nitrogen compounds to ammonia. It is a biologically mediated degradation process.

3.1.1.7 Chemical Reactions—pH and Precipitation

Chemical reactions are common in wastewater processes and occur between compounds produced from biological reactions and/or between compounds in the wastewater and chemicals added in the wastewater treatment process. These chemical reactions commonly include changes in pH through alkalinity production (denitrification) or alkalinity consumption (nitrification) or the effect of added chemicals and precipitation reactions such as phosphorus precipitation in anaerobic digesters. Because chemical precipitation is common in wastewater treatment, these reactions are commonly included in mathematical models, describing, for example, chemical precipitation of phosphorus with alum or ferric chloride.

3.1.2 Physical Reactions

Physical reactions are another important part of modeling wastewater treatment processes. Examples of physical reactions described herein include mass transport of pollutants through processes such as dispersion and diffusion in biofilms, gas–liquid gas transfer (aeration and stripping), and mixing. Physical reactions described herein are incorporated in process models, together with the biological reactions previously described in this chapter, to provide an overall representation of the treatment processes in facilities.

3.1.2.1 Gas–Liquid Mass Transfer

Mass transfer of compounds between gas and liquid phases is important in wastewater treatment. Several examples include aeration (i.e., oxygen transfer from air to mixed liquor), ammonia stripping (i.e., removal of ammonia from liquid solution), and carbon dioxide equilibrium.

Henry's law states that, at equilibrium, the ratio of the concentration of a volatile solute in a solvent to the solute's partial pressure in the vapor phase above the solvent–solute mixture is a constant. Henry's law is commonly used in mathematical models to describe the behavior of compounds that partition between liquid and gas phases (e.g., oxygen and methane).

Biological wastewater treatment systems require some type of aeration to drive aerobic biological processes. Aeration is the transfer of oxygen from air into the liquid phase so that bacteria can use oxygen dissolved in the liquid phase for aerobic growth. The rate of flux of oxygen from the gas to the liquid is given by

$$r_{O_2} = K_L a \, (C_{sat} - C) \tag{2.3}$$

where

r_{O_2} = the oxygen-transfer rate per unit volume, mg/L·s
$K_L a$ = the volumetric mass-transfer rate coefficient, L/s
C_{sat} = the saturation concentration or liquid oxygen concentration (mg/L) in equilibrium
      with the bulk gas phase (governed by Henry's law constant)
C = the concentration in the bulk liquid, mg/L

The mass transfer rate coefficient, $K_L a$, is system-specific and depends on many factors, primarily the type of aeration system (surface aeration, fine bubble diffusers, coarse bubble diffusers, etc.), size and shape of the aeration basins, among others. Other factors that need to be accounted for in wastewater aeration systems are the clean water vs wastewater factors, which affect oxygen transfer efficiency.

3.1.2.2 Diffusion

Diffusion is a physical process whereby mass is transported via gradients in concentration, temperature, or pressure gradients. In wastewater treatment processes, we are mainly concerned with diffusion because of concentration gradients in systems such as biofilms. Fick's second law governs mass transport caused by concentration gradients, as follows:

$$\frac{\partial C}{\partial t} = D\frac{\partial^2 C}{\partial x^2} \tag{2.4}$$

where

   C = the concentration in the bulk liquid, mg/L
   D = the diffusion coefficient, m^2/s
   t = time, s
   x = the distance from the base of the biofilm, m

   Biofilms are matrices of microorganisms and their extrapolymeric substances that grow attached to a fixed surface, such as rocks, plastic media, and so on. Concentration gradients are established within the biofilm because of the need to transport material throughout the biofilm. Mathematical models for biofilm systems have to capture the existence of these concentration gradients to predict mass fluxes into and out of the biofilm.

3.1.2.3 Adsorption

Adsorption is a physical process in which substances adhere to a surface, resulting in the removal of such substances from the bulk liquid. Common examples of the use of adsorption processes in wastewater treatment include the use of granular or powdered activated carbon to remove residual chlorine, synthetic dyes, or metals from wastewater. Adsorption is also a factor in the removal of colloidal particulates by activated sludge and in chemical coagulation processes.

3.1.3   Solids Separation

Solids–liquid separation processes play an important role in modeling wastewater treatment systems and have a direct effect on the prediction of effluent quality. Solids–liquid separation processes include sedimentation (primary and secondary) and size exclusion filtration via media filtration and membranes.

3.1.3.1 Sedimentation

In terms of modeling, primary and secondary sedimentations present different situations. In primary sedimentation, particle concentration is low enough that the particles typically do not have significant interactions. Moreover, the system can be modeled using Stoke's law, which governs the settling velocity of particles as a function of the densities of the particles and the fluid and the size of the particle. Primary sedimentation is typically modeled using an assumed solids capture efficiency, sometimes based on statistical correlations between solids capture efficiency and surface overflow rate.

In secondary sedimentation, mixed liquor with a relatively high concentration of particles needs to settle in the clarifiers to produce clarified effluent. Because of the high concentration of particles in the mixed liquor and interactions between particles, Stoke's law does not apply anymore. Rather, a concept called *zone settling* is often used to model secondary sedimentation, where the velocity of the sludge relative to the liquid is a function of the local concentration. At high particle concentrations toward the bottom of the clarifier, flocs change size and shape and a phenomenon known as *compaction* occurs. In the compaction zone, water has to be displaced so that the dense sludge can settle downward, producing a complex phenomenon to represent mathematically. Available clarifier models range from simple two-compartment models to complex computational fluid dynamic models suited for a range of modeling objectives. The development of different clarifier model equations is described in Chapter 1 and their use is described in Chapter 3.

3.1.3.2 Filtration

Solids separation by size-exclusion filtration applies to filtration of secondary effluent or filtration of mixed liquor in membrane bioreactors. Filtration is based on size exclusion. Particles of a certain cutoff size will be retained in the filter, producing a clear effluent. Modeling filtration in wastewater treatment processes is typically limited to simple models involving an assumed solids capture efficiency or an assumed effluent quality. Backwash from filtration processes can represent a significant flow component and, therefore, it is desirable to include it in process models.

3.1.4   Reactor Hydraulics

Biological and chemical reactions in facilities take place in engineered reactors designed to provide the desired hydraulic conditions to achieve treatment objectives. Such reactors use different mixing regimes, and the overall arrangement and mixing regime of the reactors affects the overall rate and extent of biological and chemical conversions that occur during treatment.

Based on mixing regimes, there are two ideal mixing conditions that cover most reactors used in wastewater treatment: perfect mixing and ideal plug flow. In reality, mixing characteristics typically fall within these two ideal conditions, although it is important to understand how these ideal mixing conditions are used in reactor modeling. Simulators have different types of reactors available with certain mixing regime assumptions, and the simulator user needs to judge which reactor type to use and understand the effects of reactor configurations.

3.1.4.1 Perfect Mixing—Continuously Stirred Tank Reactor

Perfect mixing in a continuous-flow reactor occurs when the concentration of a compound inside the reactor is perfectly uniform. This means that the concentration of a compound inside a continuous-flow reactor is exactly the same as the concentration in the reactor outlet and that any particle entering a perfectly mixed reactor is instantaneously dispersed within the perfectly mixed volume. These types of reactors are commonly known as *continuously stirred tank reactors* (CSTRs).

3.1.4.2 Ideal Plug Flow

Ideal plug flow occurs when the velocity throughout the reactor is steady and constant across the flow section and there is no molecular diffusion or mixing of the compound being described in the model in the axial direction. Ideal plug flow also implies complete mixing in the cross-sectional plane perpendicular to the direction of flow. Particles entering a continuous plug flow reactor (PFR) travel through the reactor and exit at exactly the theoretical hydraulic detention time. Particles entering the reactor do not mix along the length of the reactor.

3.1.4.3 Tanks-in-Series Model

In practice, the conditions for perfect mixing (CSTR) and ideal plug flow are not achieved for flow-through systems. Reactors used in wastewater treatment typically fall somewhere in between these two ideal conditions. Tracer studies using conservative substances can be used to mathematically characterize reactor systems using hydraulic models that fall in between perfect mixing and ideal plug flow.

A commonly used mathematical model is the tanks-in-series model. This model assumes that the reactor studied can be considered as an arrangement of CSTR tanks placed one after the other (N) (each tank with a volume equal to V_{total}/N) such that the total volume of the reactor being modeled is equal to the sum of the volumes of the equally sized CSTR tanks.

The residence time density function for this model is given by the following equation (Clark, 1996):

$$f(t) = \frac{N^N \cdot t^{N-1}}{\theta_H{}^N (N-1)!} \exp\left(-\frac{N \cdot t}{\theta_H}\right) \tag{2.5}$$

where
  N = the number of CSTRs in series
  θ_H = the hydraulic residence time (s) in the overall system ($\theta_H = V/Q$) where V = the system volume (L) ($V = V_1 + V_2 + V_3 + \ldots + V_N$) and Q = the flow into and out of the system (m^3/s)

When considering $N = 1$, this model gives the solution for the ideal CSTR, and as N approaches infinity, the solution approaches the solution for the ideal PFR. In this model, the parameters are the number of CSTRs in series (N) and the hydraulic residence time (θ_H). Using Eq. 2.5, it can be shown that plug flow conditions are approached for $N \geq 3$. Using more than 3 reactors in series provides little benefit in terms of approaching ideal plug flow conditions.

3.1.4.4　Axial Dispersion Model

The axial dispersion model is based on the PFR configuration, but assumes that the contribution of dispersion in the axial direction of the reactor to the total mass transport is relatively important compared to advection such that an ideal PFR cannot be assumed. The governing equation for this model, considering a conservative tracer (no reaction), is as follows (Clark, 1996):

$$\frac{\partial C}{\partial t} + u \frac{\partial C}{\partial x} = D \frac{\partial^2 C}{\partial x^2} \tag{2.6}$$

where
u = the velocity (m/s) of the water ($u = Q/A$)
D = the dispersion coefficient, m²/s

The model parameters are the dispersion coefficient, D, and water velocity, u. In this model, as D approaches infinity, the solution approaches the one for the ideal CSTR, while, as D approaches zero, the solution resembles the ideal PFR.

3.1.4.5　Sequencing Batch Reactor

A sequencing batch reactor (SBR) operates in fill-react-settle-decant-idle cycles. As opposed to continuous-flow systems where biochemical reactions occur over space (i.e., along the length of a reactor or through several tanks in series), reactions in an SBR are carried out over time. The batch operation of an SBR results in a system that operates like an ideal PFR, but with time instead of space as the independent variable of the system.

3.2　State and Combined Variables

Process engineers are accustomed to using influent parameters that are commonly and easily measured, such as BOD, COD, and TSS. These parameters are indicators of the wastewater strength, which is important for facility design and operation, but they do not provide sufficient details of the wastewater composition for many models. While models do not require the exact chemical makeup of raw wastewater, they do require more information than just BOD and TSS to predict the growth of microorganisms and the fate of pollutants within the model.

Model equations are based on basic building blocks called *state variables*. State variables are the simplest constituent parts within a model and are divided into two general categories: soluble and particulate (IWA nomenclature designation of "S" and "X", respectively; see Corominas et al. [2010] for further reference). In addition, some models use state variables for colloidal material, which is difficult to define as truly soluble, or particulate, which constitutes large molecules that may or may not be removed by filtration depending on the filter pore size. Examples of soluble state variables include soluble biodegradable substrate (expressed as COD), ammonia (nitrogen [N]), and orthophosphate (PO_4) (phosphorus [P]). When several state variables are combined, they constitute composite variables such as BOD, COD, TSS, total Kjeldahl nitrogen, and total phosphorus. An example of the manner in which models combine state variables into composite variables is shown in Figure 2.2 (for COD), Figure 2.3

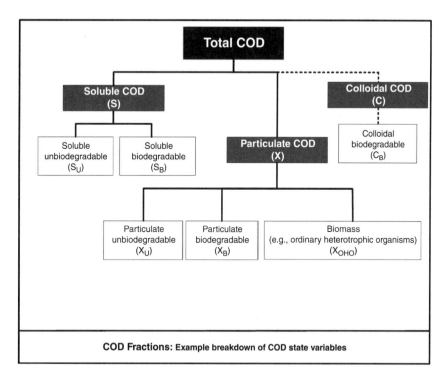

FIGURE 2.2 Example state variables that compose the composite variable, COD.

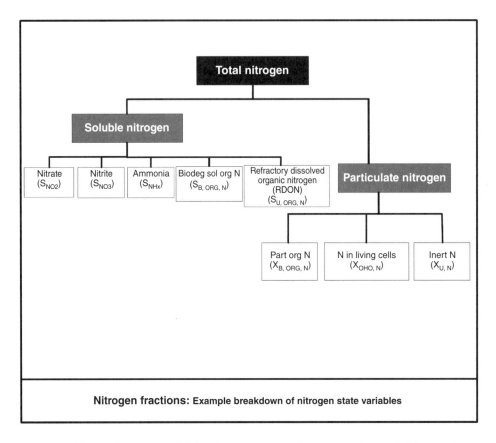

Nitrogen fractions: Example breakdown of nitrogen state variables

FIGURE 2.3   Example state variables that compose the composite variable, total nitrogen.

(for total nitrogen), and Figure 2.4 (for total phosphorus). The state variables are shown in white, while the black and gray boxes refer to composite variables. The examples shown for state and composite variables are typical for many wastewater models, although the exact state variables used are specific to the model selected. Chemical oxygen demand, rather than BOD, is used as the basis for describing carbonaceous substrate in the IWA models.

The intent of this section is to provide a general overview of state variables and explain what they represent as an introduction to the more detailed discussion of models presented in Chapter 3. Different mathematical models use similar state

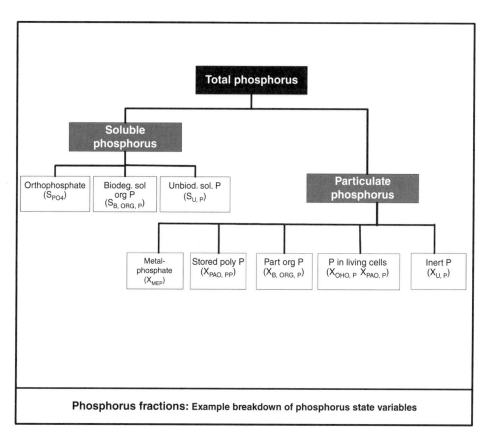

FIGURE 2.4   Example state variables that compose the composite variable, total phosphorus.

variables, although with some differences that correspond to their specific model structure. Table 2.1 illustrates the range of state variables considered in commonly used activated sludge models.

3.3 Kinetic Expressions

Kinetic expressions describe the rate of change of state variables in a model. It is beyond the scope of this manual to give a detailed explanation of the different kinetic expressions used in process models. Instead, a brief description is given for the

TABLE 2.1 Range of state variables considered in commonly used activated sludge models.

Description	Parameter	Units	ASM1 Symbol	ASM2d Symbol	ASM3 Symbol	ASM3 Bio-P Symbol	Barker and Dold Symbol	UCTPHO+ Symbol	ASM2d + TUD Symbol
COD soluble									
Soluble biodegradable organics	S_B	g COD.m-3	S_S		S_S	S_S			
Fermentable organic matter	S_F	g COD.m-3		S_F			S_{BSC}	S_F	S_F
Fermentation product (volatile fatty acids)	S_{VFA}	g COD.m-3		S_A			S_{BSA}	S_A	S_A
Soluble undegradable organics	S_U	g COD.m-3	S_I	S_I	S_I	S_I	S_{US}	S_I	S_I
Dissolved oxygen	S_{O2}	- g COD.m-3	S_O	S_{O2}	S_O	S_O	S_O	S_{O2}	S_O
COD particulate and colloidal									
Particulate biodegradable organics	X_C	Bg COD.m-3	X_S	X_S	X_S	X_S	S_{ENM}	X_{ENM}	X_S
Adsorbed slowly biodegradable substrate	X_{Ads}	g COD.m-3						X_{ADS}	
Particulate undegradable organics	X_U	g COD.m-3		X_I	X_I	X_I			X_I
Particulate undegradable organics from the influent	$X_{U,Inf}$	g COD.m-3	X_I				S_{UP}	X_I	
Particulate undegradable endogenous products	$X_{U,E}$	g COD.m-3	X_P				Z_E	X_E	
Nitrogen									
Ammonium and ammonia nitrogen ($NH_4 + NH_3$)	S_{NHx}	g N.m-3	S_{NH}	S_{NH4}	S_{NH}	S_{NH}	N_{H3}	S_{NH4}	S_{NH}

(*continued*)

Description	Symbol	Units							
Nitrate and nitrite (NO_3 + NO_2) (considered to be NO_3 only for stoichiometry)	S_{NOx}	g N.m-3	S_{NO}	S_{NO3}	S_{NO}	S_{NO}	N_{O3}	S_{NO3}	S_{NO}
Dissolved nitrogen gas	S_{N2}	g N.m-3		S_{N2}	S_{N2}	S_{N2}			S_{N2}
Particulate biodegradable organic N	$XC_{B,N}$	g N.m-3	X_{ND}			N_{BP}			
Soluble biodegradable organic N	$S_{B,N}$	g N.m-3	S_{ND}			N_{BS}			
Soluble inert organic N	$S_{U,N}$	g N.m-3					N_{US}		
Phosphorus									
Soluble inorganic phosphorus	S_{PO4}	g P.m-3		S_{PO4}	S_{PO4}	S_{PO4}	P_{O4}	S_{PO4}	S_{PO}
Biomass									
Ordinary heterotrophic organisms (OHOs)	X_{OHO}	g COD.m-3	$X_{B,H}$	X_H	X_H	X_H	Z_H	X_H	X_H
Autotrophic nitrifying organisms (NH_4 to NO_3^-)	X_{ANO}	g COD.m-3	$X_{B,A}$	X_{AUT}	X_A	X_A	Z_A	X_{NIT}	X_A
Phosphorus accumulating organisms (PAOs)	X_{PAO}	g COD.m-3	X_{PAO}	X_{PAO}	X_{PAO}	X_{PAO}	Z_P	X_{PAO}	X_{PAO}
Organisms (biomass)	X_{Bio}	g COD.m-3							
Internal cell products									
Storage compound in OHOs	$X_{OHO,Stor}$	g COD.m-3			X_{STO}	X_{STO}			
Storage compound in PAOs	$X_{PAO,Stor}$	g COD.m-3	X_{PHA}	X_{PHA}	X_{PHA}	X_{PHA}	S_{PHB}	X_{PHA}	

TABLE 2.1 Range of state variables considered in commonly used activated sludge models. (*continued*)

Description	Parameter	Units	ASM1 Symbol	ASM2d Symbol	ASM3 Symbol	ASM3 Bio-P Symbol	Barker and Dold Symbol	UCTPHO+ Symbol	ASM2d + TUD Symbol
Stored poly-β-hydroxyalkanoate in PAOs	$X_{PAO,PHA}$	g COD.m-3							X_{PHA}
Stored glycogen in PAOs	$X_{PAO,Gly}$	g COD.m-3							X_{GLY}
Stored polyphosphates in PAOs	$X_{PAO,PP}$	g P.m-3		X_{PP}		X_{PP}		X_{PP}	X_{PP}
Releasable stored polyphosphates in PAOs (low molecular weight)	$X_{PAO,PPLo}$	g P.m-3					$P_{PP\text{-}LO}$		
Nonreleasable stored polyphosphates in PAOs (high molecular weight)	$X_{PAO,PPHi}$	g P.m-3					$P_{PP\text{-}HI}$		
Inorganics									
Metal hydroxide compounds	X_{MeOH}	g TSS.m-3		X_{MeOH}					
Metal phosphate compounds	X_{MeP}	g TSS.m-3		X_{MeP}					
Alkalinity (HCO_3^-)	S_{Alk}	mol HCO_3^--m-3	S_{ALK}	S_{ALK}	S_{ALK}	S_{HCO}			S_{HCO}
TSS									
Total suspended solids	X_{TSS}	g TSS.m-3		X_{TSS}	X_{SS}	X_{SS}			X_{TSS}

kinetic expression used for the basis of many biological models, including those for the activated sludge process, namely the Monod equation, which was first proposed by Monod (1942) to describe the growth of bacteria. In Eq. 2.7, the Monod equation is first described in words and then in its simple algebraic form using common notations and units used in wastewater treatment. It should be noted that the form of this equation is identical to the Michaelis–Menten (Michaelis and Menten, 1913) formula used to describe enzyme reactions, although, unlike the Michaelis–Menten formula, the Monod equation does not have any fundamental biological or chemical basis but has proven to be a useful, simple expression to describe biological activity with a significant experience base. The Monod equation is as follows:

Bacterial growth = Maximum rate × Substrate concentration/
(Constant + Substrate concentration) × Biomass concentration

or

$$dX/dt = \mu \times S/(K_s + S) \times X \qquad (2.7)$$

where
 X = biomass concentration, mg/L
 μ = maximum specific growth rate, h^{-1}
 S = substrate concentration, mg/L
 K_s = half-saturation constant (concentration at which the specific growth rate is half the maximum rate), mg/L
 t = time, h

Figure 2.5 shows the effect that substrate concentration has on the biomass growth rate using a Monod expression. The process designer should be aware of the following salient features of this curve:

- At high substrate concentrations, the rate approaches its maximum rate (μX), and the kinetics are effectively "zero-order", which means the rate is not dependent on the substrate concentration.

- At zero substrate concentration, the rate is zero.

- At a substrate concentration of K_s, the growth rate is exactly half of the maximum rate, hence the name of the K_s constant, *half-saturation constant*.

Other kinetic expressions have also been used to describe transformations in process models, including first-order kinetics, where the rate is always directly

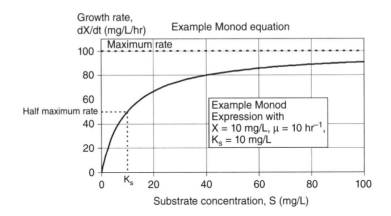

FIGURE 2.5   Example growth rate versus substrate using a Monod equation.

proportional to a substrate concentration, and zero-order kinetics, where the rate is constant, irrespective of the substrate concentration.

In addition to using Monod expressions to describe how organisms use substrate to grow, similar equations can be used to adjust the rate based on environmental factors. These expressions are termed *switching functions* and have a value between 0 and 1, depending on the environmental condition and the concentration of the variable that defines that environmental condition. There are two common forms of switching functions: one that decreases the growth rate as the variable decreases and one that decreases the growth rate as the variable increases. The equation form and curve shape for the first type is identical to the Monod equation previously described. The second form, which reduces the rate as the variable concentration increases, can be termed a *reverse Monod*, and its curve shape is shown in Figure 2.6. Figure 2.6 also shows that the rate is at half its maximum at the half-saturation concentration (K_s); however, when the variable concentration is zero, the rate is at its maximum, and, at high concentrations, the rate approaches zero. Switching functions is commonly used to switch rates on and off, depending on the presence of compounds such as dissolved oxygen and nitrate (NO_3) to define if conditions are aerobic, anoxic, or anaerobic. These formulas are more straightforward to handle in numerical solvers than functions that simply turn processes on or off at a particular threshold concentration. Care should be taken when making adjustments to switching function parameters because changing them slightly can significantly affect model outputs.

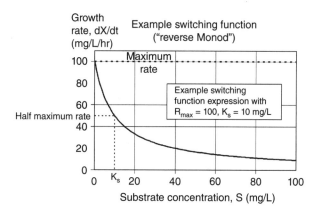

FIGURE 2.6   Example switching function.

3.4   Stoichiometry

All models contain stoichiometric parameters that are used to define mass relationships between state variables. Examples of stoichiometric parameters include biomass yields, nutrient fractions in biomass, and the ratio of COD to VSS for biomass or influent wastewater. Not to be confused with influent stoichiometric parameters, which are selected based on sampling (see Chapter 8), model yields and biomass fractions typically should not be adjusted from values reported in the literature as they describe fundamental relationships within the model. For example, a typical microorganism is a volatile solid that can be expressed as $C_5H_7O_2N$ (molecular weight = 113) and will yield a typical COD/VSS ratio of 1.42 based on the following equation:

$$C_5H_7O_2N + 5O_2 \rightarrow 5CO_2 + 2H_2O + NH_3 \qquad (2.8)$$

So, looking at the stoichiometric coefficients of the aforementioned equation, the oxygen, O_2 (or COD), equivalent for the microorganism is as follows:

$$5 \times 32 \text{ g } O_2 1/113 \text{ g VSS} = 1.42 \text{ g } O_2 (\text{g VSS})^{-1} \qquad (2.9)$$

3.5   Gujer Matrices

Models are often presented by means of "Gujer matrices" (Henze et al., 2000) (previously referred to as "Petersen matrices", [Petersen, 1965]), which graphically describe all the interactions of system components. The Gujer matrix for a given model can be

Not simplified but cons till S//ow.

used by the process designer to understand how state variables are affected by different process rates.

The Gujer matrix consists of the following three parts:

- Stoichiometric matrix—This part of the matrix defines the changes in state variables for each of the processes included in the model.
- Composition matrix—This matrix defines the COD, nitrogen, phosphorus, and charge in each state variable included in the model.
- Kinetic rate expressions—Kinetics for each process included in the model is defined by a process rate equation.

Appendix A presents a simplified example for ASM1 heterotrophic bacteria that are growing by using a soluble substrate and autotrophic bacteria growing on ammonia. The following sections explain the components of the Gujer matrix for the simplified example presented in Appendix A.

3.5.1   Stoichiometric Matrix Example

Two fundamental processes occur with the heterotrophic biomass: the concentration of biomass increases by cell growth and decreases by decay. As a consequence of cell growth, coupled with the system stoichiometry, substrate removal also occurs and, in aerobic systems, oxygen is used. In the matrix, three state variable components (biomass, substrate, and oxygen) are listed across the top and two processes (growth and decay) are identified in the left-most column of the matrix.

3.5.2   Kinetic Rates Example

The kinetic expressions or rate equations for each process (in this example, a Monod equation for the growth and a first-order decay with respect to biomass) are recorded in the right-most column. Parameters used in the kinetic expressions are defined in the lower right corner of the matrix.

The Gujer matrix can be used to easily identify which state variables (substrate and biomass, in this example) are involved in which reactions. The reaction term for a particular state variable within the boundaries of the system is obtained by summing the products of the stoichiometric coefficients in the column for the state variable and the process rate expressions. For example, the rate of reaction for biomass, X_{OHO}, at a point of the system would be as follows:

$$r_{XOHO} = \mu \times S_B/(K_{SB} + S_B) \times X_{OHO} - b \times X_{OHO} \tag{2.10}$$

Similarly, the rate expression for substrate utilization would be as follows:

$$r_{SB} = -(1/Y_{OHO}) \times \mu \times S_B/(K_{SB} + S_B) \times X_{OHO} \qquad (2.11)$$

In addition, the rate expression for oxygen utilization would be as follows:

$$r_{SO_2} = (1 - (1/Y_{OHO})) \times \mu \times S_B/(K_{SB} + S_B) \times X_{OHO} \qquad (2.12)$$

It should be noted that the value shown in the matrix is the true maximum specific growth rate, which cannot be measured directly using laboratory testing as decay also occurs during the measurement. An observed or "net" maximum specific growth rate, μ_{net}, that is measured at high substrate concentrations in a laboratory test, where it is ensured that the rate is not limited by substrate, includes the decay process and must be converted using the following equation:

$$\mu = \mu_{net} + b \qquad (2.13)$$

If a cell within the matrix does not contain anything, it is an indication that the state variable is not changed by the particular process. For the aforementioned example, the substrate (S_s) and oxygen (S_{O_2}) are both not used in the decay process; thus, the rate expression for each of these state variables does not include the decay rate term. This is a simple example of a matrix; more complex matrices can be used to describe other models with more state variables and processes.

3.5.3   Composition Matrix Example

Elements shown within the matrix are the stoichiometric coefficients, which set out the mass relationships in consistent units between the components in the individual processes. For example, growth of biomass (+1) occurs at the expense of soluble substrate ($-1/Y_{OHO}$).

4.0   IMPLEMENTATION OF MODELS IN SIMULATION TOOLS

Models are implemented in simulators to produce a facility model. The intent of this chapter is to differentiate models from simulators and to discuss how they are linked together.

4.1   Simulator Versus Model

A common mistake in process modeling is to use the term *model* to describe the software package on which the model is run. The correct term for the software is

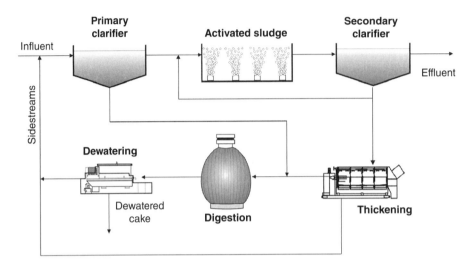

FIGURE 2.7 Example model flow schematic showing process flow linkages between unit processes.

simulator, as the term *model* is actually the set or sets of equations that are solved within the simulator. A facility model can be made up of many different unit process models that are linked together; this is often represented in a simulator by a schematic showing how process units are connected (e.g., Figure 2.7).

Each unit process model, in turn, may be made up of several submodels that describe different facets of the unit process operation and performance. For example, an activated sludge model may include a kinetic model, which describes biokinetic reactions, and an aeration system model, which describes the mass transfer of oxygen to the activated sludge. Each model is made up of a set of equations that may be linked in series or may occur in parallel and be unlinked.

Models that describe different unit processes will use some variables that are common (e.g., flow and temperature), but also some that are only applicable to a particular unit process. Examples of this are models of the activated sludge and anaerobic digestion processes. The rate equations and parameters used to describe the transformations in activated sludge are different than those used for anaerobic digestion. Two approaches are commonly used to handle the interaction between different process models, and they have traditionally been termed the *supermodel* approach and the *interface* approach. In the supermodel approach, the equations for all kinetic reactions and the variables used in them are available in all unit processes. For those reactions that do not occur in a process (e.g., nitrification in an anaerobic digester), the reaction rate simply becomes very low or zero and, as such, does not have a significant effect on model behavior. In the interface approach, unit process models only contain equations to describe the reactions that occur within that unit process model (e.g., no nitrification rate equations in the anaerobic digester). In this approach, output that is a constituent from one unit process may not necessarily be an input to another unit process model or it may be described using a different term; thus, an "interface" module is used to define how the constituents are passed between model types.

The main advantage of the supermodel approach is that it completely describes all unit processes, and errors are not introduced in the conversion of constituents between process types. The supermodel approach can also pick up on transformations in a unit process that were not expected to occur, but do actually happen. The disadvantage of this approach is that it requires a large number of equations that must be handled in all unit processes, many of which may be redundant. This can slow down simulation times and make it more difficult for numerical solvers to find solutions. Analysis by Grau et al. (2009) provides further discussion on the merits of the two approaches and describes a "tailored" supermodel approach that attempts to draw on the benefits of a supermodel using a reduced matrix that matches the model requirements without requiring unnecessary processes to be modeled. The approach most commonly used in process simulators is the interface approach.

4.2 Numerical Solvers

When simulating any conditions, all the differential equations of the model must be solved simultaneously by an integration algorithm. This means that the derivatives must be evaluated and passed to the integrator. Ensuring that numerical integration routines are adequate is crucial for most systems (Olsson and Rosén, 2005), although most users regard numerical integration of the ordinary differential equations as a black-box routine because it is carried out within the simulator software at the simple click of a button, and, for most process design applications, the default solvers in the simulator are good enough. The choice of numerical solver depends on whether the model is steady state or dynamic.

Steady-state models describe mechanisms of wastewater treatment processes as algebraic equations (time is not accounted for), and the solution is found by iterative calculations of the equations. Some examples of steady-state solving routines are the Newton–Raphson method, golden section search, secant method, and Broyden's method (Chapra and Canale, 2009). Alternatively, a steady-state solution can be found by dynamically running the simulation over a long period of simulated time (typically, approximately 3 times the simulated sludge age) until all concentrations become steady.

In dynamic models, mechanisms are described by differential equations. To solve these models, explicit solvers such as Runge–Kutta or Adams–Bashforth are used for nonstiff models and implicit solvers such as Adams–Moulton (Chapra and Canale, 2009), GEAR (Hindmarsh, 1974), or CVODE (https://computation.llnl.gov/casc/sundials/main.html), are used for stiff models. (A system is called *stiff* when the range of the time constants within the model is large. This would occur when some of the systems react quickly and some are significantly slower. This reaction time difference makes the simulation of such a system challenging; as such, special methods have been developed to handle these situations.) Some solvers in commercial wastewater treatment simulators are called *stiff solvers* and, consequently, are capable of solving stiff systems. However, a problem common to all stiff solvers is the difficulty handling dynamic input. The more stochastic or random an input variable is, the more problematic it is for a stiff solver.

4.3 Static and Dynamic Modeling

4.3.1 Steady-State Modeling

The terms *steady-state model*, or *static model*, are used in wastewater process modeling to refer to mathematical models describing equilibrium relationships between model variables independent of time. When used for wastewater process modeling, static models tend to be used to represent average conditions over a relatively long period of time. The definition of *steady state* is intrinsically linked to time scales. For fast processes (i.e., chemical reactions and gas-transfer reactions), time scales to achieve steady-state conditions are typically on the order of minutes. For the slow processes involved in biological wastewater treatment (i.e., bacterial growth), time scales to achieve steady state are typically on the order of days or weeks. Therefore, steady-state modeling in wastewater process modeling typically is used to represent average conditions over long time scales of at least a few weeks or even several months.

Steady-state facility models are useful for design and capacity evaluation purposes. However, model users are cautioned to recognize that actual facility flows, concentrations, and operating conditions vary during the course of the day and from week to week. Consequently, projections from a steady-state model must not be expected to accurately replicate individual samples taken on any particular day. However, these model predictions can be compared with a reasonable estimate of average concentrations seen over a period of time for which the evaluation is being done. When targeting high levels of removal, steady-state modeling of a system subject to diurnal loadings generally tends to be optimistic. A significant deviation between steady-state and dynamic modeling results can occur, depending on the magnitude of diurnal variations and the characteristics of the system (i.e., conservatism of the design).

4.3.2   Dynamic Modeling

The term *dynamic modeling* refers to mathematical models that use differential equations as a function of time. For example, truly dynamic models can provide the expected range of contaminant concentrations in the effluent stream from facilities during the course of a day. Dynamic models use time-varying inputs, such as changing flowrates and influent concentrations, and predict the dynamic response of the process. Dynamic models can also take into account changes in process operation such as internal pumping rates or aeration patterns varying over time. Evaluations using dynamic models are particularly useful for specific issues related to diurnal or daily variations in a facility, such as variations in the required airflow, pumping rates, or expected variations in effluent characteristics. One of the benefits of dynamic modeling is that it allows for the prediction of the behavior of biological systems comprised of bioreactors and final clarifiers under wet weather conditions.

A dynamic model can be used to generate steady-state solutions. When a dynamic model is solved numerically or run for a long time with constant inputs (influent flow, concentrations, return rate, etc.), a steady-state solution is obtained.

5.0   SUMMARY

Models are an approximation of reality. Wastewater treatment processes include biological, chemical, and physical reactions that can be represented with mathematical equations to explain and predict the behavior of a system under a given set of conditions. Therefore, it is critical that model users understand the assumptions and limitations of any model being used to avoid using them outside of their intended scope of conditions. It is also important that model users look for a balance between the degree of complexity of mathematical models and use of model results. The most practical application of wastewater process models is to use the simplest model that can provide the answers needed.

Mathematical models use state variables, kinetic expressions, and stoichiometric relationships to define processes. The wastewater treatment modeling community has "standardized" the use of Gujer matrices to summarize stoichiometry and kinetics associated with each model. Simulation tools are used to implement and solve models, either under steady-state or dynamic conditions. This chapter presented a general overview and background of modeling fundamentals to provide a framework for details presented in subsequent chapters of this manual.

6.0 REFERENCES

Batstone, D. J.; Keller, J.; Angelidaki, I.; Kalyuzhnyi, S.; Pavlostathis, S. G.; Rozzi, A.; Sanders, W.; Siegrist, H.; Vavilin, V. (2002) *Anaerobic Digestion Model No. 1 (ADM1)*; IWA Publishing: London, U.K.

Chapra, S.; Canale, R. (2009) *Numerical Methods for Engineers*, 6th ed.; McGraw-Hill: New York.

Corominas, Ll.; Rieger, L.; Takacs, I.; Ekama, G.; Hauduc, H.; Vanrolleghem, P. A.; Oehmen, A.; Gernaey, K. V.; van Loosdrecht, M. C. M.; Comeau, Y. (2010) New Framework for Standardized Notation in Wastewater Treatment Modeling. *Water Sci. Technol.*, **61** (4), 841–857.

Grady, L., Jr.; Daigger, G. T.; Lim, H. C. (1999) *Biological Wastewater Treatment*, 2nd ed.; Marcel Dekker: New York.

Grau, P.; Copp, J. B.; Vanrolleghem, P. A.; Takács, I.; Ayesa, E. (2009) A Comparative Analysis of Different Approaches for Integrated WWTP Modeling. *Water Sci. Technol.*, **59** (1), 141–147.

Henze, M.; Gujer, W.; Mino, T.; van Loosdrecht, M. (2000) *Activated Sludge Models ASM1, ASM2, ASM2d, and ASM3*; IWA Scientific and Technical Report No. 9; International Water Association: London, U.K.

Hindmarsh, A. C. (1974) *GEAR: Ordinary Differential Equation Solver*, 3rd ed.; National Technical Information Service: Alexandria, Virginia.

International Water Association Task Group on Good Modeling Practice (2008) Plant Model Set-Up and Calibration. *Proceedings of the International Water

Association Task Group on Good Modeling Practice Course on Modeling Activated Sludge Plants; Vienna, Austria, Sept 5–6.

Michaelis, L.; Menten, M. (1913) Die Kinetik der Invertinwirkung. *Biochem. Z.,* **49**, 333–369.

Monod, J. (1942) *Recherches sur la Croissance des Cultures Bactériennes;* Librairie Scientifique: Paris, France.

Olsson, G.; Rosén, C. (2005) *Industrial Automation Applications, Structures, and Systems;* Lund University: Lund, Sweden.

Petersen, E. E. (1965) *Chemical Research Analysis;* Prentice-Hall: Englewood Cliffs, New Jersey.

Wanner, O.; Eberl, H.; Morgenroth, E.; Noguera, D.; Picioreanu, C.; Rittman, B.; Van Loosdrecht, M. (2006) *Mathematical Modeling of Biofilms;* IWA Task Group on Biofilm Modeling, Scientific and Technical Report 18; IWA Publishing: London, U.K.

Water Environment Federation (2010) *Nutrient Removal;* WEF Manual of Practice No. 34; McGraw-Hill: New York.

Chapter 3

Unit Process Model Descriptions

(continued)

1.0 INTRODUCTION

The activated sludge process is the most widely researched and modeled unit process in wastewater treatment. There are many activated sludge models and much information is available to model this process. However, activated sludge models (also known as *suspended growth models*) are not the only process models available to the process designer. This section describes various models that are available for different unit processes, including biofilm models, and their advantages and disadvantages and key inputs and outputs. Detailed descriptions of the processes are not presented. Instead, the reader is directed to relevant sources such as *Design of Municipal Wastewater Treatment Plants* (WEF and ASCE/EWRI, 2010).

2.0 ACTIVATED SLUDGE (SUSPENDED GROWTH) PROCESS MODELS

Carbon, nitrogen, and phosphorus removal processes for biological wastewater treatment have all been incorporated in mathematical models for activated sludge, which describe the reactions involved in bacterial metabolism. While many mathematical models are available in the literature, this section primarily focuses on International Water Association (IWA) models and some of the most used derivations from these models, plus a brief mention of metabolic models.

The IWA Task Group on Mathematical Modeling for Design and Operation of Activated Sludge Processes has published a series of mathematical models that are widely used within the activated sludge modeling community. These models are summarized in Table 3.1, grouped by the significant processes included in each model.

TABLE 3.1 Summary of IWA activated sludge models.

Process / model	Activated Sludge Model No. 1	Activated Sludge Model No. 2	Activated Sludge Model No. 2d	Activated Sludge Model No. 3
Particle hydrolysis	X	X	X	X
Fermentation		X	X	
Carbon storage		X (for biological phosphorus removal [Bio-P] process only)	X (for Bio-P process only)	X
Carbon oxidation	X	X	X	X
Nitrification	X	X	X	X
Denitrification	X	X	X	X
Biological phosphorus removal		X	X	
Biological phosphorus removal with denitrification			X	

The quality of simulation results can be affected by various types of errors in the published model (typing, inconsistencies, or conceptual errors) and/or in the underlying numerical model description. A systematic approach to verify models by tracking typing errors and inconsistencies in model development and software implementation was proposed by Hauduc et al. (2010). In this work, stoichiometry and kinetic rate expressions were checked for seven of the most commonly used activated sludge models. Errors found in these models were reported in spreadsheet format with corrected matrices with the calculations of all stoichiometric coefficients for the discussed biokinetic models. The spreadsheet is presented in Appendix A.

2.1   Activated Sludge Model No. 1

The first model released by the IWA Task Group was Activated Sludge Model No. 1 (ASM1), which incorporated significant processes of interest to water resource recovery facilities (WRRFs), namely carbon oxidation and nitrogen removal (nitrification and denitrification). The model is based on Monod kinetics and well-established biological reactions describing bacterial growth. The model maintains an oxygen balance through the use of chemical oxygen demand (COD). Therefore, the influent is defined in terms of soluble, particulate, biodegradable, and unbiodegradable fractions of the total influent COD. This influent definition based on COD is common to all the IWA activated sludge models. In ASM1, particulate substrates (COD and organic nitrogen) are hydrolyzed and converted to soluble components. Bacterial growth is assumed to occur on soluble substrates only, under aerobic and anoxic conditions.

A spreadsheet with the ASM1 Gujer matrix (model equations) is presented in Appendix A. Bacterial decay is modeled using a death–regeneration approach, assuming that a fraction of the decayed bacterial mass is transformed into inert material and another fraction is transformed into particulate substrate, which, in turn, is hydrolyzed into soluble substrate and then can be used by the active bacterial mass.

2.2   Activated Sludge Model No. 2

Activated Sludge Model No. 2 (ASM2) was later released to incorporate a mathematical description of the biological phosphorus removal process using ASM1 as a base model. The ASM2 includes carbon oxidation, nitrification, denitrification, and biological phosphorus removal. The metabolism of phosphorus accumulating organisms (PAOs) was described using internal storage products, namely Poly-β-hydroxyalkanoate (PHA) and polyphosphate.

While glycogen is believed to play an important role in the metabolism of PAOs, it was not included in the model structure. The ASM2 considered growth of PAOs under aerobic conditions only and did not incorporate the denitrifying metabolism of PAOs. Similar to ASM1, ASM2 used a death–regeneration approach to describe bacterial decay. To minimize the number of state variables, ASM2 and, subsequently, Activated Sludge Model No. 2d (ASM2d) and Activated Sludge Model No. 3 (ASM3) do not have state variables for particulate nitrogen or phosphorus; instead, they use a simplification for describing particulate nitrogen and phosphorus as a ratio of the particulate COD state variables. This can cause difficulties that must be worked around if the influent nitrogen/COD and phosphorus/COD change significantly during of the simulation process (i.e., in dynamic simulations). As a result, ASM2 is no longer used by wastewater treatment practitioners.

The ASM2d (Henze et al., 1999) was later developed to incorporate the denitrifying metabolism of PAOs and is more commonly used today. Growth of PAOs under anoxic conditions was described assuming one population of PAOs growing at a reduced rate under anoxic conditions, similar to the mathematical description of aerobic and anoxic growth of heterotrophic bacteria in ASM1. A spreadsheet with the ASM2d Gujer matrix (model equations) is presented in Appendix A.

2.3   Activated Sludge Model No. 3

The ASM3 does not include the biological phosphorus removal process, but does incorporate models for carbon oxidation and nitrogen removal. The carbon removal process in ASM3 is conceptually different from ASM1. In ASM3, a bacterial growth metabolism is considered based on the internal storage of carbon. Particulate substrate is hydrolyzed (i.e., transformed into soluble substrate) and then stored as an internal product by the heterotrophic bacteria. Bacterial growth is based on the stored substrate only. This approach effectively separates primary hydrolysis (i.e., hydrolysis of particulate substrate in the wastewater) from secondary hydrolysis (i.e., hydrolysis of decayed biomass).

Bacterial decay in ASM3 is also described differently than in ASM1. An endogenous respiration approach is used in ASM3. In this approach, a fraction of the active biomass is transformed into inert material, and an additional oxygen demand is assumed to occur. Unlike in ASM1, decay of biomass in ASM3 does not produce any usable substrate.

A spreadsheet with ASM3 Gujer matrix (model equations) is presented in Appendix A. Arguably, ASM1 remains the most widely used model among this family of models published by IWA (currently, ASM1 remains the most cited of all IWA models.).

2.4   Dold's General Model

Wentzel et al. (1989) developed a dynamic model for PAOs. That model was based on data from "enhanced cultures" developed in continuously flowing biological nutrient removal laboratory-scale systems at the University of Cape Town (UCT) (Cape Town, South Africa). Dold (1991) merged the PAO model with an ASM1-type model (ordinary heterotrophic organisms [OHOs] and one-population nitrifying biomass).

The Barker and Dold (1997) model is commonly referred to as *Dold's General Model* or, simply, the *General Model*. The General Model is based on 19 state variables and 36 process expressions covering the following functional categories:

- Growth and decay of OHOs under aerobic and anoxic conditions on two substrates (readily biodegradable COD [RBCOD] and short chain volatile fatty acids [SCVFA]), that is, BOD removal and denitrification. The model introduces the concept of "reduced yield" for anoxic growth. In addition, growth processes were duplicated for using either ammonia or nitrate (if ammonia was depleted) as the nitrogen (N) source for synthesis.

- Fermentation of complex RBCOD to SCVFA under anaerobic conditions.

- Enmeshment of slowly biodegradable COD (SBCOD) and hydrolysis to RBCOD under aerobic, anoxic, and anaerobic conditions. The stoichiometry for anaerobic hydrolysis included a COD loss component to model reduced sludge production in enhanced biological phosphorus removal (EBPR) systems.

- Hydrolysis of organic nitrogen and ammonification for conversion of soluble organic nitrogen to ammonia.

- Growth and decay of a single-population nitrifier biomass, that is, one-step oxidation of ammonia to nitrate with inorganic carbon for synthesis of biomass (fixing carbon dioxide [CO_2]).

- Growth and decay of PAOs with (a) anaerobic release of phosphate from stored polyphosphate and uptake of volatile fatty acids (VFAs) and storage as PHA and (b) use of stored PHA with uptake of phosphate under both anoxic and aerobic conditions.

The General Model formed the basis for the activated sludge model used in early versions of BioWin simulation software (EnviroSim Associates Ltd., Flamborough, Ontario, Canada; http://www.envirosim.com). Since 1997, the activated sludge part of the General Model has been expanded extensively and refined. The activated sludge biokinetic model has been integrated with an anaerobic digestion model, all within a single model matrix, to allow full-scale facility modeling. Hence, the number of state variables and processes has been extended. Modeling of pH was also included, enabling the effect of pH on process rates to be incorporated, as well as modeling of several chemical precipitation reactions. Gas–liquid mass-transfer behavior for six gases was incorporated to the model. Additions to the activated sludge part of the model include

- Growth and decay of methylotrophs, that is, denitrification using methanol.
- Modification of fermentation processes with acetate, propionate, and dissolved hydrogen as possible products. This allowed for a more fundamental approach to accounting for observed COD loss.
- Considering nitrification as a two-step process with growth and decay of ammonia oxidizing biomass (AOB) oxidizing ammonia to nitrite and growth and decay of nitrite oxidizing biomass (NOB) converting nitrite to nitrate.
- Processes for assimilative reduction of nitrate or nitrate to provide nitrogen for synthesis of biomass if ammonia is depleted.
- Allowing both nitrate and nitrite to act as terminal electron acceptors in OHO denitrification.
- Growth and decay of anaerobic ammonia oxidizers (Anammox), that is, conversion of ammonia and nitrite to nitrogen gas and nitrate using the energy to synthesize organic material from inorganic carbon (fixing carbon dioxide).

The anaerobic digestion part of the model applied in the BioWin simulator is described in Section 6.3 of this chapter.

2.5 New General Model

The New General model, based on Dold's General Model, is used in GPS-X Version 6.1 (Hydromantis, Inc., Hamilton, Ontario, Canada) and has been evaluated by Hauduc et al. (2010).

Among some of the differences, the New General Model includes anoxic decay of PAOs, resulting in anoxic lysis of PHA and stored polyphosphate (releasable) being modeled in a similar way to the aerobic lysis of these variables in the General Model. To account for the lysis of stored polyphosphate, the New General Model includes three new processes: one for aerobic lysis, one for anoxic lysis, and one for anaerobic lysis.

The New General Model includes hydrolysis under anoxic and anaerobic conditions and uses these processes as a sink for "COD loss". In the New General Model, the fermentation of readily biodegradable (soluble) substrate to VFAs is associated with the growth of heterotrophic organisms and, hence, a portion of the fermented COD winds up as new biomass.

For a complete description of the New General Model, the reader is referred to Appendix A.

2.6   Mantis2 Model

Mantis2 (Hydromantis, Inc.) is an activated sludge biological model available in the GPS-X simulation platform. The model includes processes for carbon, nitrogen, and phosphorus removal and anaerobic treatment. The most important features of the models are

- A single set of state variables for both the liquid (activated sludge) and solid (anaerobic digestion) processes

- Two-step nitrification and denitrification processes

- Methylotrophic denitrification on external substrates such as methanol

- Anaerobic ammonium oxidation process

- Precipitation of common precipitates of aluminum, iron, calcium, magnesium, and orthophosphate in the liquid and sludge streams

- The pH estimation in both the liquid and solid streams

- Elemental mass balance for COD, carbon, nitrogen, and phosphorus and other inorganic components like calcium, magnesium, and potassium

- Gas transfer of soluble gases oxygen, carbon dioxide, hydrogen, methane, and nitrogen gas

The underlying structure of the comprehensive model is developed by consistent integration of the processes, kinetics, and stoichiometric descriptions in ASM2d

(Henze et al., 1999), UCTADM1 (Sotemann et al., 2005), the New General Model (Barker and Dold, 1997), ADM1 (Batstone et al., 2002), and the integrated physical/chemical process model of Musvoto et al. (2000). The models for two-step nitrification and two-step denitrification are based on the review provided in a discussion by Sin et al. (2008). The process model for Anammox is based on the stoichiometry and kinetic information provided in a study by Strous et al. (1998). The mass balance in the comprehensive model is extended to COD, carbon, nitrogen, phosphorus, calcium, magnesium, potassium, and charge. The model uses a unified set of 48 state variables (21 soluble plus 27 particulate) and 56 biological, chemical, and physical reactions. Algebraic equations for estimating pH and alkalinity are implemented in the model. The chemical precipitation reactions of precipitation of calcium carbonate ($CaCO_3$), magnesium hydrogen phosphate ($MgHPO_4$), calcium phosphate ($CaPO_4$), aluminum phosphate ($AlPO_4$), iron phosphate ($FePO_4$), and struvite are also included in the model.

2.7 Metabolic Models

Metabolic models have also been proposed to describe bacterial metabolism in activated sludge processes (Roels, 1983). Metabolic models take into account the formation of intermediate compounds in substrate removal pathways and use fixed stoichiometric relationships to describe substrate conversions. Intermediate compounds that are produced, such as adenosine triphosphate (ATP) and nicotinamide adenine dinucleotide (NAD), are considered in metabolic models to derive such stoichiometric relationships. In contrast, the IWA activated sludge models are based on observable reactions, although not much consideration is given to the formation of intermediate compounds along the substrate conversion pathways. Metabolic models are still not as widely used as the IWA family of activated sludge models, but several examples can be found in the literature for both carbon removal (Beun, Paletta, van Loosdrecht, and Heijnen, 2000; Beun, Verhoef, van Loosdrecht, and Heijnen, 2000; Filipe et al., 2001) and EBPR (Murnleitner et al., 1997; Pramanik et al., 1999; Smolders et al., 1995).

3.0 BIOFILM MODELS

Several different types of biofilm models exist including pseudo-analytical, analytical, one-dimensional numerical, two-dimensional numerical, and three-dimensional numerical biofilm models. Consistent with the IWA task group who composed

Scientific and Technical Report Number 18, *Mathematical Modeling of Biofilms* (Wanner et al., 2006), a single utilitarian mechanistic biofilm model is not recognized by the authors of this chapter because selecting a steady-state or dynamic pseudo-analytical, analytical, or numerical biofilm model is objective specific. However, there is general agreement that one-dimensional biofilm models are sufficient for the planning, evaluation, process design, and optimization of proposed or existing biofilm reactors. Unfortunately, there is no consensus as to which one-dimensional biofilm model should be applied for the design of biofilm reactors (Boltz, Morgenroth, and Sen, 2010). In addition, only a framework exists for the calibration of biofilm (reactor) model (Boltz et al., 2012). The selection and use of a biofilm model for process design is dependent on modeling objectives and the capability of the individual modeler. Considering a variety of modeling objectives and modeler capabilities, only existing one-dimensional biofilm models will be discussed in this chapter. Multidimensional biofilm models are presently considered to be tools used exclusively for research.

Biofilm discretization has a considerable effect on simulation results when simulating mixed-culture biofilms (Boltz et al., 2011). However, it cannot be overemphasized that the state variables, transformation processes, kinetic expressions, and stoichiometry that make up the Gujer matrix that is commonly associated with IWA activated sludge models is also an essential component of biofilm models. The Gujer matrix, which is used to mathematically describe biochemical transformation processes inside the biofilm, is incorporated with a model that describes soluble substrate mass-transfer (with Fick's law) and transport processes such as particle movement inside the biofilm and particle attachment and detachment from the biofilm surface. The one-dimensional biofilm models described in this chapter vary in complexity. Pseudo-analytical and analytical biofilm models can be implemented in a commercially available spreadsheet and, when applied to describe mixed-culture biofilms, the commercially available spreadsheet's Solver function may be used as an optimization tool. These simple biofilm models will allow less experienced model users to evaluate the models and understand how variables interact and influence one another. Numerical one-dimensional biofilm models require coding for execution, and, therefore, the modeler must have an additional skill set. However, commercially available software, namely AQUASIM™ (Reichert 1998a, 1998b), exists that will allow biofilm modelers who are unable to code to implement these models.

Essentially, the one-dimensional biofilm models described in this chapter require the same basic information to execute a simulation. The data required include diffusion coefficients of each soluble substrate i in water (e.g., $D_{W,i}$) and inside the biofilm (e.g., $D_{F,i}$), kinetic expressions (e.g., ρ_j), biokinetic parameter values (e.g., q_{max}, K, b, and η), and stoichiometric coefficients (e.g., $v_{j,i}$, $i_{c,i}$, f, and Y). Bioreactor geometric data are needed along with biofilm surface area (A_F), bulk-liquid volume (V_B), mass-transfer boundary layer thickness (L_L), and the influent water flowrate (Q). Information required for biofilm particulate characterization depends on the method of biofilm compartmentalization and discretization. The biofilm model user will need to know the volume fraction of the biofilm liquid phase ($\varepsilon_{S,i}$) and the particulate component concentration ($\rho_{S,i}$) if the model is compartmentalized such that it accounts for biofilm liquid/solid void fractions. If the biofilm model does not account for liquid/solid void fractions, biofilm biomass density is commonly notated as "X_F". In addition, expressions for the attachment (u_{att}) and detachment (u_{det}) velocities of particulate components are required.

3.1 Motivation for Modeling Biofilms

Biofilm reactors and activated sludge systems (or suspended growth reactors) can achieve similar treatment objectives with respect to organic matter removal, nitrification, and denitrification. The same types of microorganisms, exposed to the same local environmental conditions such as electron donor and acceptor, pH, and temperature, are responsible for the governing biochemical transformation processes. Mathematical modeling of biofilms must account for mass transfer because of the effect it has on biofilm structure and function and the resulting effect on biofilm reactor design and operation (Boltz and Daigger, 2010). Microbial competition within the biofilm is based not only on the availability of substrate at the biofilm–liquid interface, but also on detachment and the location of the different groups of bacteria within the biofilm. Bacteria closer to the biofilm surface have the advantage of more direct access to substrates in the bulk of the liquid. On the other hand, bacteria growing further away from the biofilm surface (i.e., nearer the substratum) are given secondary access to substrates in the bulk phase; however, these bacteria are offered more protection from detachment and washout (of the biofilm).

3.2 What Are Biofilms?

Biological treatment processes have the following two conditions in common: (1) active microorganisms have to be concentrated within the system, and (2) microorganisms have to be removed from the treated effluent (WEF et al., 2009). Biofilms consist of microorganisms immobilized in a dense layer growing attached to a fixed or free-moving solid surface. Maintaining a high active biomass concentration in the biofilm (reactor) is independent of settler underflow recycle. Bacteria in biofilms are offered some protection from system washout and can grow in locations where their food supply remains abundant. Whether or not a biofilm will play a dominant role in a system depends on the rate of washout (or the solids retention time [SRT]) of suspended biomass. If the washout rate of suspended biomass is larger than the growth rate of a particular group of microorganisms, then these microorganisms will grow preferentially in a biofilm. With low washout rates, the need to overcome mass-transfer resistances associated with the biofilm may provide less of an incentive for microorganisms to develop inside the biofilm.

Mass transfer of soluble substrates, namely electron donors and electron acceptors, within the biofilm is primarily based on molecular diffusion, which is typically a much slower process than the biochemical transformation of the soluble substrate (of type i). Consequently, soluble substrate concentration gradients develop inside the biofilm. A consequence of the concentration gradients resulting from resistance to mass transfer is that substrate removal inside biofilms is often limited by the rate of mass transfer and not the rate of biochemical transformation. Concentration gradients inside biofilms are far more substantial than those in (suspended biomass) flocs. Therefore, bacteria growing inside the biofilm do not have the same accessibility to soluble substrates (S) or particulate (X) components as the bacteria growing inside the (suspended biomass) flocs. On the other hand, substrate gradients also allow for the development of different ecological niches within the biofilm with different local substrate and electron acceptor concentrations and some protection inside the biofilm from toxic or inhibitory substances (Morgenroth, 2008). Understanding the interactions between mass transfer and substrate conversion processes is necessary to understand the overall performance of biofilm systems.

3.3   Why Model Biofilms?

Mathematical modeling can be used to describe certain features of a biofilm system (e.g., a bioreactor) by selecting and solving mathematical expressions. The following questions are relevant for biofilm reactor design and operation:

- What is the rate of soluble substrate removal as a function of concentrations in the bulk of the liquid at any predetermined location within a biofilm reactor?

- To what extent does the available biofilm surface area (A_F), total amount of biofilm biomass (X_F), or specific microorganism presence within the biofilm control particulate (X) and soluble substrate (S) removal?

This section will evaluate mathematical modeling approaches that can help address some of these questions, particularly in the context of WRRF simulation and biofilm reactor design.

3.4 Existing Biofilm Models

Biofilm models are more complex than process models used to describe the suspended growth compartment. Three approaches to modeling biofilm systems are a graphical technique, empirical and semi-empirical models, and mathematical one-dimensional biofilm models. The mathematical one-dimensional biofilm models summarized in this section include pseudo-analytical (Rittmann and McCarty, 1980a and 1980b), analytical (e.g., Harremoës [1978]), and numerical one-dimensional biofilm (Wanner and Reichert, 1996) models.

Biofilm models can be generally categorized by their objective, that is, mechanistic biofilm models that are used primarily as research tools or those used for engineering practice (Noguera et al., 2004). Wanner et al. (2006) described five significant biofilm model classes: pseudo-analytical, analytical, one-dimensional, two-dimensional, and three-dimensional numerical. Harremoës (1978) presented analytical solutions to the diffusion-reaction equations based on an assumed homogeneous biofilm biomass distribution with the biofilm thickness, L_F, and biomass density, X_F, known a priori. Sáez and Rittmann (1992) developed a pseudo-analytical biofilm model that described the flux of a soluble substrate across the surface of a biofilm ($J_{LF,i}$) of any thickness (L_F). Subsequently, dynamic one-dimensional biofilm models that account for biofilm heterogeneity and rely on complex numerical methods were advanced by the work of Kissel et al. (1984), Wanner and Gujer (1986), and Wanner and Reichert (1996). Dynamic biofilm models account for the development of microbial species over the entire biofilm thickness (L_F) in response to local substrate availability and microbial growth rates. Advanced mathematical descriptions of biofilms include cellular automata (Noguera et al., 2004; Picioreanu et al., 1998) and individual-based (Kreft et al., 2001) multidimensional biofilm models. Multidimensional biofilm models have been used to estimate the influence of biofilm structure on local fluid dynamics and mass transport external to the biofilm (e.g., Eberl et al. [2000]).

A trend of increasing model complexity, longer computational period, and significant deviation from the mass-based approach inherent to IWA activated sludge models limits the present usefulness of multidimensional numerical biofilm models singularly to research. The models presented in this progression (i.e., pseudo-analytical to individual-based and deterministic cellular automata approaches) follow an inductive

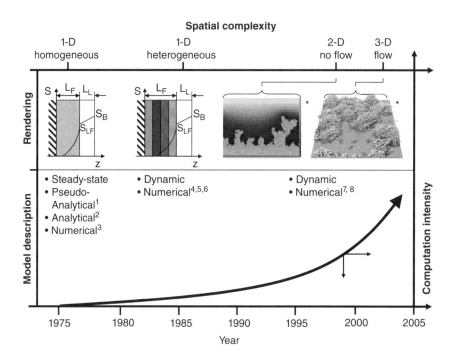

FIGURE 3.1   Biofilm models have become more complex and more computationally intensive during the past 30 years. http://www.biofilms.bt.tudelft.nl/material .html, [1]Williamson and McCarty (1976), [2]Harremoës (1978), [3]Harris and Hansford (1978), [4]Kissel et al. (1984), [5]Wanner and Gujer (1986), [6]Wanner and Reichert (1996), [7]Picioreanu et al. (1998), [8]Picioreanu et al. (2003) (Boltz, Morgenroth, and Sen [2010]).

approach to biofilm modeling, with the general direction being from most simplistic to complex mathematical models. Figure 3.1 conceptually illustrates the developmental sequence of increasingly complex biofilm models. During the past 40 years, numerous advances in mathematical modeling of biofilms have expanded the fundamental understanding of biofilms. In addition, the inductive development of increasingly complex biofilm models has made possible the deductive analysis of more simplistic one-dimensional biofilm models. Morgenroth et al. (2000) used such a deductive approach by comparing three-dimensional and one-dimensional numerical biofilm models. More recently, Pérez et al. (2005), with corrigendum by Gapes et al. (2006), compared an analytical biofilm model based on composite first- and zero-order kinetics to complex biofilm models including a numerical one-dimensional biofilm model implemented in the AQUASIM™ platform. These deductive modeling studies have demonstrated that one-dimensional biofilm models are capable of accurately simulating the flux of a soluble substrate i across the surface of a biofilm ($J_{LF,i}$).

Wuertz and Falkentoft (2003) stated that the use of biofilm models in engineering practice required the creation of simplified models, that is, biofilm models that are based on simplifying assumptions characteristic of the biofilm system to be simulated. However, model simplicity is less important than developing biofilm models that are sufficiently complex (in terms of accounting for the processes and variables that are a required to completely evaluate parameters of interest to process designers) and capable of answering questions relevant to practicing engineers. The biofilm model must capture essential fundamental mechanisms in the simplest fashion possible, yet account for full-scale process diversity and complexity that is typically associated with biofilm-reactor-based full-scale WRRFs. Comprehensive simplified biofilm (reactor) models have been developed and used in engineering design (see Boltz, Johnson, Daigger, and Sandino [2009]; Boltz, Johnson, Daigger, Sandino, and Elenter [2009]; Rauch et al. [1999]; and Sen and Randall [2008a and 2008b]). In addition to comprehensive simplified biofilm (reactor) models, alternative biofilm reactor modeling techniques exist including graphical techniques and empirical and semi-empirical models.

3.5   Graphical Biofilm Models

Graphical procedures can be used to ascertain only a fraction of the information that may be gleaned from mechanistic mathematical biofilm models. For example, graphical procedures can be used to determine the total hydraulic load (THL) required to decrease a substrate concentration and, by definition, the biofilm area, A_F, required for a system to treat influent wastewater until the desired concentration of a soluble substrate i remains in the effluent stream. These items can be determined directly. The graphical procedure can be used to determine effluent substrate concentration from any series of continuously flowing stirred tank reactors (CFSTRs), but a stepwise procedure must be used when a series of CFSTRs is analyzed. Additional information describing the graphical procedure summarized here can be found elsewhere (see Antoine [1976]; WEF et al. [2009]; and Grady et al. [2011]). This graphical procedure is valid for any biofilm-based CFSTR; however, should the CFSTR be comprised of multiple stages with different characteristics, it will be necessary to establish different flux-response curves to describe each stage of the CFSTR series.

This procedure requires a graphical representation of soluble substrate i flux ($J_{LF,i}$) as a function of the bulk-phase concentration of substrate i ($S_{B,i}$). The process designer should recognize that the relationship between flux and the bulk-liquid concentration of substrate i is based on a specific system and location. Therefore, the flux-response curve required to implement the graphical procedure may not be obtained from, or correlate well with, values reported in the literature or from different systems. As a result, the process designer should carefully consider the conditions under which the flux-response curve was developed before applying results. Ultimately, the flux-response curve incorporates system-specific mass-transfer characteristics, operating mode, and environmental conditions under which the specific system was operated. Therefore, the flux-response curve may not be representative of different biofilm reactor types designed to meet the same treatment objective nor may it be representative of the same biofilm reactor type if it is designed for different operating conditions. However, a flux-response curve generated for the same biofilm reactor type, but under different operating conditions, may still offer some indication of expected performance in the absence of a system-specific numerical simulation or pilot- or full-scale test system performance observations.

When using the graphical procedure to evaluate pilot-scale facility observations, fluxes should be compared to rates in full-scale systems. Pilot or experimental systems may promote a greater flux than expected. Therefore, any flux that deviates significantly from those reported in published studies for similar biofilm reactors should be used only after careful consideration. The basis for the graphical procedure is a material balance on a soluble substrate i in a biofilm-based CFSTR, as follows:

$$0 = \underbrace{Q \cdot S_{\text{in},i}}_{\substack{\text{mass per} \\ \text{time input}}} - \underbrace{Q \cdot S_{\text{B},i}}_{\substack{\text{mass per} \\ \text{time output}}} - \underbrace{J_{\text{LF},i} \cdot A_{\text{F}}}_{\substack{\text{biofilm} \\ \text{transformation rate}}} - \underbrace{r_{\text{B},i} \cdot V_{\text{B}}}_{\substack{\text{suspended growth} \\ \text{transformation rate}}} \tag{3.1}$$

where

Q = flowrate through the system, m^3/d

$S_{\text{in},i}$ = influent concentration of soluble substrate i, g/m^3

$S_{\text{B},i}$ = effluent, or bulk-liquid, concentration of soluble substrate i, g/m^3

$J_{\text{LF},i}$ = flux of soluble substrate i across the biofilm surface ($\text{g}/\text{m}^2 \cdot \text{d}$)

A_{F} = biofilm surface area, m^2

$r_{\text{B},i}$ = rate of soluble substrate i conversion because of suspended biomass, $\text{g}/\text{m}^3 \cdot \text{d}$

V_{B} = bulk-liquid volume, m^3

Assuming that transformation occurring in the bulk liquid is negligible, the "suspended growth transformation rate" (in Eq. 3.1) can be neglected.

Rearranging Eq. 3.1 yields Eq. 3.2, which provides rationale for the graphical procedure, as follows:

$$J_{LF,i} = \underbrace{\frac{Q}{A_F} \cdot S_{in,i}}_{\text{constant}} - \underbrace{\frac{Q}{A_F} \cdot S_{B,i}}_{\text{slope}}$$

(3.2)

The slope, or $\left(-\dfrac{Q}{A}\right)$, is referred to as the *operating line* and represents the THL on each stage. Figure 3.2 illustrates the graphical method. The flux curves have been created based on observations in the first and second stage of a postdenitrification biofilm reactor. The ordinate represents nitrate-nitrogen flux and the abscissa nitrate-nitrogen concentration remaining in the effluent stream. The graphical procedure depends on the substrate flux curve(s). The method requires development of multiple flux curves if the performance characteristics of respective stages vary significantly.

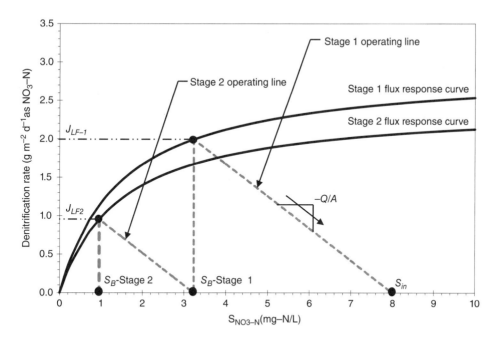

FIGURE 3.2 Example graphical procedure for describing the response of a biofilm reactor to defined conditions (see WEF and ASCE/EWRI [2009] for design example), including (1) first- and second-stage operating lines and (2) flux curves based on observations at a pilot-scale denitrification biofilm reactor.

When using pilot-scale facility data to generate a flux-response curve, appropriate scale considerations must be given when designing the pilot unit and experiments.

3.6 Empirical and Semi-Empirical Models

Empirical models can be implemented easily by hand or by using a spreadsheet, but they have limited applicability because of their "black box" consideration of system parameters. Because environmental conditions and bioreactor configuration affects biofilm reactor performance, a system can respond differently from the description provided by an empirical model. Significant sources of variability in values include differences in biofilm carrier type and configuration, the extent of concentration gradients external to the biofilm surface, and biofilm composition. Empirical models can produce results that vary 50% to 100% of actual system performance.

Empirical models are inadequate for describing complex processes such as the explicit evaluation of two-step ammonium oxidation first to nitrite by ammonia oxidizing bacteria and then to nitrate by nitrite oxidizing bacteria. Therefore, empirical models have limited application in defining the conditions that either promote or deter complex processes in biological systems.

Historically, biofilm reactors have been designed using empirical criteria and design formulations (or empirical and semi-empirical models), but this trend is changing. One should recognize that the coefficients in empirical models describing biofilm reactors include system and, often, location-specific mass-transfer resistances (Grady et al., 2011). For this reason, the values typically differ from apparent or intrinsic values reported in the literature. Once a suitable flux value ($J_{\mathrm{LF},i}$) has been determined, Eq. 3.1 can be rearranged, neglecting bulk-phase conversion processes, to calculate the material concentration remaining in the effluent stream, which is expressed as follows:

$$S_{\mathrm{B},i} = S_{\mathrm{in},i} - J_{\mathrm{LF},i} \cdot \frac{A_{\mathrm{F}}}{Q} \qquad (3.3)$$

If sufficient data exists to allow for the development of parameter values and mathematical relationships capable of describing a complete range of conditions expected when treating municipal wastewater, then empirical models can be used reliably. The addition of model components to account for specific phenomena encroaches on the premise of mechanistic mathematical model development. For this reason, a distinction is made here between empirical and semi-empirical models. Gujer and Boller (1986) and Sen and Randall (2008a and 2008b) provide examples of semi-empirical models describing nitrifying trickling filters and moving bed biofilm reactors/integrated fixed-film activated sludge systems, respectively.

3.7   Mathematical Biofilm Models

Mathematical modeling can be used to describe certain features of a biofilm or biofilm system (such as a bioreactor) by selecting and solving mathematical expressions. Mathematical biofilm models have been used extensively as research tools for more than 40 years, but only recently (i.e., during the past 10 years) have mechanistic biofilm models been used in engineering practice.

3.7.1   Approaches to Mathematical Modeling of Biofilms

A difficult aspect of using biofilm models in engineering practice is selecting the appropriate level of model complexity. The following are summaries of different modeling approaches (WEF, 2010):

- Zero-dimensional biofilm—One aspect of modeling biofilms is that bacteria are retained in the system and are not washed out with the effluent stream. The simplest approach for biofilm modeling would be to assume that all biomass in the reactor is exposed to dissolved and particulate substrate concentrations in the bulk of the liquid, neglecting the effect of mass-transfer limitations (i.e., a zero-dimensional biofilm model). In full-scale wastewater treatment scenarios, biofilms may be relatively thick and are typically mass-transfer limited. Thus, the zero-dimension modeling approach that neglects mass-transfer limitations is not useful except in limited special cases.

- One-dimensional homogeneous biofilm (with a single limiting substrate)—This approach takes into account mass-transfer limitations in the biofilm and the corresponding effects on concentration profiles and substrate flux into the biofilm. It is assumed that active bacteria are homogeneously distributed over the biofilm thickness. This approach is valid only if calculations are performed for the rate-limiting substrate that has to be determined a priori by the user, as described in Morgenroth (2008). The flux of the nonlimiting substrates can be calculated based on reaction stoichiometry.

- One-dimensional homogeneous biofilm (multiple substrates and multiple biomass components)—One key aspect of modeling biofilms is to evaluate the competition and coexistence of different bacterial groups and local environmental conditions. Local process conditions can be accurately determined by calculating penetration depths for different soluble substrates. Based on the fluxes, growth of individual groups of bacteria can be determined. To simplify calculations, it can be assumed that all bacterial groups are homogeneously distributed over the biofilm thickness (Boltz, Johnson, Daigger, and Sandino, 2009; Rauch et al., 1999).

- One-dimensional heterogeneous biofilm—Different bacterial groups compete in a biofilm, not only for substrate but also for space, where bacteria toward the surface are less influenced by mass-transfer limitations. Bacteria growing toward the base of the biofilm are often rate-limited by substrate availability resulting from mass-transfer limitations. On the other hand, these bacteria are better protected from (surface) detachment. These one-dimensional heterogeneous biofilm models must keep track of local growth and decay of the different bacterial groups and detachment to calculate biomass distributions over the biofilm thickness (L_F).

- Two-dimensional and three-dimensional biofilm models—From a practical standpoint, biofilms are not as smooth and flat as is assumed in one-dimensional biofilm models. Mathematical models have been created that predict the development of biofilms in two or three dimensions, the influence of the heterogeneous structure on fluid flow, and, ultimately, the combination of fluid flow and biofilm structure on substrate availability and removal inside the biofilm. Multidimensional biofilm models are not necessary for most questions related to biofilm reactors. However, it is important for model users to recognize that biofilm structure influences system hydrodynamics and external mass transport (through the mass-transfer boundary layer, J_{MTBL}), which are simultaneously affected by biofilm reactor appurtenances and mode of operation. Such interactions are not accounted for in existing one-dimensional biofilm models because of the rigid segregation of the bulk-liquid mass-transfer boundary layer and biofilm (which is assumed to have a uniform thickness and smooth surface). Multidimensional biofilm models have been used to quantify the influence of biofilm structure on local fluid dynamics and external mass transport (Eberl et al., 2000).

Other than the fact that one-dimensional biofilm models are sufficient for biofilm reactor analyses in engineering practice, no simple or general recommendations can be given as to what approach is the most appropriate for describing biofilm reactors. In *Mathematical Modeling of Biofilms*, Wanner et al. (2006) provide a detailed description of different modeling approaches and a discussion of how the modeling approaches compare to different modeling scenarios. Many commercially available WRRF simulators used for biofilm reactor design and evaluation take into account multiple substrates and biomass fractions in either a heterogeneous or homogeneous one-dimensional biofilm.

3.7.2 Basic Equations Governing Mathematical Biofilm Models

Submerged and completely mixed biofilm reactors facilitate the application of state-of-the-art knowledge and understanding of biofilm mechanics. Moreover, modern biofilm reactors have characteristics that are amenable to simulation with existing one-dimensional biofilm models (Boltz and Daigger, 2010; Boltz et al., 2011). As a result, most existing WRRF simulators now include biofilm reactor modules that are based on the mathematical description of a one-dimensional biofilm. Table 3.2 lists software packages and references for the incorporated biofilm model that constitutes the biofilm reactor module.

Biofilm models are a quantitative representation of what is known about biofilms. Proper use of mechanistic biofilm models for reactor design is dependent on a user understanding model structure and components, governing equations, and the input that is required to obtain accurate biofilm model output that may be applied to biofilm reactor design and operation. A mechanistic biofilm model user should understand the model basis, its underlying assumptions, and limitations inherent to the modeling approach before applying the model to research or biofilm reactor design.

A biofilm schematic is illustrated in Figure 3.3. The schematic illustrates diffusion and reaction occurring inside a one-dimensional biofilm. In addition, concentration gradients external to the biofilm surface are illustrated in the manner that they are modeled, namely as an external mass-transfer resistance through the "mass-transfer boundary layer". The partial differential equation describing molecular diffusion, substrate use inside a biofilm, and the dynamic accumulation of a single rate-limiting soluble substrate i is presented as follows:

$$\underbrace{\frac{\partial S_{F,i}}{\partial t}}_{\text{accumulation}} = \underbrace{D_{F,i} \frac{\partial^2 S_{F,i}}{\partial x^2}}_{\text{diffusion}} - \underbrace{r_{F,i}}_{\substack{\text{biochemical} \\ \text{reaction}}} \tag{3.4}$$

where
$S_{F,i}$ = concentration of soluble substrate i in the biofilm, g/m^3
x = distance from the biofilm surface, m
t = time, d
$D_{F,i}$ = soluble substrate i diffusion coefficient inside the biofilm, m^2/d
$r_{F,i}$ = rate of soluble substrate i conversion per biofilm volume, $g/m^3 \cdot d$

TABLE 3.2 Water resource recovery facility simulators—biofilm model characteristics and citations (modified WEF and ASCE/EWRI, 2009).

Software	Company	Biofilm model type and biomass distribution*	Reference
AQUASIM™	EAWAG, Swiss Federal Institute of Aquatic Science and Technology, Dübendorf, Switzerland (www.eawag.ch/index_EN)	1-D, DY, N, Heterogeneous	Wanner and Reichert (1996) (modified)
AQUIFAS™	Aquaregen, Mountain View, California (www.aquifas.com)	1-D, DY, SE, N, Heterogeneous	Sen and Randall (2008a, 2008b)
BioWin™	EnviroSim Associates Ltd., Flamborough, Canada (www.envirosim.com)	1-D, DY, N, Heterogeneous	Wanner and Reichert (1996) (modified), Takács et al. (2007),
GPS-X™	Hydromantis Inc., Hamilton, Ontario, Canada (www.hydromantis.com)	1-D, DY, N, Heterogeneous	
Pro2D™	CH2M HILL Inc., Englewood, Colorado (www.ch2m.com/corporate)	1-D, SS, N(A), Homogeneous (constant L_F)	Boltz, Johnson, Daigger, and Sandino (2009); Boltz, Johnson, Daigger, Sandino, and Elenter (2009)
Simba™	ifak GmbH, Magdeburg, Germany (www.ifak-system.com)	1-D, DY, N, Heterogeneous	Wanner and Reichert (1996) (modified)
STOAT™	WRc, Wiltshire, U.K. (www.wateronline.com/storefronts/wrcgroup.html)	1-D, DY, N, Heterogeneous	Wanner and Reichert (1996) (modified)
WEST™	MIKE by DHI, Hørsholm, Denmark (www.mikebydhi.com)	1-D, DY, N(A), N, Homogeneous, Heterogeneous	Rauch et al. (1999), Wanner and Reichert (1996) (modified)

*1-D = one dimensional; DY = dynamic; N = numerical; and N(A) = numerical solution using analytical flux expressions.

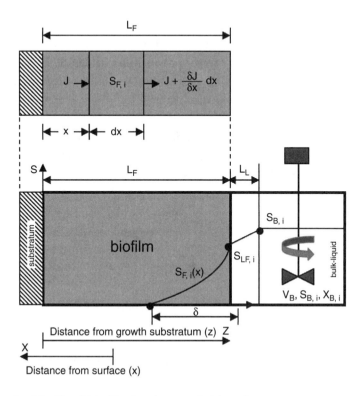

FIGURE 3.3 An idealized biofilm having a soluble substrate concentration of component i ($S_{F,i}$) is moved and may be limited by molecular diffusion (top). This transport process occurs in an idealized one-dimensional biofilm with an assumed homogeneous biomass distribution (bottom). The biofilm has a uniform thickness (L_F). A soluble substrate concentration profile of component i is illustrated with a completely mixed bioreactor's bulk-liquid substrate concentration decreasing through a mass transfer boundary layer of thickness, L_L, until reaching the liquid–biofilm interface. Here, the interfacial soluble substrate concentration of component i ($S_{LF,i}$) exists. Then, the concentration of soluble substrate component i decreases in one-dimension (perpendicular to the growth substratum) through the biofilm where the soluble component i concentration is defined at any point inside the biofilm as $S_{F,i}(x)$. The conceptual homogeneous biofilm is grown in a submerged, completely mixed biofilm reactor that is characterized by a bulk-liquid volume (V_B), a soluble substrate concentration of component i ($S_{B,i}$), and a particulate concentration of component i ($X_{B,i}$).

It should be emphasized that the basis for a mathematical description of the one-dimensional biofilm, as described with Eq. 3.4, is simultaneously occurring molecular diffusion and biochemical reaction. Molecular diffusion is based on Fick's second law of diffusion. A hyperbolic saturation-type kinetic expression such as the Monod equation is typically applied to describe the biochemical transformation rate ($r_{F,i}$).

Two boundary conditions are used to derive constants that are required to solve the second-order partial differential equation (i.e., Eq. 3.4). Boundary condition 1 (BC1) and boundary condition 2 (BC2) are expressed mathematically by Eq. 3.5 and Eq. 3.6, respectively, as follows:

$$BC1: \frac{dS_{F,i}}{dx} = 0 \quad \text{when} \quad x = L_F \tag{3.5}$$

and

$$BC2: S_{F,i} = S_{LF,i} \text{ when } x = 0 \tag{3.6}$$

The flux of any soluble substrate i at a point inside the biofilm $[J(x)]$ is proportional to the concentration gradient at a specific location (x) inside the biofilm, according to the following equation:

$$J_{F,i}(x) = -D_{F,i} \frac{dS_{F,i}(x)}{dx} \tag{3.7}$$

Using Eq. 3.7, the flux of any soluble substrate i through the biofilm surface can be calculated by the following equation:

$$J_{LF,i} = -D_{F,i} \frac{dS_{F,i}}{dx} \text{ at } x = 0 \tag{3.8}$$

The flux of a soluble substrate i across the biofilm surface ($J_{LF,i}$), as presented in Eq. 3.8, will be used in material balances for the overall biofilm reactor. Equation 3.8 is dependent on knowing the soluble substrate i concentration at the biofilm-liquid interface. Soluble substrate i concentration(s) gradients are known to exist in the vicinity of the biofilm surface. Figure 3.4 illustrates this condition for dissolved oxygen concentration gradients in the vicinity of the surface of a biofilm grown in a flow chamber.

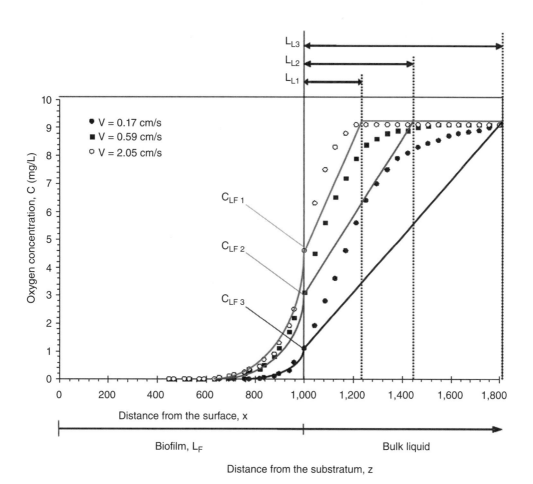

FIGURE 3.4   Dissolved oxygen concentration microprofiles measured inside biofilms grown in a flow chamber subjected to varying flow conditions. Increasing the chamber volumetric flowrate increased liquid velocity in the vicinity of the biofilm surface. Consequently, the concentration gradient external to the biofilm surface decreased. The least flow velocity resulted in the least dissolved component i concentration at the liquid–biofilm interface. The straight lines illustrate a modeling trend, which simulates the concentration gradient external to the biofilm surface as a straight line (WEF and ASCE/EWRI, 2009; raw data by Zhang and Bishop, 1994).

The concentration gradient external to the biofilm surface is not explicitly modeled. Rather, it is modeled as a mass-transfer resistance according to the following equation:

$$J_{\text{MTBL},i} = \frac{1}{R_{\text{L}}}(S_{\text{B},i} - S_{\text{LF},i})$$

(3.9)

where

$J_{\text{MTBL},i}$ = soluble substrate i flux through the mass-transfer boundary layer, g/m²·d

R_{L} = mass-transfer resistance external to the biofilm

Additional methods for modeling external mass-transfer resistances are summarized by Kissel (1986). Defining a conceptual mass-transfer boundary layer thickness provides a more intuitive understanding when compared with the mass-transfer resistance (R_{L}). Resistance to mass transfer and the mass-transfer boundary layer thickness (L_{L}) are related according to the following equation:

$$R_{\text{L}} = \frac{L_{\text{L}}}{D_{\text{W},i}}$$

(3.10)

where

L_{L} = mass-transfer boundary layer thickness, m

$D_{\text{W},i}$ = soluble substrate i diffusion coefficient in water, m²·d⁻¹

The soluble substrate i flux through the mass-transfer boundary layer is linked to its flux across the biofilm surface. This provides an additional boundary condition, as follows, that is required to calculate the unknown value of the substrate concentration at the liquid–biofilm interface ($J_{\text{LF},i}$):

$$\text{BC3: } J_{\text{MTBL},i} = J_{\text{LF},i}$$

(3.11)

3.8   Mass Balances—Particulate and Dissolved Components in the Bulk Phase

The simplest application of biofilm models assumes that the simulated biofilm grows in a CFSTR (i.e., the biofilm is completely submerged and the bulk of the liquid is completely mixed) and bulk-liquid dissolved oxygen concentration (in an aerobic biofilm reactor) is kept constant by aeration. The following mass balance can be used to describe the fate of particulate and soluble substrates:

$$\text{Accumulation} = \text{Input} - \text{Output} - \text{Substrate utilization} \qquad (3.12)$$

Under steady-state conditions, the "accumulation" term is equal to zero. Accounting for biochemical transformation processes resulting from both suspended growth and biofilm compartments, a general mass balance on particulate and soluble substrates in a submerged, completely mixed (biofilm) reactor is represented by Eq. 3.13 and Eq. 3.14, respectively, as follows:

$$\underbrace{\frac{d}{dt}(V_B \cdot S_{B,i})}_{\text{accumulation}} = \underbrace{Q \cdot S_{in,i}}_{\text{input}} - \underbrace{Q \cdot S_{B,i}}_{\text{output}} + \underbrace{V_B \cdot (r_{AS,i} + r_{F,i} \cdot \varsigma)}_{\text{dissolved subsrate utilization}} \qquad (3.13)$$

and

$$\underbrace{\frac{d}{dt}(V_B \cdot X_{B,i})}_{\text{accumulation}} = \underbrace{Q \cdot X_{in,i}}_{\text{input}} - \underbrace{Q \cdot X_{B,i}}_{\text{output}} + \underbrace{V_B \cdot (r_{AS,i} + r_{F,i} \cdot \varsigma)}_{\text{particulate substrate utilization}} \qquad (3.14)$$

where

Q = is the liquid volumetric flowrate, m^3/d
V_B = bulk-liquid volume which biochemical transformation processes occur, m^3
$S_{in,i}$ = influent concentration of the soluble substrate i, g/m^3
$S_{B,i}$ = bulk-liquid concentration of the soluble substrate i, g/m^3
$r_{AS,i}$ = substrate i suspended biomass utilization rate, g COD/m^3 bioreactor·d
$r_{F,i}$ = substrate i biofilm utilization rate, g COD/m^3 bioreactor·d
$X_{in,i}$ = particulate component i influent concentration, g/m^3
$X_{B,i}$ = particulate component i bulk-liquid concentration, g/m^3
ς = empty-bed volume biofilm carrier fraction (in the range 0 to 1) (dimensionless)

The biofilm carrier volume displacement factor (Φ) accounts for a reduction in the bulk-liquid (or reaction) volume because of empty-bed volume displacement by the biofilm carrier (dimensionless), and can be calculated as follows:

$$\Phi = \frac{V_R - V_D}{V_R} \qquad (3.15)$$

Here, V_R is the bulk-liquid volume in the empty bioreactor (m^3) and V_D is the bulk-liquid volume displaced by the biofilm carrier(s) (m^3). Similarly, bulk-liquid volume may be further displaced by the biofilm itself. The total bulk-liquid volume displaced by the biofilm carrier(s) and the biofilm itself can be calculated as follows:

$$V_B = V_R - \underbrace{\Phi \cdot V_R}_{\substack{\text{bulk-liquid volume} \\ \text{displaced by carriers}}} - \underbrace{A \cdot L_F}_{\substack{\text{bulk-liquid volume} \\ \text{displaced by biofilm}}} \qquad (3.16)$$

where
  A = biofilm surface area, m^2
  L_F = biofilm thickness, m
  Φ = biofilm carrier volume displacement factor

Biofilm carriers are characterized by their bulk specific surface area, net specific surface area, bulk-liquid volume displacement, and net liquid volume displacement. These terms are defined as follows:

- Bulk specific surface area—biofilm area per unit volume of biofilm carriers or

$$\left(\frac{\text{m}^2 \text{ of biofilm}}{\text{m}^3 \text{ of biofilm carrier}} \right) \qquad (3.17)$$

- Net specific surface area—biofilm area per unit bioreactor volume or

$$\left(\frac{\text{m}^2 \text{ of biofilm}}{\text{m}^3 \text{ of reactor volume}} \right) \qquad (3.18)$$

- Bulk-liquid volume displacement—liquid volume displaced per unit volume of biofilm carriers or

$$\left(\frac{\text{m}^3 \text{ of liquid displaced}}{\text{m}^3 \text{ of biofilm carrier}} \right) \qquad (3.19)$$

- Net-liquid volume displacement—liquid volume displaced per unit bioreactor volume or

$$\left(\frac{\text{m}^3 \text{ of liquid displaced}}{\text{m}^3 \text{ of reactor volume}} \right) \qquad (3.20)$$

The net specific surface area is defined as

$$a \left(= \frac{A_F}{V_R} \right) \left(\text{m}^2_{\text{biofilm}} \ \text{m}^{-3}_{\text{bioreactor}} \right) \qquad (3.21)$$

Applying the specific surface area and converting the biofilm reaction rate to more explicit terms results in the following definition:

$$V_{B} \cdot r_{F,i} = J_{LF,i} \cdot A_{F} \tag{3.22}$$

where $J_{LF,i}$ is the flux of soluble substrate i across the biofilm surface $(g_{COD}/m_{F}^{-2} \cdot d^{-1})$. A bulk-liquid reaction volume-based biofilm rate expression is obtained by applying the active biofilm growth media specific surface area, a $(m_{biofilm}^{2}/m_{bioreactor}^{3})$, rather than the biofilm surface area, A_{F} $(m_{biofilm}^{2})$. As a result, Eq. 3.13 and Eq. 3.14 can be modified for dissolved and particulate components, respectively, into Eq. 3.23 and Eq. 3.24, as follows:

$$\underbrace{\frac{d}{dt}(V_{B} \cdot S_{B,i})}_{\text{accumulation}} = \underbrace{Q \cdot S_{in,i}}_{\text{input}} - \underbrace{Q \cdot S_{B,i}}_{\text{output}} + \underbrace{(V_{B} \cdot r_{AS,i})}_{\text{dissolved substrate utilization}} + \underbrace{J_{LF,i} \cdot A_{F} \cdot \varsigma}_{\text{dissolved substrate utilization}} \tag{3.23}$$

and

$$\underbrace{\frac{d}{dt}(V_{B} \cdot X_{B,i})}_{\text{accumulation}} = \underbrace{Q \cdot X_{in,i}}_{\text{input}} - \underbrace{Q \cdot X_{B,i}}_{\text{output}} + \underbrace{(V_{B} \cdot r_{AS,i})}_{\text{particulate substrate utilization}} + \underbrace{J_{LF,i} \cdot A_{F} \cdot \varsigma}_{\text{particulate substrate utilization}} \tag{3.24}$$

3.9 Pseudo-Analytical Biofilm Models

Closed-form analytical solutions are only available for simplified situations including homogeneous biofilm biomass distribution (i.e., uniform X_{F}), known biofilm thickness (L_{F}), and explicitly analytical solutions evaluating the flux of a single soluble substrate i according to first-order or zero-order kinetics as defined by the model user. A pseudo-analytical biofilm model is a simple alternative when one or more of the simplifications must be eliminated (Wanner et al., 2006). Pseudo-analytical solutions consist of a small set of algebraic equations that can be solved by hand or by using a spreadsheet. Solution outputs include the flux of a soluble substrate i ($J_{LF,i}$) when inputting the bulk-liquid concentration of soluble substrate i ($S_{B,i}$). The relative ease of using a pseudo-analytical biofilm model makes the set of equations particularly useful for routine application in process design and as a teaching tool (Grady et al., 2011; Rittmann and McCarty, 2001).

Pseudo-analytical biofilm model solutions first evolved from work by Atkinson and Davies (1974) and Atkinson and How (1974), who found a pseudo-analytical solution for instances when nonlinear Monod kinetics represent rate limitation by a single dissolved component i. Rittmann and McCarty (1980) included a biomass balance so that the pseudo-analytical biofilm model predicts a steady-state biofilm thickness. Rittmann and McCarty (1980) incorporated an external mass-transfer resistance to Monod kinetics. Finally, Sáez and Rittmann (1992) presented the latest and most accurate pseudo-analytical solution for steady-state biofilms that can be applied to describe multispecies biofilms. Table 3.3 lists a step-by-step set of instructions for using the pseudo-analytical biofilm model of Sáez and Rittmann (1992). Appendix B also provides a comprehensive description of the pseudo-analytical biofilm model of Sáez and Rittmann (1992).

3.10   Analytical Biofilm Models

The analytical approach is the simplest solution to the general biofilm model. The analytical solution to the biofilm model is unique in that it is obtained by mathematical derivation without any numerical techniques. This allows the model user to independently analyze the effects of each term, variable, and parameter (e.g., mass-transfer and biokinetic expressions). A disadvantage inherent to the analytical biofilm model is that simulating the effect of multiple components in the kinetic expression, complex spatial biofilm geometries, and temporal dynamics is difficult, if not impossible, to account for with an analytical biofilm model (Wanner et al., 2006). However, the analytical biofilm model may be modified by including numerical techniques to overcome, at least to some extent, some of the aforementioned limitations inherent to the analytical biofilm model. Basic equations resulting from the analytical solution(s) of Eq. 3.4 are summarized in Table 3.4. Appendix C presents an in-depth analysis of analytical biofilm models.

3.11   Numerical One-Dimensional Biofilm Models

The numerical one-dimensional biofilm model is remarkably flexible with regard to the number of dissolved and particulate components, biokinetics, and, to a certain extent, the physical and geometrical properties of a biofilm that are described (Wanner et al., 2006). This model is commonly used in both research and design. The numerical one-dimensional biofilm model treats its variables differently in that attached particulate components form the biofilms solid matrix and suspended particulate and dissolved components exist in the bulk phase. Particulate components, for example, include active microbial species, extracellular polymeric substances, and organic and inorganic substrates. Dissolved components, for example, include metabolites, products, the hydrogen ion, and organic and inorganic substrates. According to Wanner et al. (2006), model output includes spatial profiles of particulate components, accumulation of particulate components and loss of mass from the biofilm, spatial profiles of dissolved components, flux and concentration remaining in the effluent stream of dissolved components, and biofilm thickness as a function of

TABLE 3.3 Step-by-step instructions for using the pseudo-analytical model of Saéz and Rittmann (1992).

Step	Descriptions
Step 0: Parameters needed a priori	q_{max} = maximum specific rate of dissolved component utilization $(g/g_X \cdot d)$
	K_S = half-saturation concentration (g/m^3)
	Y = true yield coefficient (g_X/g_s)
	b = biomass loss coefficient $(1/d)$
	b_{ina} = biomass inactivation coefficient $(1/d)$
	b_{res} = biomass respiration coefficient $(1/d)$
	b_{det} = biomass detachment coefficient $(1/d)$
	D_w = diffusion coefficient of dissolved component i in water (m^2/d)
	D_F = diffusion coefficient of dissolved component i in the biofilm matrix (m^2/d)
	L_L = mass-transfer boundary layer thickness (m)
	X_F = biofilm biomass density, or concentration (g/m^3)
	$S_{B,i}$ = dissolved component I concentration in the bulk phase (g/m^3)
Step 1: Calculate dimensionless parameters, $\tilde{K}$ and $S_{min,i}$	$\tilde{S}_{min,i} = \dfrac{b}{Y \cdot q_{max} - b}$; $\tilde{K} = \dfrac{D}{L_L}\left[\dfrac{K_S}{q_{max} \cdot X_F \cdot D_F}\right]^{\frac{1}{2}}$
Step 2: Calculate dimensionless bulk-phase dissolved component i concentration, $\tilde{S}_{B,i}$	$\tilde{S}_{B,i} = \dfrac{S_{B,i}}{K_S}$
Step 3: Calculate coefficients: α, β	$\left.\begin{array}{l}\alpha = 1.5557 - 0.4117 \cdot \tanh\,(\log_{10}\tilde{S}_{min}) \\ \beta = 0.5035 - 0.0257 \cdot \tanh\,(\log_{10}\tilde{S}_{min})\end{array}\right\} \tanh = \dfrac{(e^x - e^{-x})}{(e^x + e^{-x})}$
Step 4: Iterate to find $\tilde{S}_{B,i}$	$\tilde{S}_{LF,i} = \tilde{S}_{B,i} - \dfrac{\tanh\left[\alpha \cdot \left(\dfrac{\tilde{S}_{LF,i}}{\tilde{S}_{min,i}} - 1\right)^{\beta}\right] \cdot \{2 \cdot [\tilde{S}_{LF,i} - \ln(1 + \tilde{S}_{LF,i})]\}^{\frac{1}{2}}}{\tilde{K}}$
Step 5: Calculate the dimensionless flux $\tilde{J}_{LF,i}$	$\tilde{J}_{LF,i} = \tilde{K} \cdot (\tilde{S}_{B,i} - \tilde{S}_{LF,i})$
Step 6: Calculate steady-state flux, $\tilde{J}_{LF,i}$	$J_{LF,i} = \tilde{J}_{LF,i} \cdot (K_S \cdot q_{max} \cdot X_F \cdot D_{F,i})^{\frac{1}{2}}$
Step 7: Calculate steady-state biofilm thickness, L_F	$L_F = \dfrac{J_{LF,i} \cdot Y}{X_F \cdot b}$

TABLE 3.4 Summary of analytical biofilm models (modified from Morgenroth [2008])*.

Biofilm kinetics	Substrate flux across the biofilm surface (J = g/m²·d)	Degree of biochemical transformation rate limited by mass transfer	Concentration profile over the biofilm thickness	Appendix equation-number
First order	$$J_{LF,i}^1 = \frac{q_{max,i} \cdot X_{F,k} \cdot L_F \cdot S_{LF,i}}{K_i}$$ $$\cdot \frac{\tanh\beta}{\beta};$$ $$\beta = \sqrt{\frac{q_{max,i} \cdot X_{F,k} \cdot L_F^2}{D_{F,i} \cdot K_i}}$$	$$J_{LF,i}^1 = \underbrace{\varepsilon}_{\substack{\text{Mass transport}\\\text{limitation}}} \cdot q_{max,i} \cdot X_{F,k} \cdot L_F \cdot S_{LF,i}$$ $$\varepsilon = \frac{\tanh\left(\dfrac{L_F}{L_{crit}}\right)}{\dfrac{L_F}{L_{crit}}}$$	$$S_{F,i}(K) = \frac{\cosh\left(\dfrac{L_F - x}{L_{crit,i}}\right)}{\cosh\left(\dfrac{L_F}{L_{crit,i}}\right)} \cdot S_{LF,i}$$ $$L_{crit,i} = \sqrt{\frac{D_{F,i}}{q_{max,i} \cdot X_{F,k}}}$$	(B7)
Zero-order (0th), biofilm partially penetrated by the rate-limiting substrate (β ≤ 1)	$$J_{LF,i}^{0,pp} = \sqrt{2 \cdot D_{F,i} \cdot q_{max,i} \cdot X_{F,k}} \cdot \sqrt{S_{LF,i}}$$	$$J_{LF,i}^{0,pp} = \underbrace{\beta}_{\substack{\text{Mass transport}\\\text{limitation}}} \cdot q_{max,i} \cdot X_{F,k} \cdot L_F$$ $$\beta = \sqrt{\frac{2 \cdot D_{F,i} \cdot S_{LF,i}}{q_{max,i} \cdot X_{F,k} \cdot L_F^2}}$$	$$S_{F,i,0,pp}(x) = S_{LF} - \left(x \cdot \beta \cdot L_F - \frac{x^2}{2}\right)$$ $$\cdot \frac{q_{max,i} \cdot X_{F,k}}{D_{F,k}}$$ $$S_{F,i,0,pp}(x) = S_{LF} \cdot \left[1 - \left(\frac{2 \cdot x}{\beta \cdot L_F} - \frac{x^2}{(\beta \cdot L_F)^2}\right)\right]$$	(B5)

Zero-order (0ᵗʰ), biofilm completely penetrated by the rate-limiting substrate (β ≥ 1)

$$J_{LF,i}^{0,cp} = q_{max,i} \cdot L_F \cdot X_{F,k}$$

Corresponds to a biofilm where substrate flux is not influenced by resistance to mass transfer (full penetration and β ≥ 1)

$$S_{F,i,0,cp}(x) = S_{LF} - \left(x \cdot L_F - \frac{x^2}{2} \right) \cdot \frac{q_{max,i} \cdot X_{F,k}}{D_{F,i}}$$

$$S_{F,i,0,cp}(x) = S_{LF} \cdot \left\{ 1 - \left[\frac{2 \cdot x}{\beta^2 \cdot L_F} - \frac{x^2}{(\beta \cdot L_F)^2} \right] \right\} \tag{B6}$$

*J = mass flux normal to biofilm surface; q = maximum specific conversion rate for a substrate i; $X_{F,k}$ = concentration of the particulate biomass component k in the biofilm or biomass type k per unit biofilm volume; L_F = biofilm thickness; $L_F > L_{crit}$ (L_{crit}) biofilm thickness separating mass-transfer and non-mass-transfer rate limited biofilms (e.g., biofilm with $L_F > L_{crit}$ is mass-transfer-limited); $S_{LF,i}$ = concentration of component i at the biofilm surface; $S_{F,i}$ = concentration of component i at a point x inside the biofilm; K = Monod half-saturation constant of component i; $D_{F,i}$ = molecular diffusion coefficient of a soluble component i inside the biofilm; ε = first-order biofilm efficiency factor; α = the biofilm efficiency factor; α = the biofilm constant; and x = distance from the biofilm surface to a specified point inside the biofilm.

the production and decay of particulate material in the biofilm and of the attachment and detachment of cells and particles at the biofilm surface and inside the biofilm. For each of the aforementioned parameters, time-development (and steady-state) solutions can be calculated. Processes considered in the model include biochemical transformation processes, advection and diffusion of attached particulate components in the biofilm solid matrix, attachment and detachment of particulate components at the biofilm surface and inside the biofilm, diffusion of suspended particulate and dissolved components in the biofilm liquid phase and in the mass-transfer boundary layer, and complete mixing of suspended particulate and dissolved components in the bulk phase.

Special characteristics of the numerical one-dimensional biofilm model include physical and geometrical biofilm properties that may change with time and space. A variable volume fraction of the biofilm liquid phase can be used to reproduce the observation that the biofilm density changes with time and space (Lewandowski and Boltz, 2011). Variable diffusion coefficients can be used to reproduce the observation that the transport of suspended and dissolved particulate components in the biofilm can be enhanced by advection and turbulent diffusion reaching into the biofilm (Zhang and Bishop, 1994).

The one-dimensional biofilm model is comprised of a system of stiff, nonlinear partial differential equations. Because of the stiffness of the equation system, integration methods designed for stiff systems must be used or the dissolved and particulate components must be treated separately. Furthermore, biofilm growth (i.e., the displacement of the biofilm-water interface) creates a moving boundary problem (Kissel et al., 1984; Wanner and Gujer, 1986). The commercially available program AQUASIM™ (Reichert 1998a, 1998b) is commonly used when numerically modeling one-dimensional biofilms because of its ability to handle the associated stiff equation system and moving boundary problem.

The numerical one-dimensional biofilm model is described completely in articles by Wanner and Reichert (1996) and Reichert and Wanner (1997). The equations presented in these manuscripts are complex and rigorous beyond practicality (at least in many cases). Consistent with Scientific and Technical Report Number 18, *Mathematical Modeling of Biofilms* (Wanner et al., 2006), only the most important definitions and equations of the model are described. Furthermore, for the sake of simplicity, only those processes required for practical significance are included. Mathematical expressions that describe the numerical one-dimensional biofilm model are summarized in Appendix D.

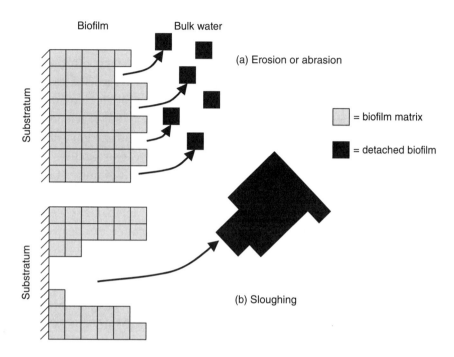

FIGURE 3.5 (a) Erosion or abrasion and (b) sloughing biofilm detachment processes (Morgenroth [2003]; reprinted with permission from IWA publishing).

3.12   Biofilm Detachment Models

Developing biofilms accumulate bacterial cells, but all biofilms eventually loose particulate components. The loss of particulate components from the biofilm and their introduction to the bulk of the liquid is called *detachment*. The detachment process may range from the removal of small biofilm fragments to larger biofilm segments (Morgenroth and Wilderer, 2000). While operating biofilm reactors, detachment is characteristic of a well-operating system. In fact, detachment is required to prevent clogging caused by excessive biofilm growth. Bryers (1988) described four biofilm detachment processes: abrasion, erosion, sloughing, and predator grazing. Abrasion and erosion are illustrated in Figure 3.5. Abrasion, which is initiated by particle collision, and erosion, initiated by hydrodynamic shear near the biofilm surface, represent the removal of small groups of cells. Predatory higher life forms such as macro- and micro-fauna graze biofilms (Boltz et al., 2008). Abrasion, erosion, and, to a certain extent, fauna grazing are associated with well-operating biofilm reactors.

Sloughing and excessive predation is detrimental to biofilm reactor performance. Avoiding excess predatory fauna accumulation and promoting continuous detachment of small biofilm fragments results from proper thickness control, which occurs in a stable environment not subject to excessive mass-transfer resistances.

When developing and using mathematical biofilm models one must include a quantitative description of detachment regardless of the "state of science" offering a limited understanding of detachment processes. The rate and category of detachment can have a significant influence on the results of biofilm reactor models (Morgenroth, 2003). The inherent problems of detachment modeling, lack of understanding of the fundamental mechanisms of detachment, and the inability of existing biofilm detachment models to predict the location of biofilm biomass detachment were identified by Kissel et al. (1984). The importance of detachment location can be exemplified by the case of sloughing because the composition of the remaining biofilm may be different from the biofilm that was originally present (Morgenroth, 2003). Table 3.5 lists several biofilm detachment models and demonstrates that numerous expressions have been proposed for specific experimental applications. The abundance of detachment models listed in Table 3.5 partly indicates the failure of any one rate expression to model biofilm detachment for a wide range of biofilm reactor operating conditions (Peyton and Characklis, 1993).

3.13   Using Mechanistic Biofilm Models to Describe a Biofilm Reactor

Models have been successfully applied both in research and practice, and serve as the benchmark for new or expanded activated sludge models (Morgenroth et al., 2000). Mathematical modeling of biofilm reactors is more complicated than suspended growth reactor modeling because of increased complexity in describing the fate of wastewater constituents (i.e., particulate vs soluble substrate), differences in biofilm reactor configuration and operation, the effect of bulk-liquid hydrodynamics, and biofilm diffusional resistances. Until recently, the complexity of available mechanistic biofilm models has led to their limited adoption in engineering practice. Consequently, biofilm reactor design has been based on empirical criteria and design formulations, which can provide a good basis for design when applied to the conditions for which empirical data were collected, but are inadequate for describing complex processes and fail to provide a mechanistic understanding of redox zone formation and how the different redox zones affect biofilm reactor performance (e.g., simultaneous nitrification and denitrification, deammonification, and greenhouse gas emission) unless data are collected and supporting empirical models are developed for that purpose.

TABLE 3.5 Detachment rate expressions (modified from Morgenroth [2008, 2003]; Peyton and Characklis [1993]; and Tijhuis et al. [1995]).

Mechanism of detachment relationship	Detachment rate expression r_{det} [=] g/m^2 · d*	Reference
None specified	0 $\dfrac{dL_F}{dx} = 0$ (constant biofilm thickness)	Kissel et al. (1984) Fruhen et al. (1991) Wanner and Gujer (1984)
Biofilm thickness	$k_d \, \rho_F \, L_F^2$ $k_d \, \rho_F \, L_F$	Truler and Characklis (1982) Bryers (1984) Wanner and Gujer (1986) Kreikenbohm and Stephan (1985) Chang and Rittmann (1987) Rittmann (1989)
Shear	$k_d \, \rho_F \, \tau$ $k_d \, \rho_F \, L_F \, \tau^{0.58}$	Bakke et al. (1984) Rittmann (1982)
Growth rate or substrate utilization rate	$L_F \, (k_d' + k_d'' \cdot q_{max})$ $k_d \cdot r_S \cdot L_F$	Speitel and DiGiano (1987) Robinson et al. (1984) Peyton and Characklis (1993) Tijhuis et al. (1995)
Backwashing down to a predicted base biofilm thickness	$\begin{cases} 1 \cdot k_d' & \text{normal operation} \\ 2 \cdot k_d'' \end{cases} (L_F - L_{\text{base thickness}}) \quad \text{backwashing}$	Rittmann et al. (2002)

*k_d', k_d'' = detachment rate coefficients; ρ_F = biofilm volumetric mass density (g/m^3); L_F = biofilm thickness (microns); $L_{\text{base thickness}}$ = predefined biofilm thickness after backwashing (microns); q_{max} = specific growth rate (d^{-1}); r_s = substrate utilization rate (g/m^2·d); and τ = shear stress (g/m·d^2).

Values of key parameters such as bulk-liquid volume, biofilm surface area, and bulk-liquid substrate concentrations can be identified in modern biofilm reactors (e.g., moving bed biofilm reactor). Furthermore, the continuously flowing biofilm reactor systems are typically configured in a series of completely mixed reactors that are analogous to a common method for their pseudo two-dimensional simulation. These factors contribute to the fact that a majority of modern biofilm reactors are amenable to simulation with existing biofilm models. As a result, each of the WRRF simulators listed in this chapter has been expanded to include a biofilm reactor module based on a revision or expansion to an existing mathematical model describing a one dimensional biofilm (Boltz, Morgenroth, and Sen, 2010).

4.0 WATER CHEMISTRY MODELS

Modeling water chemistry in conjunction with biological processes presents some challenges, mainly because the underlying components and reactions are faster and more complex. Water chemistry models discussed in this section include alkalinity, pH, and chemical precipitation.

4.1 Alkalinity

Alkalinity is a state variable in many activated sludge models (e.g., ASM1, ASM2d, and ASM3) and is assumed to be all of the carbonate alkalinity that is available and completely dissociated for acid neutralization. Current models calculate alkalinity by including the effect of all of the ionic species at the current system state.

The approach, to date, in IWA activated sludge models has been to track alkalinity changes instead of calculating complex pH reactions. Thus, alkalinity is used as a pseudo-indicator of potential pH instability problems. This approach assumes that the pH remains approximately constant and is in a region where it does not affect biological activity.

The following are some limitations of using alkalinity as a surrogate measurement of pH: alkalinity cannot be used for modeling physical–chemical precipitation and alkalinity is not a good indicator of steady pH conditions when modeling acid fermenters or digesters or in systems where significant gas transfer may occur.

4.2 pH

Modeling pH is an important factor in simulating the performance of biological wastewater treatment processes, including activated sludge and anaerobic digestion. The pH affects the species distribution of the weak acid systems (carbonate, ammonia, phosphate, acetate, propionate, etc.) present in the process. This, in turn, dictates the rate of many of the biological and physico-chemical phenomena occurring in these systems such as biological activity that can be severely limited outside an optimal pH range, chemical precipitation reactions when metal salts such as alum or ferric chloride are added for chemical phosphorus removal, spontaneous precipitation of magnesium and calcium phosphates (struvite, hydroxydicalcium phosphate [HDP] and hydroxyapatite [HAP]), and stripping of ammonia at high pH.

Modeling pH is difficult because the underlying components and reactions are fast and complex. The pH calculations include several processes that can be described by kinetic approaches (Musvoto et al., 2000; Vavilin et al., 1995). However, the rates of many of the reactions involved are typically 4 to 20 orders of magnitude faster than typical biological rates. As a result, calculation of a kinetic-based model will significantly reduce simulation speed.

Most current models for pH, such as that presented by Fairlamb et al. (2003), use a mixed kinetic/equilibrium-based approach to minimize the negative effect on simulation speed.

The pH models are typically based on the following reactions:

- Equilibrium modeling of the phosphate, carbonate, ammonium, volatile fatty acid systems and typical strong ions in wastewater

- Incorporation of activity coefficients based on the ionic strength of the solution

- Chemical precipitation reactions

- Gas–liquid transfer of ammonia and carbon dioxide

- Biological activity affecting compounds included in the model

4.3 Chemical Precipitation

Modeling chemical precipitation provides considerable insight into the behavior of a WRRF. Some chemical precipitation models are based on the stoichiometry of the basic metal-ion reactions. Other chemical precipitation models use an equilibrium/kinetic approach. The following are common chemical precipitation reactions:

- Chemical precipitation by alum or ferric

- Struvite precipitation, which includes magnesium, nitrogen, and phosphorus removal from the liquid phase and struvite precipitation during degassing of carbon dioxide (Musvoto et al., 2000)

- Calcium phosphate precipitation, including HDP and HAP processes (Maurer et al., 1999)

5.0   SOLID–LIQUID SEPARATION MODELS

Solid/liquid separation models, mainly thickening/clarification (settler) models, are frequently used in process engineering and design work. The first dynamic settler models, used in combination with ASM1, enabled complete modeling of the activated sludge process (Nolasco, 1989; Takacs et al., 1991; Vitasovic, 1989). Over time, settler models used in combination with a library of process models enabled full-scale facility modeling by interconnecting various types of processes (Daigger and Nolasco, 1995).

In the general context of facility-wide modeling, solids–liquid phase separation is used not only in the activated sludge stage, but also in primary and tertiary clarification, sludge thickening, and other processes. This section provides a background of the most common mathematical models used to describe secondary clarifiers. The settler models described here also can be used for primary and tertiary clarifiers and gravity thickeners, although they require different parameter ranges to be used within the model. Settler models can be categorized according to modeling objectives. Modeling objectives define how much detail the model should contain about the physical setup of the actual clarifier structure and, potentially, the reactions that occur in the clarifier. The four most frequently used clarifier representations, shown in Figure 3.6, are

- Point settlers
- Simple (ideal) clarifiers
- Flux-based one-dimensional models
- Computational fluid dynamic (CFD) (two-dimensional or three-dimensional) models

5.1   Point Settler (Simple Mass Balance) Model

The point settler (simple mass balance) model represents a clarifier without volume, area, or depth. The sole purpose of the model often is to retain the mixed liquor suspended solids (MLSS) in the system. The model is based on an instantaneous mass balance around the clarifier.

The return activated sludge (RAS) concentration can be expressed from the mass balance if the flows and the operating MLSS concentration are known. Most early activated sludge simulation software (e.g., SSSP) set the effluent solids to zero because their focus was on biological performance and soluble components in the effluent. However, more recent implementations allow the user to set the effluent

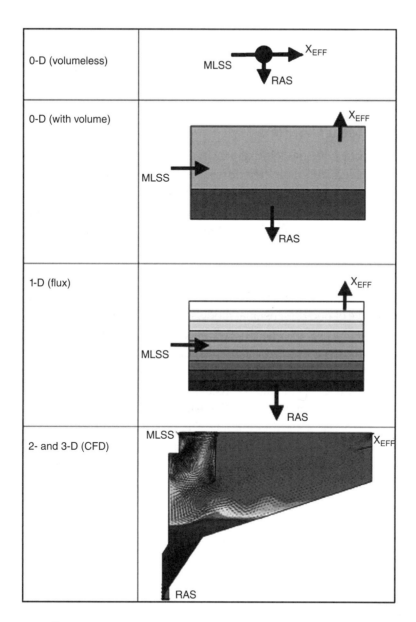

FIGURE 3.6 Different representations of the same clarifier (CFD model credit: BioMath, Ghent University, Ghent, Belgium).

solids concentration using simple empirical approaches, typically linked to input MLSS (percent of MLSS removed) or applied hydraulic or solids flux. In this instance, solids lost through the effluent must be included in the mass balance used to determine the RAS concentration.

5.2   Simple, Ideal Clarifiers

Simple, ideal clarifiers are models that are similar in concept to the point settler models previously discussed, although they do account for the liquid and (fixed) sludge volume contained in the clarifier. An advantage of simple, idea clarifiers is their simplicity. When biochemical reactions are calculated within the fixed sludge volume contained in the clarifier, these models can be used to approximate anoxic and anaerobic reactions occurring in the sludge blanket.

5.3   One-Dimensional Flux Model

The main objective of one-dimensional models is an estimation of the effluent and RAS total suspended solids (TSS) concentration because the model responds to changes in hydraulic and solids loading. The model predicts the solids stored in the settler.

When combined with a mechanistic model of the reactions that occur within the aeration basin in an activated sludge process, sludge mass stored in the clarifier and the varying mass distribution between the reactors and the clarifier can also be simulated using these types of clarifier models. The layered one-dimensional flux model represents the clarifier as a stack of horizontal layers. Horizontal movement of liquid is not considered. Circular and rectangular tanks are not distinguished in one-dimensional models because the model assumes that all solids enter the clarifier and are distributed evenly throughout one layer of the clarifier.

The basis of flux theory used in secondary clarifiers is the Vesilind settling function (Vesilind, 1968), which links the settling velocity of the solids to the solids concentration (Figure 3.7). Bulk flow and settling flux-based dynamic mass balances are implemented in each layer, and the output of the model is a vertical solids profile, with one concentration calculated for each layer, as shown in Figure 3.8 (Nolasco, 1989). The sludge blanket height can also be estimated.

Effluent solids are simulated using a modification of the Vesilind settling function to account for discrete settling (i.e., double-exponential model) (Nolasco, 1989; Takacs et al., 1991). Conceptually, the mixed liquor solids entering the final clarifier can be thought to consist of three fractions: unsettleable, slowly settling, and rapidly

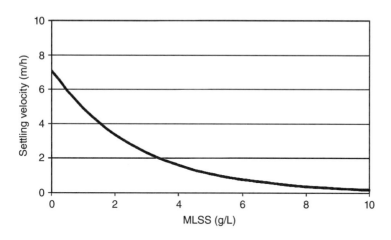

FIGURE 3.7   Example Vesilind settling velocity function.

settling fraction. The double exponential settler model accounts for the three settling fractions. The settling velocity (V_s) for a given suspended solids concentration (X) in this model is given by the following formula:

$$V_s = V_{max}\left[e^{-r_{floc}(x-x_{min})} - e^{-r_{coll}(x-x_{min})}\right]$$ (3.25)

where
$X_{min} = Fns \times X_{in}$ = minimum attainable effluent suspended solids concentration, where Fns = unsettleable influent suspended solids fraction and X_{in} = influent suspended solids concentration (i.e., MLSS concentration)
r_{floc} = settling parameter for the floc particles (rapidly settling solids fraction)
r_{coll} = settling parameter for the slowly settling solids fraction

It is important to note the following four ranges of suspended solids concentrations (X_i) within a layer i (Figure 3.9):

- The settling velocity equals zero as the solids attain the minimum attainable concentration $(X_i = X_{min})$.

- The settling velocity is dominated by the flocculating nature of the particles; thus, the settling velocity is sensitive to the r_{floc} parameter.

- The settling velocity has become independent of solids concentration (particles have reached their maximum size).

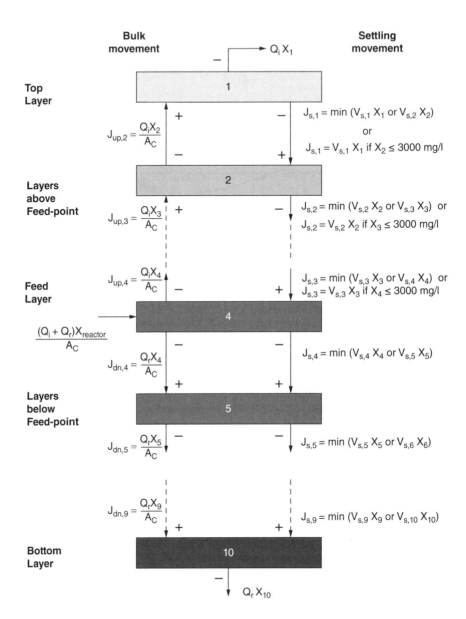

FIGURE 3.8   Bulk flow and settling flux in settling model (Nolasco, 1989).

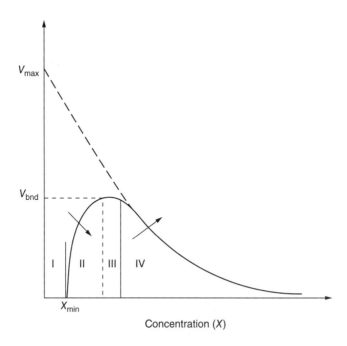

FIGURE 3.9 Double exponential settling velocity (Hydromantis, Inc.).

- The settling velocity is affected by hindrance and becomes dependent on the r_{coll} parameter (the model reduces to the Vesilind equation).

One-dimensional dynamic models play an important role in activated sludge and facility-wide process predictions. Because of their simple structure, they do not add a significant computational load to the process model, and they can reasonably predict the three main functionalities of secondary settlers: clarification, thickening, and sludge storage. The reasonable prediction of secondary effluent TSS concentrations resulting from the application of these models enable better prediction of effluent quality, even under rainstorm events when solids washouts tend to occur.

Model prediction of underflow solids is useful for sludge wastage estimates, which, in turn, influences SRT calculations and modeling of the sludge treatment train. Estimation of sludge storage within the secondary clarifier through dynamic sludge blanket predictions allows for a better assessment of overall solids inventory within the activated sludge process.

Under certain environmental conditions, biological or chemical reactions such as denitrification and/or phosphorus release may occur in secondary settlers. These reactions may have considerable effects on overall facility performance. Prediction of solids inventory within the clarifier by one-dimensional models allows for modeling of these reactions by turning each settling zone into a reactor with all biochemical reactions, which is generally referred to as *reactive settler models*.

5.4 Reactive Settler Models

Reactive settler (clarifier) models are used to model both clarification/thickening (as described in the previous section) and biochemical reactions within the tank. Taking advantage of the solids inventory capabilities of the clarifier model, reactive settler models are capable of modeling biological activity taking place within each layer of a one-dimensional flux clarifier model (generally, 10 layers are used).

Generally, the same biochemical model used in the aeration tank of an activated sludge system is applied to the clarifier model, resulting in 10 additional reactors for each clarifier being modeled. This increase in reactors slows down the speed at which the overall model runs. For this reason, reactive settler models are recommended only in those instances when the practitioner considers that the estimation of what goes in the settler biochemically is of importance to the overall objective of the simulation exercise.

5.5 Computational Fluid Dynamic Models (Two-Dimensional and Three-Dimensional)

The main objective of CFD models in clarifiers is their detailed structural design. One-dimensional layered models, while providing effluent and underflow TSS predictions, cannot be used to investigate details of clarifier structures (i.e., tank geometry or baffle placement). On the other hand, two- or three-dimensional CFD models can be used for this purpose. Standard commercial simulation software, generally used for process design, does not include CFD capabilities. Specialized software, such as the following, is required to carry out CFD modeling: Fluent (Lebanon, New Hampshire); CFX (ANSYS, Inc., Canonsburg, Pennsylvania); and SolidWorks (Dassault Systèmes SolidWorks Corp., Concord, Massachusetts).

Computational fluid dynamic models are based on conservation of fluid mass (continuity), momentum, solids mass (transport of suspended solids), and enthalpy (heat balance). To achieve a stable numerical solution, the settler has to be discretized

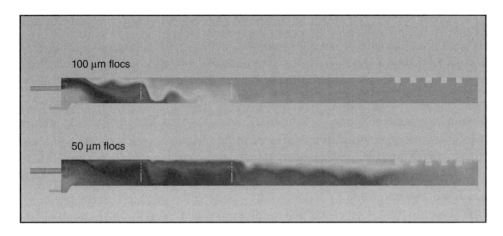

100 µm flocs

50 µm flocs

FIGURE 3.10   Detailed profile of particle contours resulting from a CFD model of a rectangular clarifier (courtesy of Black & Veatch, Kansas City, Missouri).

into a fine grid, often using tens of thousands of grid elements. Under these computational conditions, it is possible to account for fine physical details, such as baffles, and their exact geometry, placement, and angle. The equations in such a model are subsequently solved at each node and at each time step.

Although CFD modeling demands a significant computational load, it can result in a detailed picture of solids distribution and flow patterns within the clarifier. In the example shown in Figure 3.10, the fate of 50- and 100-µ flocs are simulated along the cross section of a rectangular clarifier structure. Density waves and the effect of baffles on flow patterns are also made visible in this two-dimensional example. Adding the third dimension increases complexity and execution time for these models. Therefore, three-dimensional modeling should only be performed in special instances.

Use of CFD models has accelerated significantly in recent years as a result of advances in hydrodynamic modeling, model calibration techniques, and computational speed. The main driving force behind the use of CFD models is their increased capability (compared to one-dimensional models) of predicting clarifier behavior (particularly, effluent concentration) for different internal tank configurations (e.g., tank geometry, baffle arrangements, weir layout, etc.). However, cost of implementation remains one of the main obstacles to the use of this type of clarifier model.

6.0   ANAEROBIC PROCESS MODELS

Digestion models are important in wastewater process modeling, not only because they can be used for the design and evaluation of digestion systems, but because they have a direct effect on the predicted performance of the activated sludge process in whole- facility model simulations. Information that is typically searched for in anaerobic digestion models includes the volatile solids destruction rate and biogas production and composition. Nitrogen and phosphorus concentrations in recycle streams are particularly important in biological nutrient removal systems. In addition, struvite precipitation in biological phosphorus removal facilities with anaerobic digestion is also important.

Historically, mathematical models for digestion have incorporated biochemical and physical processes to anaerobic digestion with varying degrees of complexity, resulting in a wide array of models published in the literature. The main processes involved in anaerobic digestion include hydrolysis, fermentation, and methane production (methanogenesis). Early models only considered the reactions occurring in the methanogenic phase of anaerobic digestion. Examples of early models include the work of Andrews (1969) and Andrews and Graef (1971). After the work of Mosey (1983), models began to include the fermentation process and the effect of hydrogen gas and higher organic acids on methanogenesis. Some examples of recent models include the work of Bagley and Brodkorb (1999), Costello et al. (1991), and Massé and Droste (2000). The work of Pavlostathis and Giraldo-Gomez (1991) was critical to the development of anaerobic digestion models. Because of the wide range of models available in the literature and the difficulty in sampling full-scale digesters, anaerobic digestion models have not been validated as extensively as suspended growth activated sludge models.

6.1   Anaerobic Digester Model No. 1

Similar to the development of activated sludge models, IWA formed the Task Group for Mathematical Modeling of Anaerobic Digestion Processes to develop the Anaerobic Digestion Model No. 1 (ADM1) (Batstone et al., 2002). The task group's purpose was to create and publish a generic model that would serve as a common platform for dynamic simulations of anaerobic digestion processes.

In developing ADM1, the task group tried to establish common nomenclature, units, and model structure consistent with existing anaerobic modeling literature and most activated sludge models. The structure was devised to encourage specific extensions or modifications where required, but still maintain a common platform. Currently,

ADM1 has found wide application in the anaerobic digestion modeling community and is becoming a benchmark model for the simulation of anaerobic digestion processes.

In ADM1, complex particulate waste is broken down via a disintegration step to carbohydrates, proteins, fats, inert particulates, and inert soluble-state variables. This disintegration step includes a variety of processes involved in digestion and separates them from hydrolysis. Hydrolysis of carbohydrates, proteins, and fats produces sugars, amino acids, and long-chain fatty acids. Different groups of microorganisms carry out acidogenesis from sugars and amino acids, and acetogenesis from long-chain fatty acids, propionate, butyrate, and valerate.

Methanogenesis from acetate and hydrogen gas is the final step in the digestion model, resulting in methane and carbon dioxide formation. Decayed biomass is incorporated to the complex particulate waste pool. Physical–chemical reactions in the model include acid–base chemistry reactions of the most important liquid–liquid acid and base pairs and gas–liquid exchange reactions for carbon dioxide, methane, and hydrogen. The ADM1 does not include precipitation reactions.

6.2 Dold's General Model—Anaerobic Digestion

As discussed in previous sections, Dold's General Model describes the biological processes occurring in activated sludge and anaerobic digestion systems, several chemical precipitation reactions, and the gas–liquid mass-transfer behavior for six gases. The anaerobic digestion model within the General Model contains the following three main functional categories:

- Heterotrophic growth through fermentation—Two pathways are used to simulate the fermentation of readily biodegradable (complex) substrate to acetate, propionate, carbon dioxide, and hydrogen. The dominant pathway is governed by the dissolved hydrogen concentration. These processes are mediated by ordinary heterotrophic organisms. The base rate expression for the fermentation growth process is the product of the maximum specific growth rate constant, the heterotrophic biomass concentration, and a Monod expression for the readily biodegradable (complex) substrate. This base rate is modified to account for nutrient limitations (ammonia, phosphate, and other cations and anions) and pH inhibition. In activated sludge vessels, there is an anaerobic growth factor applied. Ammonia is used as a nitrogen source for cell synthesis.

- Growth and decay of propionic acetogens—These two processes describe the growth and decay of propionic acetogens, converting propionate to acetate, carbon dioxide, and hydrogen. The nitrogen source for cell synthesis is ammonia. The base rate expression for the growth process is the product of the maximum specific growth rate, the propionic acetogen biomass concentration, and a Monod expression for propionate. This base rate is modified to account for environmental conditions (growth remains off unless the environmental conditions are anaerobic; growth is inhibited by hydrogen and acetate), nutrient limitations (nitrogen, phosphate, other cations and anions), and pH inhibition. The model uses ammonia as a nitrogen source for cell synthesis. The decay process has a rate that varies according to the electron acceptor environment.

- Growth and decay of methanogens—This category consists of six processes describing the growth and decay of two of the principal groups of obligate anaerobic microorganisms: acetoclastic methanogens converting acetate (or methanol) to methane and carbon dioxide and hydrogenotrophic methanogens converting carbon dioxide (or methanol) and hydrogen to methane and water. The base rate expression for each of the four growth processes is the product of the maximum specific growth rate constant, the appropriate biomass concentration, and a Monod expression for each of the substrates. This base rate is modified to account for nutrient limitations (ammonia, phosphate, and other cations and anions) and pH inhibition. Ammonia is used as a nitrogen source for cell synthesis. For both populations, the decay rate varies according to the electron acceptor environment.

6.3 Upflow Anaerobic Sludge Blanket Reactors' Modeling

Upflow anaerobic sludge blanket (UASB) reactors' modeling is generally accomplished by a combination of anaerobic digester models and solids/liquid separation models. To date, no industry standards (i.e., well-established mathematical models) have been proposed for UASB reactors. Nevertheless, available commercial simulation platforms present alternative ways of modeling UASBs and expended granule sludge blanket reactors.

7.0   MODELS FOR OTHER UNIT PROCESSES
7.1   Gas-Transfer Modeling

There are at least six gas–liquid mass-transfer processes modeled in unit processes, namely oxygen, carbon dioxide, methane, nitrogen, ammonia, and hydrogen. Supply of oxygen constitutes a significant operating cost for biological wastewater treatment systems. A number of theories have been proposed to explain the phenomenon of mass transfer in gas–liquid systems. These theories describe the behavior under the following two groups: laminar or turbulent flow conditions.

The mass-transfer theories proposed suggest several possible mechanisms controlling the phenomenon of mass transfer in bubble aerated aqueous systems. Determination of which mechanism most closely represents the actual transfer process must be based on experimental observations. Because direct measurement of the liquid-phase mass-transfer coefficient typically is not possible, an average liquid-phase mass-transfer coefficient is generally inferred from experimental determinations of the overall mass-transfer coefficient.

The overall mass-transfer coefficient (kLa$_L$) may be determined from a mass balance on the experimental data. Simultaneous mass balances for a component should be conducted on both the gas and liquid phases.

7.2   Membrane Bioreactor Modeling

Even though the behavior of membranes can be the subject of complex analysis, MBR models commonly used in practice are relatively simple. For example, overall solids retention (typically 99.9% and higher) and the colloidal material capture rate are the main model parameters required.

The solids percent retention defines the percentage of the incoming solids mass stream that is retained by the MBR and is returned to the mixed liquor stream, with the remaining solids (1% or less) coming out in the permeate stream. Under steady-state conditions, the mass coming out of the MBR will be the same as the mass entering the MBR.

The colloidal percent retention defines the percentage of the incoming unabsorbed colloidal COD that is retained by the MBR and is returned to the mixed liquor stream. This setting can be used to decouple solids retention and permeate COD to a degree if there is unadsorbed colloidal material in the influent to the MBR.

7.3   Sidestream Treatment Process Modeling

A key concern in WRRFs that need to meet ammonia or total nitrogen effluent criteria is the amount of nutrient release in the digesters and, therefore, the additional loading on the liquid train caused by the return of solids processing sidestreams (Newbigging et al., 1995). Similarly, phosphorus released in anaerobic digester returns to the liquid train. Phosphorus returned in recycle streams (sidestreams) is even greater in biophosphorus facilities because of the storage of excess phosphorus in the biological phosphorus removal process. Struvite precipitation reactions in anaerobic digesters, although not desirable under uncontrolled conditions, may reduce phosphorus and nitrogen loadings in sidestreams.

A number of "named" sidestream biological processes have been developed to treat the ammonia component of the reject water before returning it to the liquid train (e.g., InNitri, Babe, SHARON, Anammox, Canon, Demon, etc.). These systems involve one or more (or a combination) of the following biological transformations:

- Nitritation mediated by ammonia oxidizing biomass (i.e., conversion of ammonia to nitrogen dioxide) or partial nitration (i.e. converting a portion of the ammonia to nitrogen dioxide).

- Nitratation mediated by nitrite oxidizing biomass (i.e., conversion of nitrite to nitrate).

- Denitritation mediated by heterotrophic bacteria where nitrite serves as an electron acceptor on the addition of organic substrate with production of nitrogen gas.

- Denitratation mediated by heterotrophic bacteria where nitrate serves as an electron acceptor on the addition of organic substrate with production of nitrite.

- Nitrogen removal by autotrophic Anammox bacteria. The process converts ammonia directly into nitrogen gas in unaerated conditions, using nitrite as an electron acceptor.

Several models for a number of sidestream processes have been reported. For example, Volcke (2006) developed a two-step nitrification and denitrification model to represent the SHARON process. Wett and Rauch (2003) developed a two-step nitritation/denitritation model based on detailed data from two full-scale reject water SBR treatment processes. Van Hulle (2005) incorporated Anammox reactions in a two-step nitrification/denitrification model.

7.4   Filtration

In a manner similar to thickening and dewatering, filtration can be modeled as a simple solids-separation device in which the designer specifies solids capture efficiency and, in this case, a reject water quantity or quality. Alternatively, more complex filtration models are available that account for the accumulation of solids on or within the filter media. These models use equations such as the Kozeny–Carman equation (Carman, 1937).

7.5 Disinfection

Several models are available for the various disinfection processes. For disinfection using chlorine or another oxidizing agent, an empirical exponential equation can be used to represent the dose–response curve commonly used to relate kill rate to chemical concentration and retention time. Models are also available for UV disinfection, including the Water Environment Research Foundation (Alexandria, Virginia) UV model (WERF, 1995), which relates UV dose to coliform concentration. The disinfection chapter of *Design of Municipal Wastewater Treatment Plants* (WEF and ASCE/EWRI, 2009) describes disinfection kinetics and reactor design, including the use of CFDs, in more detail.

7.6 Thickening/Dewatering

Three options are available for thickening and dewatering. The simplest and most commonly used approach for both thickening and dewatering processes is a mass balance based on the solids capture efficiency of the unit and the thickness of the sludge or cake produced. For whole- facility modeling, this typically is the most useful approach. For gravity thickening and dissolved air flotation thickeners, process models are available that use the same approach as that used for clarifiers using one-dimensional flux models of sedimentation or flotation. Finally, there are empirical dewatering models available, such as that developed by American Society of Civil Engineers Task Committee on Belt Filter Presses (1988), which link capture efficiency to optimal polymer dose in belt filter presses. Models are also available to model centrifuge performance. The reader is referred to *Design of Municipal Wastewater Treatment Plants* (WEF and ASCE/EWRI, 2009) for additional information on thickening and dewatering.

7.7 Pretreatment—Grit Removal, Screening, and so on

Mathematical models for pretreatment processes (grit removal; fats, oils, and grease removal; screening; etc.) are not frequently part of commercial software used by WRRF designers. Even though some level of modeling for these processes may be available in simulators, these models are far from mechanistic and do not have the level of research and development observed in models for other unit processes described in this chapter. For these reasons, pretreatment process modeling is not further presented in this chapter. For design purposes, the reader is referred to design manuals such as the aforementioned *Design of Municipal Wastewater Treatment Plants* (WEF and ASCE/EWRI, 2009).

7.8 Greenhouse Gas Emissions

There is gaining interest in the water sector to understand and minimize the carbon footprint of WRRFs. Modeling tools are needed to gain insight to the potential carbon footprint effect of different design, operational, and planning scenarios, and ultimately for developing mitigation strategies. As with any modeling effort, the greenhouse gas modeling approach should be based on the specific goals of the wastewater utility.

In general, commercial simulation software available today provides the capability to account for indirect (electric) carbon dioxide emissions from aeration data, methane (CH_4) production from anaerobic digestion, and carbon dioxide emissions from biomass. In some cases, embedded GHG emissions using life-cycle assessment methods can also be predicted.

However, nitrous oxide emissions, having 300 times the global warming potential of carbon dioxide in a 100-year cycle (IPCC, 2007), are yet to be developed to the same level of certainty as other greenhouse gases emitted at WRRFs. Generic emission factors such as those recommended by the U.S. Environmental Protection Agency (2010) and the Intergovernmental Panel on Climate Change (2007) could be used for nitrous oxide, but these factors may not provide an accurate representation of actual emissions from a particular facility as evidenced by the wide variability in nitrous oxide emissions from a nationwide survey of various facilities (Ahn et al., 2010; IPCC, 2006).

Nitrous oxide emissions reported by Chandran et al. (2011), Hiatt and Grady (2008), and Kampschreur et al. (2009) relate nitrous oxide production to the presence of anoxic/oxic zones, high nitrous oxide concentrations, and variable influent ammonia loading (Ahn et al., 2010).

Nitrous oxide models developed as an extension of activated sludge models (Chandran et al., 2011; Houweling et al., 2011; Mampaey et al., 2011; Yu et al., 2010) could be used to predict nitrous oxide emissions and develop mitigation strategies for a particular facility. However, to date, no consensus has been reached on actual nitrous oxide emission mechanisms and current models have yet to be calibrated and validated.

8.0 REFERENCES

Ahn, J. H.; Kim, S.; Park, H.; Rahm, B.; Pagilla, K.; Chandran, K. (2010) N2O Emissions from Activated Sludge Processes, 2008-2009: Results of a National Monitoring Survey in the United States. *Environ. Sci. Technol.*, **44**, 4505–4511.

American Society of Civil Engineers Task Committee on Belt Filter Presses (1988) Belt Filter Press Dewatering of Wastewater Sludge. *J. Environ. Eng.*, **114** (5), 991–1007.

Andrews, J. F. (1969) Dynamic Model of the Anaerobic Digestion Process. *Am. Soc. Civ. Eng. J. Sanit. Eng. Div.*, **95** (1), 95–116.

Andrews, J. F.; Graef, S. P. (1971) Dynamic Modeling and Simulation of the Anaerobic Digestion Process. In *Anaerobic Biological Treatment Processes*; American Chemical Society: Washington, D.C.

Antoine, R. L. (1976) *Fixed Biological Surfaces–Wastewater Treatment*; CRC Press: Cleveland, Ohio.

Atkinson, B.; Davies, I. J. (1974) The Overall Rate of Substrate Uptake (Reaction) by Microbial Films. Part I. A Biological Rate Equation. *Trans. Inst. Chem. Eng.*, **52**, 248–259.

Atkinson, B.; How, S. Y. (1974) The Overall Rate of Substrate Uptake (Reaction) by Microbial Films. Part II. Effect of Concentration and Thickness with Mixed Microbial Films. *Trans. Inst. Chem. Eng.*, **52**, 260–572.

Bagley, D. M.; Brodkorb, T. S. (1999) Modeling Microbial Kinetics in an Anaerobic Sequencing Batch Reactor Model Development and Experimental Validation. *Water Environ. Res.*, **71**, 1320–1332.

Bakke, R.; Trulear, M. G.; Robinson, J. A.; Characklis, W. G. (1984) Activity of Pseudomonas Aeruginosa in Biofilms: Steady State. *Biotechnol. Bioeng.*, **26**, 1418–1424.

Barker, P. S.; Dold, P. L. (1997) General Model for Biological Nutrient Removal Activated Sludge Systems: Model Presentation. *Water Environ. Res.*, **69**, 969–984.

Batstone, D. J.; Keller, J.; Angelidaki, I.; Kalyuzhnyi, S.; Pavlostathis, S. G.; Rozzi, A.; Sanders, W.; Siegrist, H.; Vavilin, V. (2002) *Anaerobic Digestion Model No. 1 (ADM1)*; IWA Publishing: London, U.K.

Beun, J. J.; Paletta, F.; van Loosdrecht, M. C. M.; Heijnen, J. J. (2000) Stoichiometry and Kinetics of Poly-Beta-Hydroxybutyrate Metabolism in Aerobic, Slow Growing, Activated Sludge Cultures. *Biotechnol. Bioeng.*, **67** (4), 379–389.

Beun, J. J.; Verhoef, E. V.; van Loosdrecht, M. C. M.; Heijnen, J. J. (2000) Stoichiometry and Kinetics of Poly-Beta-Hydroxybutyrate Metabolism Under

Denitrifying Conditions in Activated Sludge Cultures. *Biotechnol. Bioeng.*, **68** (5), 496–507.

Boltz, J. P.; Daigger, G. T. (2010) Uncertainty in Bulk-Liquid Hydrodynamics and Biofilm Dynamics Creates Uncertainties in Biofilm Reactor Design. *Water Sci. Technol.*, **61** (2), 307.

Boltz, J. P.; Goodwin, S. G.; Rippon, D.; Daigger, G. T. (2008) A Review of Operational Control Strategies for Snail and other Macrofauna Infestations in Trickling Filters. *Water Practice*, 2 (4).

Boltz, J. P.; Johnson, B. R.; Daigger, G. T.; Sandino, J. (2009) Modeling Integrated Fixed Film Activated Sludge (IFAS) and Moving Bed Biofilm Reactor (MBBR) Systems I: Mathematical Treatment and Model Development. *Water Environ. Res.*, **81** (6), 555–575.

Boltz, J. P.; Johnson, B. R.; Daigger, G. T.; Sandino, J.; Elenter, D. (2009) Modeling Integrated Fixed Film Activated Sludge (IFAS) and Moving Bed Biofilm Reactor (MBBR) Systems II: Evaluation. *Water Environ. Res.*, **81** (6), 576–586.

Boltz, J. P.; Morgenroth, E.; Brockmann, D.; Bott, C.; Gellner, W. J.; Vanrolleghem, P. A. (2011) Systematic Evaluation of Biofilm Models for Engineering Practice: Components and Critical Assumptions. *Water Sci. Technol.*, **64** (4), 930–944.

Boltz, J. P.; Morgenroth, E.; Brockmann, D.; Daigger, G. T.; Henze, M.; Rittmann, B.; Sørensen, K. H.; Takács, I.; Vanrolleghem, P. A.; van Loosdrecht, M. C. M. (2012) Framework for Biofilm Model Calibration Protocol. *Proceedings of the 3rd International Water Association/Water Environment Federation Wastewater Treatment Modeling Seminar (WWTmod2012)*; Mont-Sainte-Anne, Québec, Canada, Feb 26–28.

Boltz, J. P.; Morgenroth, E.; Sen, D. (2010) Mathematical Modeling of Biofilms and Biofilm Reactors for Engineering Design. *Water Sci. Technol.*, **62** (8), 1821–1836.

Bryers, J. D. (1984) Biofilm Formation and Chemostat Dynamics: Pure and Mixed Culture Considerations. *Biotechnol. Bioeng.*, **26**, 948–958.

Bryers, J. D. (1988) Modeling Biofilm Accumulation. In *Physiological Models in Microbiology*; Bazin, M., and Prosser, J. J., Eds.; CRC Press: Boca Raton, Florida; pp. 109–144.

Carman, P. (1937) Fluid Flow Through Granular Beds. In *Chemical Engineering Research and Design,* Vol. 15a; Institution of Chemical Engineers: Rugby, U.K.; pp. 150–166.

Chandran, K.; Stein, L.; Klotz, M.; van Loosdrecht, M. (2011) Nitrous Oxide Production by Lithotrophic Ammonia-Oxidizing Bacteria and Implications for Engineered Nitrogen-Removal Systems. *Biochem. Soc. Trans.*, **39**, 1832–1837.

Chang, H. T.; Rittmann, B. E. (1987) Mathematical Modeling of Biofilm on Activated Carbon. *Environ. Sci. Technol.*, **21**, 273–280.

Costello, O. J.; Greenfield, P. F.; Lee, P. L. (1991) Dynamic Modeling of a Single-Stage High-Rate Anaerobic Reactor I. Model Derivation. *Water Res.*, **25**, 847–858.

Daigger, G. T.; Nolasco, D. (1995) Evaluation and Design of Full-Scale Wastewater Treatment Plants Using Biological Process Models. *Water Sci. Technol.*, **31** (2), 245–255.

Dold, P. L. (1991) Modification du Modele General des Boues Activees pour Tenir Compte de la Dephosphatation. *Sciences et Techniques de L'eau*, **24**, 229–243.

Eberl, H. J.; Picioreanu, C.; Heijnen, J. J.; van Loosdrecht, M. C. M (2000) A Three-Dimensional Numerical Study on the Correlation of Spatial Structure, Hydrodynamic Conditions, and Mass Transfer Conversion in Biofilms. *Chem. Eng. Sci.*, **55**, 6209–6222.

Fairlamb, M.; Jones, R.; Takács, I.; Bye, C. (2003) Formulation of a General Model for Simulation of pH in Wastewater Treatment Processes. *Proceedings of the 76th Annual Water Environment Federation Technical Exposition and Conference* [CD ROM]; Los Angeles, California, Oct 11–15; Water Environment Federation: Alexandria, Virginia.

Filipe, C. D. M.; Daigger, G. T.; Grady, C. P. L. (2001) A Metabolic Model for Acetate Uptake Under Anaerobic Conditions by Glycogen Accumulating Organisms: Stoichiometry, Kinetics, and the Effect of pH. *Biotechnol. Bioeng.*, **76** (1), 17–31.

Fruhen, M.; Christan, E.; Gujer, W.; Wanner, O. (1991) Significance of Spatial Distribution of Microbial Species in Mixed-Culture Biofilms. *Water Sci. Technol.*, **23** (7-9), 1365–1374.

Gapes, D.; Pérez, J.; Picioreanu, C.; van Loosdrecht, M. C. M. (2006) Corrigendum to "Modeling Biofilm and Floc Diffusion Processes Based on Analytical Solution of Reaction-Diffusion Equations". *Water Res.*, **40**, 2997–2998.

Grady, L. E., Jr.; Daigger, G. T.; Love, N. G., Filipe, C. D. M. (2011) *Biological Wastewater Treatment*, 3rd ed.; Taylor and Francis: New York.

Gujer, W.; Boller, M. (1986) Design of a Nitrifying Trickling Filter Based on Theoretical Concepts. *Water Res.*, **20**, 1353.

Harremoës, P. (1978) *Biofilm Kinetics in Water Pollution Microbiology*, Vol. 2; Michell, R., Ed.; Wiley & Sons: New York.

Harris, N. P.; Hansford, G. S. (1978) A Study of Substrate Removal in a Microbial Film Reactor. *Water Res.*, **10** (11), 935–943.

Hauduc, H.; Rieger, L.; Takács, I.; Héduit, A.; Vanrolleghem, P. A; Gillot, S. (2010) Systematic Approach for Model Verification—Application on Seven Published Activated Sludge Models. *Water Sci. Technol.*, **61** (4), 825–839.

Henze, M.; Gujer, W.; Mino, T.; Matsuo, T.; Wentzel, M. C.; Marais, G. V. R.; van Loosdrecht, M. C. M. (1999) Activated Sludge Model No. 2d, ASM2d. *Water Sci. Technol.*, **39** (1), 165–182.

Hiatt, W. C.; Grady, C. P. L., Jr. (2008) An Updated Process Model for Carbon Oxidation, Nitrification, and Denitrification. *Water Environ. Res.*, **80**, 2145–2156.

Houweling, D.; Dold, P.; Wunderlin, P.; Joss, A.; Siegrist, H. (2011) N2O Emissions: Impact of Process Configuration and Diurnal Loading Patterns. *Proceedings of the International Water Association/Water Environment Federation Nutrient Recovery and Management Conference*; Miami, Florida, Jan 9–12.

Intergovernmental Panel on Climate Change (2006) IPCC Guidelines for National Greenhouse Gas Inventories; Prepared by the National Greenhouse Gas Inventories Programme; Eggleston, H. S., Buendia, L., Miwa, K., Ngara, T., Tanabe, K., Eds.; IGES: Hayama, Japan.

Intergovernmental Panel on Climate Change (2007) The Physical Science Basis-Contribution of Working Group I to the Fourth Assessment Report of the Intergovernmental Panel on Climate Change; Solomon, S., Qin, D., Manning, M., Chen, Z., Marquis, M., Averyt, K. B., Tignor, M., Miller, H. L., Eds.; Cambridge University Press: New York.

Kampschreur, M. J.; Temmink, H.; Kleerebezem, R.; Jetten, M. S. M.; van Loosdrecht, M. C. M. (2009) Nitrous Oxide Emission during Wastewater Treatment. *Water Res.*, **43** (17), 4093–4103.

Kissel, J. C. (1986) Modeling Mass Transfer in Biological Wastewater Treatment Processes. *Water Sci. Technol.*, **18** (6), 35–45.

Kissel, J. C.; McCarty, P. L.; Street, R. L. (1984) Numerical Simulation of Mixed-Culture Biofilm. *J. Environ. Eng.*, **110** (2), 393–411.

Kreft, J.-U.; Picioreanu, C.; Wimpenny, J. W. T.; van Loosdrecht, M. C. M. (2001) Individual Based Modeling of Bacterial Colony Growth. *Microbiology*, **147**, 2897–2912.

Kreikenbohm, R.; Stephan, W. (1985) Application of a Two-Compartment Model to the Wall Growth of *Pelobacter Acidigallici* under Continuous Culture Conditions. *Biotechnol. Bioeng.*, **27**, 296–301.

Lewandowski, Z.; Boltz J. P. (2011) Biofilms in Water and Wastewater Treatment. In *Treatise on Water Science*, Vol. 4; Wilderer, P., Ed.; Oxford: Academic Press; pp. 529–570.

Mampaey, K. E.; Beuckels, B.; Kampschreur, M. J.; Kleerebezem, R.; van Loosdrecht, M. C. M.; Volcke, E. I. P. (2011) Modeling Nitrous and Nitric Oxide Emissions by Autotrophic Ammonium Oxidizing Bacteria. *Proceedings of the Nutrient Recovery and Management 2011 Conference*; Miami, Florida, Jan 9–12.

Massé, D. I.; Droste, R. L. (2000) Comprehensive Model of Anaerobic Digestion of Swine Manure Slurry in a Sequencing Batch Reactor. *Water Res.*, **34,** 3087–3106.

Maurer, M.; Abramovich, D.; Siegrist, H.; Gujer, W. (1999) Kinetics of Biologically Induced Phosphorus Precipitation in Waste-Water Treatment. *Water Res.*, **33** (2), 484–493.

Morgenroth, E. (2003) Detachment—An Often Overlooked Phenomenon in Biofilm Research and Modeling. In *Biofilms in Wastewater Treatment*; Wuertz, S., Wilderer, P. A., Bishop, P. L., Eds.; IWA Publishing: London; pp. 264–290.

Morgenroth, E. (2008) Modeling Biofilm Systems. In *Biological Wastewater Treatment—Principles, Modeling, and Design*; Henze, M., van Loosdrecht, M. C. M., Ekama, G., Brdjanovic, D., Eds.; IWA Publishing: London.

Morgenroth, E.; van Loosdrecht, M. C. M.; Wanner, O. (2000) Biofilm Models for the Practitioner. *Water Sci. Technol.*, **41** (4-5), 509–512.

Morgenroth, E.; Wilderer, P. A. (2000) Influence of Detachment Mechanisms on Competition in Biofilms. *Water Res.*, **34**, 417–426.

Mosey, F. E. (1983) Mathematical Modeling of the Anaerobic Digestion Process: Regulatory Mechanisms for the Formation of Short Chain Volatile Acids from Glucose. *Water Sci. Technol.*, **21,** 187–196.

Murnleitner, E.; Kuba, T.; van Loosdrecht, M. C. M.; Heijnen, J. J. (1997) An Integrated Metabolic Model for the Aerobic and Denitrifying Biological Phosphorus Removal. *Biotechnol. Bioeng.*, **54** (5), 434–450.

Musvoto, E. V.; Wentzel, M. C.; Ekama, G. A. (2000) Integrated Chemical–Physical Processes Modeling—II. Simulating Aeration Treatment of Anaerobic Digester Supernatants. *Water Res.*, **34** (6), 1868–1880.

Newbigging, M.; Nolasco, D.; DeAngelis, W.; Dobson, B. (1995) Overlooked Impact: Redirecting Solids Recycle Streams Enhances Plant Performance. *Water Environ. Technol.*, **7** (5), 50–54.

Noguera, D. R.; Pizarro, G. E.; Regan, J. M. (2004) Modeling Biofilms. In *Microbial Biofilms*; Ghannoum, M., O'Toole, G., Eds.; ASM Press: Washington, D.C.

Nolasco, D. A. (1989) Expert System—Simulation Interaction for Control of the Activated Sludge Process. M. Eng. Thesis, McMaster University, Hamilton, Ontario, Canada.

Pavlostathis, S. G.; Giraldo-Gomez, E. (1991) Kinetics of Anaerobic Treatment: A Critical Review. *CRC Crit. Rev. Environ. Control*, **21** (5-6), 411–490.

Pérez, J.; Picioreanu, C.; van Loosdrecht, M. C. M. (2005) Modeling Biofilm and Floc Diffusion Processes Based on Analytical Solution of Reaction-Diffusion Equations. *Water Res.*, **39**, 1311–1323.

Peyton, B. M.; Characklis, W. G. (1993) A Statistical-Analysis of the Effect of Substrate Utilization and Shear-Stress on the Kinetics of Biofilm Detachment. *Biotechnol. Bioeng.*, **41** (7), 728–735.

Picioreanu, C.; Kreft, J.-U.; van Loosdrecht, M. C. M. (2003) Particle-Based Multidimensional Multispecies Biofilm Model. *Appl. Environ. Microbiol.*, **70** (5), 3024–3040.

Picioreanu, C.; van Loosdrecht, M. C. M.; Heijnen, J. J. (1998) Mathematical Modeling of Biofilm Structure with a Hybrid Differential Discrete Cellular Automaton Approach. *Biotechnol. Bioeng.*, **58** (1), 101–116.

Pramanik, J.; Trelstad, P. L.; Schuler, A. J.; Jenkins, D.; Keasling, J. D. (1999) Development and Validation of a Flux-Based Stoichiometric Model for Enhanced Biological Phosphorus Removal Metabolism. *Water Res.*, **33** (2), 462–476.

Rauch, W.; Vanhooren, H.; Vanrolleghem, P. A. (1999) A Simplified Mixed-Culture Biofilm Model. *Water Res.*, **33** (9), 2148–2162.

Reichert, P. (1998a) AQUASIM 2.0—User Manual; Swiss Federal Institute for Environmental Science and Technology (Eawag): Dübendorf, Switzerland.

Reichert, P. (1998b) AQUASIM 2.0—Tutorial; Swiss Federal Institute for Environmental Science and Technology (Eawag): Dübendorf, Switzerland.

Reichert, P.; Wanner, O. (1997) Movement of Solids in Biofilms: Significance of Liquid Phase Transport. *Water Sci. Technol.*, **36** (1), 321–328.

Rittmann, B. E. (1982) The Effect of Shear Stress on Biofilm Loss Rate. *Biotechnol. Bioeng.*, **24**, 504–506.

Rittmann, B. E. (1989) Detachment from Biofilms. In *Structure and Function of Biofilms*; Characklis, W. G., Wilderer, P. A., Eds.; Wiley & Sons: New York; pp. 49–58.

Rittmann, B. E.; McCarty, P. L. (1980a) Evaluation of Steady-State-Biofilm Kinetics. *Biotechnol. Bioeng.*, **22**, 2359–2373.

Rittmann, B. E.; McCarty, P. L. (1980b) Model of Steady-State-Biofilm Kinetics. *Biotechnol. Bioeng.*, **22**, 2343–2357.

Rittmann, B. E.; McCarty, P. L. (2001) *Environmental Biotechnology: Principles and Application*; McGraw-Hill: Boston, Massachusetts.

Rittmann, B. E.; Stilwell, D.; Ohashi, A. (2002) The Transient-State, Multiple Species, Biofilm Model for Biofiltration Processes. *Water Res.*, **36**, 2342–2356.

Robinson, J. A.; Trulear, M. G.; Characklis, W. G. (1984) Cellular Reproduction and Extracellular Polymer Formation by *Pseudomonas Aeruginosa* in Continuous Culture. *Biotechnol. Bioeng.*, **26**, 1409–1417.

Roels, J. A. (1983) *Energetics and Kinetics in Biotechnology*; Elsevier Biomedical Press: New York.

Sáez, P. B.; Rittmann, B. E. (1992) Accurate Pseudo Analytical Solution for Steady State Biofilms. *Biotechnol. Bioeng.*, **39**, 790–793.

Sen, D.; Randall, C. W. (2008a) Improved Computational Model (AQUIFAS) for Activated Sludge, Integrated Fixed-Film Activated Sludge, and Moving-Bed Biofilm Reactor Systems, Part I: Semi-Empirical Model Development. *Water Environ. Res.*, **80**, 439–453.

Sen, D.; Randall, C. W. (2008b) Improved Computational Model (AQUIFAS) for Activated Sludge, Integrated Fixed-Film Activated Sludge, and Moving-Bed Biofilm Reactor Systems, Part III: Analysis and Verification. *Water Environ. Res.*, **80**, 633–646.

Sin, G.; Kaelin, D.; Kampschreur, M. J.; Takacs, I.; Wett, B.; Gernaey, K. V.; Rieger, L.; Siegrist, H.; van Loosdrecht, M. C. M. (2008) Modeling Nitrite in Wastewater Treatment Systems: A Discussion of Different Modeling Concepts. *Proceedings of the First International Water Association/Water Environment Federation Wastewater Treatment Modeling Seminar*; Mont-Saint Anne, Quebec, Feb.

Smolders, G. J. F.; Van-Der-Meij, J.; van Loosdrecht, M. C. M.; Heijnen, J. J. (1995) A Structured Metabolic Model for Anaerobic and Aerobic Stoichiometry and Kinetics of the Biological Phosphorus Removal Process. *Biotechnol. Bioeng.*, **47** (3), 277–287.

Sotemann, S. W.; van Rensburg, P.; Ristow, N. E.; Wentzel, M. C.; Loewenthal, R. E.; Ekama, G. A. (2005) Integrated Chemical/Physical and Biological Processes Modeling Part 2—Anaerobic Digestion of Sewage Sludges. *Water SA*, **31** (4), 545–568.

Speitel, G. E.; DiGiano, F. A. (1987) Biofilm Shearing under Dynamic Conditions. *J. Environ. Eng.*, **113**, 464–475.

Strous, M.; Heijnen, J. J.; Kuenen, J. G.; Jetten, M. S. M. (1998) The Sequencing Batch Reactor as a Powerful Tool for the Study of Slowly Growing Anaerobic Ammonium-Oxidizing Microorganisms. *Appl Microbiol. Biotechnol.*, **50**, 589–596.

Takács, I.; Bye, C. M.; Chapman, K.; Dold, P.; Fairlamb, P. M.; Jones, R. M. (2007) A Biofilm Model for Engineering Design. *Water Sci. Technol.*, **55** (8–9), 329–336.

Takács, I.; Patry, G. G.; Nolasco, D. (1991) A Generalized Dynamic Model for the Thickening/Clarification Process. *Water Res.*, **25** (10), 1263–1271.

Tijhuis, L.; van Loosdrecht, M. C. M.; Heijnen, J. J. (1995) Dynamics of Biofilm Detachment in Biofilm Airlift Suspension Reactors. *Biotechnol. Bioeng.*, **45** (6), 481–487.

Van Hulle, S. (2005). Modeling, Simulation, and Optimization of Autotrophic Nitrogen Removal Processes. Ph.D. Thesis, Ghent University, Ghent, Belgium.

Vavilin,V. A.; Vasiliev, V. B.; Rytov, S. V.; Ponomarev, A. V. (1995) Modeling Ammonia and Hydrogen Sulfide Inhibition in Anaerobic Digestion. *Water Res.*, **29** (3), 827–835.

Vesilind, P. A. (1968) Theoretical Considerations: Design of Prototype Thickeners from Batch Settling Tests. *Water Sew. Works,* **115** (7), 302–307.

Vitasovic, Z. Z. (1989) Continuous Settler Operation: A Dynamic Model. In *Dynamic Modeling and Expert Systems in Wastewater Engineering;* Patry, G. G., Chapman, D., Eds.; Lewis Publishers: Chelsea, Michigan.

Volcke, E. I. P. (2006) Modeling, Analysis and Control of Partial Nitritation in a SHARON Reactor. Ph.D. Thesis, Ghent University, Ghent, Belgium.

Wanner, O.; Eberl, H.; Morgenroth, E.; Noguera, D.; Picioreanu, C.; Rittman, B.; van Loosdrecht, M. (2006) *Mathematical Modeling of Biofilms*; IWA Task Group on Biofilm Modeling, Scientific and Technical Report 18; IWA Publishing: London, U.K.

Wanner, O.; Gujer, W. (1984) Competition in Biofilms. *Water Sci. Technol.*, **17** (2/3), 27–44.

Wanner, O.; Gujer, W. (1986) A Multi-Species Biofilm Model. *Biotechnol. Bioeng.*, **28**, 314–328.

Wanner, O.; Reichert, P. (1996) Mathematical Modeling of Mixed-Culture Biofilms. *Biotechnol. Bioeng.*, **49** (2), 172–184.

Water Environment Federation; American Society of Civil Engineers/ Environmental & Water Resources Institute (2009) *Design of Municipal Wastewater Treatment Plants*, 5th ed.; WEF Manual of Practice No. 8, ASCE Manuals and Reports on Engineering Practice No. 76; McGraw-Hill: New York.

Water Environment Research Foundation (1995) *Comparison of UV Irradiation to Chlorination: Guidance for Achieving Optimal UV Performance*; Project 91-WWD-1 Final Report; Water Environment Research Foundation: Alexandria, Virginia.

Wentzel, M. C.; Dold, P. L.; Ekama, G. A.; Marais, G. v. R. (1989) Enhanced Polyphosphate Organism Cultures in Activated Sludge Systems. Part III: Kinetic Model. *Water SA,* **15** (2), 89–102.

Wett, B.; Rauch, W. (2003) The Role of Inorganic Carbon Limitation in Biological Nitrogen Removal of Extremely Ammonia Concentrated Wastewater. *Water Res.*, **37**, 1100–1110.

Williamson, K.; McCarty, P. L. (1976) A Model of Substrate Utilization by Bacterial Films. *J.—Water Pollut. Control Fed.*, **48**, 9–24.

Wuertz, S.; Falkentoft, C. M. (2003) Modeling and Simulation: Introduction. In *Biofilm in Wastewater Treatment*; Wuertz, S., Bishop, P., Wilderer, P., Eds.; IWA Publishing: London.

Yu, R.; Kampschreur, M.; van Loosdrecht, M.; Chandran, K. (2010) Mechanisms and Specific Directionality of Autotrophic Nitrous Oxide and Nitric Oxide Generation during Transient Anoxia. *Environ. Sci. Technol.*, **44**, 1313–1319.

Zhang, T. C.; Bishop, P. L. (1994) Density, Porosity, and Pore Structure of Biofilms. *Water Res.*, **28** (11), 2267–2277.

9.0 SUGGESTED READINGS

Beier, A. G. (1987) *Lagoon Performance in Alberta*; Environmental Protection Services, Standards and Approvals Division, Municipal Branch, Alberta Environment: Edmonton, Alberta, Canada.

Boltz, J. P.; La Motta, E. J. (2007) The Kinetics of Particulate Organic Matter Removal as a Response to Bioflocculation in Aerobic Biofilm Reactors. *Water Environ. Res.*, **79** (7), 725–735.

Bruce, A. M.; Boon, A. G. (1971) Aspects of High-Rate Biological Treatment of Domestic and Industrial Wastewaters. *J.—Water Pollut. Control, Fed.*, **70**, 487–513.

DeBarbadillo, C.; Shaw, A. R.; Wallis-Lage, C. L. (2005) Evaluation and Design of Deep-Bed Denitrification Filters: Empirical Design Parameters vs. Process Modeling. *Proceedings of the 78th Annual Water Environment Federation Technical Exposition and Conference* [CD-ROM]; Washington, D.C., Oct. 29–Nov. 2; Water Environment Federation: Alexandria, Virginia.

Eckenfelder, W. (1980) *Principles of Water Quality Management*; CBI Publishing Co.: Boston, Massachusetts.

Gujer, W.; Henze, M.; Mino, T.; van Loosdrecht, M. C. M. (1999) Activated Sludge Model No. 3. *Water Sci. Technol.*, **39** (1), 183–193.

Henze, M.; Grady, C. P. L.; Gujer, W.; Marais, G. V. R.; Matsuo, T. (1987) *Activated Sludge Model No. 1.*; IAWPRC Scientific and Technical Report No. 1; International Association on Water Pollution Research and Control: London.

Henze, M.; Gujer, W.; Mino, T.; Matsuo, T.; Wentzel, M. C.; Marais, G. V. R. (1995) *Activated Sludge Model No. 2*; IAWQ Scientific and Technical Report No. 3; International Association on Water Quality: London.

Morgenroth, E.; Eberl, H.; van Loosdrecht, M. C. M. (2000) Evaluating 3-D and 1-D Mathematical Models for Mass Transport in Heterogeneous Biofilms. *Water Sci. Technol.*, **41** (4-5), 347–356.

Rauch, W., Vanrolleghem, P. A. (1997) *A Dynamic Biofilm Model for Simulation of Wastewater Treatment Processes and Benthic Activity in Rivers*; Technical Report; BIOMATH Department, Gent University: Ghent, Belgium.

Sáez, P. B.; Rittmann, B. E. (1988) Improved Pseudo-Analytical Solutions for Steady-State Biofilm Kinetics. *Biotechnol. Bioeng.*, **32**, 379–385.

Spengel, D. B.; Dzombak, D. A. (1992) Biokinetic Modeling and Scale-Up Considerations for Rotating Biological Contactors. *Water Environ. Res.*, **64**, 223–235.

Thirumurthi, D. (1974) Design Criteria for Waste Stabilization Ponds. *J.—Water Pollut. Control Fed.*, **46**, 2094–2106.

Trulear, M. G.; Characklis, W. G. (1982) Dynamics of Biofilm Processes. *J.—Water Pollut. Control Fed.*, **54**, 1288–1301.

U.S. Environmental Protection Agency (2010) *Inventory of U.S. Greenhouse Gas Emissions and Sinks: 1990-2008*; EPA-430/R-10-006; U.S. Environmental Protection Agency, Office of Atmospheric Programs: Washington, D.C.

Velz, C. J. (1948) Basic Law for the Performance of Biological Filters. *Sew. Works J.*, **20**, 607–617.

Wanner, O. (1989) Modeling Population Dynamics. In *Structure and Functions of Biofilms*; Characklis, W. G., Wilderer, P. A., Eds.; Wiley & Sons: New York; pp. 91–110.

Water Environment Federation (2010) *Nutrient Removal*; WEF Manual of Practice No. 34; McGraw-Hill: New York.

Chapter 4

Process Modeling Tools

1.0 INTRODUCTION

The preceding chapters of this manual provide a general description of process modeling and the types of models that are available in general terms. In this chapter, various modeling tools (i.e., the actual software that can be used to develop and run models) are described. First, however, it is worth reiterating the difference between *models* and *tools*. A *model* is an equation or set of equations that attempts to describe a process observation, and *tools*, which are typically software packages, are what the designer can use to do the modeling design work. For example, Activated Sludge Model No. 1 (ASM1) (Henze et al., 2000) is a model, whereas Excel, BioWin,

117

GPS-X, and WEST are tools as defined by these criteria. That is, any one of these software tools could be used to simulate ASM1. The dedicated wastewater treatment software packages often contain proprietary models, but these software packages are not models in and of themselves. In this manual, software tools have been broadly categorized as one of the following types: spreadsheet-based, dedicated wastewater treatment simulators, and general-purpose simulators. These three software tools are briefly described in this chapter.

2.0 SPREADSHEET-BASED TOOLS

One of the most common methods of performing wastewater treatment process calculations is to use a spreadsheet. This section presents a variety of common applications for models and discusses important considerations in deciding to use a spreadsheet-based model.

2.1 Scope

Spreadsheet-based models can be developed by individuals for individual use or by large organizations for corporate-wide use. This section presents some common applications of models by individuals and corporations. It also addresses the applicability of spreadsheet models to dynamic systems.

2.1.1 Common Applications

This section presents common ways that spreadsheets are used to model process behavior. The following applications will be discussed:

- Activated sludge process (ASP) design
- Clarifier design
- Hydraulic design
- Mass balances
- Chemical addition

2.1.1.1 Activated Sludge Process Design

Activated sludge process design refers to using a spreadsheet to simulate or calculate a design unknown. For instance, design spreadsheets can be used to calculate initial estimates for tank volumes given an input mass load of waste and defined effluent quality or they can be used to predict effluent quality from a defined ASP process.

The simulated process could vary from a simple aerated design to a modified Ludzak–Ettinger process to a five-stage biological nutrient removal (BNR) process. Table 4.1 presents common input parameters that the user may want to specify.

Typically, the user will need to develop an interface (i.e., drop-down menus or separate tabs) for selecting different treatment processes with separate algorithms for different process configurations. The amount of time and expertise required to do so can quickly grow as there are many different kinds of technology that can be used.

Choosing between independent and dependent variables is complicated; therefore, thought should be given at the developmental stage about how a tool can be most useful. For instance, a user will often be looking to use the spreadsheet to develop a model that meets specific effluent targets (i.e., biochemical oxygen demand

TABLE 4.1   Common input parameters.

Influent	Primary effluent	Sidestream
Flow		Flow
Average day		
Maximum month		
Maximum day		
Peak hour		
Chemical oxygen demand (COD)	COD	COD
Biochemical oxygen demand (BOD)	BOD	BOD
Total suspended solids (TSS)	TSS	TSS
Volatile suspended solids (VSS)	VSS	VSS
Total Kjeldahl nitrogen (TKN)	TKN	TKN
Ammonia	Ammonia	Ammonia
		Nitrate
Total phosphorus	Total phosphorus	Total phosphorus
Orthophosphate	Orthophosphate	Orthophosphate
Alkalinity	Alkalinity	Alkalinity
pH		
Temperature		
Fractionation of COD, BOD, TKN, and total phosphorus		

[BOD] < 5 mg/L, total suspended solids [TSS] < 5 mg/L, ammonia < 1 mg/L, total nitrogen < 5 mg/L, and total phosphorus < 1 mg/L) within a specified factor of safety. Using this approach, the model would return a required size for each zone or process tank. However, the user/developer may prefer to adjust the size of the BNR basins to achieve a suitable result.

Another important feature of ASP modeling tools is the ability to link conceptual process volume to the concrete structures that will be built or already exist. While a model can predict the total process volume required, some engineering judgment is subsequently required to allocate that volume among a variety of basins and to develop a footprint, baffle wall placement, and so on. Redundancy must also be considered. Furthermore, for existing basins, it is important to reflect how "plug flow" the basins actually are to accurately predict their performance. Again, judgment, experience, and flexibility in model development are valuable.

2.1.1.2 Clarifier Design

Clarifier design refers to the selection of an appropriately sized clarifier for a process. A steady-state approach often used for clarifier modeling is state point analysis. State point analysis considers influent flow, return activated sludge (RAS) flow, mixed liquor suspended solids (MLSS) concentrations, surface area of the clarifiers, and settling parameters.

Whereas ASP models will typically assume that clarifiers are performing optimally, it may be desirable to incorporate a clarifier model to an ASP model to account for worst-case settling conditions and/or the storage of solids that might occur in the clarifiers.

2.1.1.3 Hydraulic Models

Spreadsheets can be an excellent tool for modeling facility hydraulics. One common approach to this is to develop modules for different hydraulic conditions that a user can cut and paste together to build a water resource recovery facility (WRRF) hydraulic profile from downstream to upstream.

Because it is simple to incorporate several flow scenarios to a model, the user may want to consider water surface elevations over a variety of flow conditions that bracket the operation of the facility. The user should take care to ensure internal recycle flows are also included such as RAS, nitrified recycle, and sidestream flows.

2.1.1.4 Mass Balances

A spreadsheet can be an easy way to develop mass balances around critical points in the process. For example, a mass balance around clarifiers can be used to assess the quality of facility data. This can be particularly useful in resolving conflicts during the calibration of dynamic models. For instance, using facility wasting records to predict the mixed liquor concentration may not result in an accurate comparison. Therefore, a mass balance can be a way to determine whether MLSS or the wasting data are unreliable.

2.1.1.5   Chemical Addition
Chemical addition reactions can be easily and reliably modeled with a spreadsheet. Two examples of chemical addition that one may want to model are precipitation reactions for orthophosphate removal and supplemental carbon addition for nitrogen removal. Typically, flexibility in these spreadsheets is achieved by allowing users the option to select from a variety of chemical alternatives. For instance, carbon addition spreadsheets may include methanol, acetic acid, ethanol, and glycerin. Chemical precipitation reactions may include alum, aluminum sulfate, and ferric chloride.

An example of a carbon addition model is shown in Table 4.2. This spreadsheet is used to predict the required dosage of carbon needed to remove a specified mass of nitrate. With a chemical precipitation spreadsheet, the objective is typically to estimate the mass of solids formed and the dose of chemical required to achieve a particular effluent goal. Total phosphorus goals are a common reason coagulants are added.

2.1.2   Steady-State and Dynamic Simulations
As stated earlier, spreadsheet models are best suited for steady-state modeling. However, it is possible to arrive at an aggregate dynamic solution by using multiple steady-state models. For example, one could simulate predicted effluent quality over a 24-hour period. The user could develop 24 similar tabs, and vary the input similar to the influent diurnal loading pattern. Then, the user could take the flow-weighted average of the effluent quality to estimate what a 24-hour composite sample would measure. This approach, however, would be laborious and inefficient for a sustained dynamic simulation (i.e., a year-long simulation).

2.2   Considerations
Numerous issues should be considered when deciding on the use of a spreadsheet platform for wastewater treatment design and optimization. One of the obvious benefits is that the user interface is typically well known to modeling practitioners,

TABLE 4.2   Carbon addition model.

Parameter	Glycerin (typical)	Methanol (typical)	Units*
Estimated influent concentration of NOx-N to denitrify:	6.0	6.0	mg/L
Typical nitrate concentration leaving the anoxic zone (default is same as influent, and enter same value as influent if using denite filters):	4.8	4.8	mg/L
Estimated flow	10	10	mgd
Estimated RAS rate (enter 0 if to filters)	50	50	% of Q in
Estimated internal recycle rate (0 if post-anoxic zone or filters)	0	0	mgd
Target effluent nitrate-N	3	3	mg/L
Estimated dissolved oxygen in aeration basin effluent	2	2	mg/L
COD of product (mg/L)	930 000	1 180 000	mg/L
Solution density (lb/gal)	10.0	6.6	lb/gal
Heterotrophic yield for substrate	0.54	0.40	lb TSS/lb COD
Cost per gallon of carbon source	2.25	1.5	$/gal
Additional mass of nitrate-N to be denitrified	225	225	lb NO_3-N/d
C:N Ratio for carbon source on COD basis	6.2	4.8	lb COD/lb nitrate-N
Mass of COD needed to denitrify only	1400	1073	lb COD/d
COD to substrate ratio	0.78	1.49	lb COD/lb (product - glycerin or methanol)
Volume product to denitrify only	181	109	gpd
Wasted COD to remove this dissolved oxygen	544	417	lb COD/d
Total lb COD required	1,944	1,490	lb/d as COD
Volume product needed (denitrify + consume dissolved oxygen)	251	151	gpd

TABLE 4.2   Carbon addition model. (*continued*)

Parameter	Glycerin (typical)	Methanol (typical)	Units*
Summary			
Pump feed rate	10.4	6.3	gph
Pump feed rate	658	398	mL/min
Number of days an 8000-gal tanker will last	32	53	
Repeat theoretical COD:N expected (no dissolved oxygen upstream)	6.2	4.8	lb COD/lb N removed
Actual COD:N expected	8.6	6.6	lb COD/lb N removed
Cost per year (chemical costs only)	$205,831	$82,914	$/year for product - glycerin or methanol

*mgd $\times$ 3785 = m³/d; lb/d $\times$ 0.45 = kg/d; gal $\times$ 3.8 = L; kg $\times$ 0.45 = lb; gpd $\times$ 0.004 = m³/d; and gph $\times$ 1.051 = mL/s.

which makes setup relatively easy. This aspect speeds the learning curve for users of these spreadsheets. The spreadsheet platform also provides flexibility and customization in achieving modeling goals. However, this same flexibility can also be detrimental if quality control is not incorporated to the development of the tool.

Spreadsheet tools can be freely distributed as needed, provided users have the spreadsheet platform already. Therefore, users typically are not required to purchase licenses or copies of the design tool. There are also some spreadsheet models that can be downloaded for free, which could alleviate some of the development effort.

Standardized spreadsheet tools also allow institutional design knowledge to be easily converted to computer code and communicated to the users of the tool. In addition, the developer of the tool will benefit from the process of building the model, although the process of building the model can be extremely time consuming, particularly with a full ASP model. Having a single corporate tool may simplify the subsequent quality control aspects of a design project if a team of people routinely uses the same approach and tool.

Use of spreadsheet models for design is clearly beneficial for repetitive tasks that are suitably performed with a steady-state tool. Prior to embarking on the development of a specific tool, users should carefully consider the goal for the tool, the time that will be required to develop the tool, and the methods that will be used to ensure error checking and quality control.

3.0   DEDICATED WASTEWATER TREATMENT SIMULATORS

This section provides an overview of design tools and dynamic models. In this chapter, a distinction is made between dedicated commercial tools for design and common commercially available dynamic simulators that support design decisions. The subsequent sections discuss these tools separately.

3.1   Introduction

The most commonly used process modeling tools are commercial simulators that have been developed during the past 20 years. These tools vary in the number and complexity of the processes that can be modeled, and the features included in each package also vary. In addition, it should be noted that the software purpose can also vary. Some simulators are running models specifically focused on wastewater treatment design, while others use dynamic models to predict the effect of a design. This is a subtle difference, but, in the former case, the simulator is used to calculate the size of a process and the required equipment given a user-defined required level of treatment. In the latter case, the model is used to predict the expected treatment behavior given user-defined process sizes and equipment. Specific design simulators have their roots in the 1970s and some use empirical design rules based on typically measured parameters such as BOD, food-to-microorganism ratio, and safety factors to come up with tank sizes. Equipment is sized based on empirical curves and a defined database of available equipment options. In contrast, dynamic simulators can be used to explore the behavior of a user-defined design (i.e., What is the expected effluent quality from a facility of a certain size and operated in a certain way? How does the effluent vary under dynamic loading conditions? How much oxygen is required for the process? How does oxygen demand vary throughout the day, week, and year?). Unlike dedicated design software (where the output might be a predictor of tank size), dynamic simulators can be used to predict process dynamics and, with sufficient data, the dynamic risk associated with a particular design (i.e., How often will the effluent ammonia exceed a given limit? Can dynamic control change that frequency?).

3.2 Design Software

Although this manual is primarily focused on dynamic model use, it is important to recognize that dedicated design software exists and that these software products are also based on models. Dedicated design software is available and can be used to help designers with preliminary tank sizing and equipment estimates. These simulators are not capable of running models dynamically, but, in some instances, are based on steady-state output from a dynamic model. Typically, users of these tools input a wastewater strength and a level of treatment and the simulator solves the various models to come up with a design solution. Not unlike the aforementioned spreadsheet (with solver) options, these simulators tend to be relatively easy to use and the models in these simulators are generally capable of calculating equipment sizes. However, the user needs to recognize that the output solution is based on a single set of user inputs. Plan-It STOAT from WRc plc (Swindon, U.K.) and CapdetWorks from Hydromantis Environmental Software Solutions, Inc., (Hamilton, Ontario, Canada) are examples of design tools.

CapdetWorks is based on the CAPDET model that was developed by the U.S. Army Corps of Engineers for the U.S. Environmental Protection Agency in 1973. The CAPDET model uses a parametric cost-estimating approach, which limits the overall utility of the model, although a revised cost-estimating procedure using both parametric and unit cost-estimating techniques has been developed. Revisions to the original model have expanded its capabilities and improved the accuracy of planning-level design, but the approach is still limited.

In contrast, Plan-It STOAT includes process models for most significant processes used at facilities and both specific design models, but also uses International Water Association ASM1, Activated Sludge Model No. 2d, and Activated Sludge Model No. 3. As with CapdetWorks, the output from Plan-It STOAT includes process volumes, chemical and energy requirements, and cost estimates.

Dedicated design software has its place in the marketplace, but it is not the panacea for designing viable solutions nor is it the industry standard approach that professional engineers use. Engineering judgment, equipment design limitations, site limitations, regulatory reliability, and maintenance standards all enter into formulation of good design. Dynamic disturbances and the ability of a facility to cope with those disturbances is also a critical part of design decisions related to equipment, tank sizes, and operation. The dynamic simulators discussed in the next section provide more insight to the expected dynamic behavior for the facility and thus provide even more knowledge that can be used to make the most of tanks, equipment, and operation.

3.3   Dynamic Simulators

The following sections outline key aspects of commercially available tools to provide an unbiased overview of things to consider when choosing a commercial software tool.

3.3.1   Understanding the Tools

Publications by Mehrotra et al. (2005), Sedran et al. (2006), and Takács et al. (2007) have attempted to discuss differences in the output of several different software products to highlight differences in model predictions. However, an issue with this type of analysis is that it must take into consideration some of the most fundamental differences in commercial products to be considered complete. Based on experience, commercially available simulation software packages will calculate precisely the same output if they are set up the same way and are simulating the same model (Copp, 2002). However, this is not often the case and, as such, comparison of out-of-the-box simulator results is problematic. In addition, it is important to recognize that model output is affected by simulator-specific features. Therefore, even if simulators are running the same model with the same parameter values, it is still possible that slightly different results may be generated.

Substantial effort went into verifying steady-state and dynamic output from several commercial products and the results were verified using BioWin™, EFOR™, GPS-X™, Simba®, STOAT™, and WEST® software (Alex et al., 1999; Pons et al., 1999). Cross platform testing provided significant insight to the simulators and the simulation process.

The results of those studies indicated that each of the simulators required simulator-specific alterations and fine-tuning to reach agreement in the output data. It is expected that the simulation software should have no effect on modeling output such that different simulators modeling the same system should give the same result. However, because of the many simulator-specific options, this is not the case, nor is it a trivial task to ensure similar results using different simulators. Although the simulators can be forced to give essentially identical steady-state results, the study found that it is not realistic for each simulator to produce the same instantaneous dynamic results. For instance, Figure 4.1 shows the dynamic output from three simulators using the same solids flux model. The differences illustrated in the figure are the result of different means used to propagate soluble components through the settler. In these three instances, the particulate components are modeled in the same way, but the soluble components are modeled differently. When using any tool,

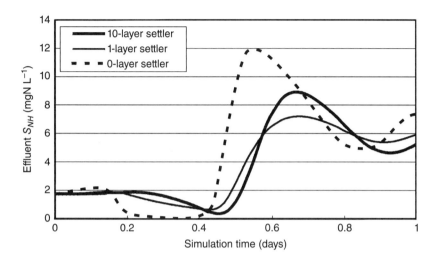

FIGURE 4.1   An example of the dynamic output produced for ammonia by three different simulators running the same 10-layer flux-based clarifier model for solids, but a different number of layers to simulate the soluble components.

care should be taken to understand the implementation of the models and the effect of that implementation.

3.3.2   *Freeware Tools*

Freeware is the generic terminology used herein to describe software that is generally available at no financial cost. Freeware simulators are typically available for download (STOAT) or are Web-based and intended to be run through a Web-based application (JASS).

In some instances, these freeware simulators were once sold commercially, but are now available for free, while others were academic exercises and have always been in the public domain. For example, STOAT is a dynamic simulator that was widely used in the U.K. prior to privatization of water companies and has a standard set of activated sludge processes and models. STOAT used to be commercially sold to clients; however, with increased competition from emerging simulators and a desire to limit development and ongoing support to clients, the developers (WRc plc) made the software available at no cost, but no longer provide support to users. By providing only a standard set of models, STOAT's capabilities are limited, but the software can be useful for simple activated sludge studies. Similar limitations are noted with other freeware options. JASS (Uppsala University, Uppsala, Sweden) is a Web-based application with limited application, but an option worth considering if the goals of the simulation exercise are consistent with the software capability (e.g., as a teaching tool or mass balance calculator).

In summary, the key aspect to consider with a freeware tool is the lack of support. Because the software is free, the developers tend to not provide product or modeling support; this could affect a user's ability to deliver a project result.

3.3.3 Commercial Tools

The most common process modeling tools used today are dynamic commercial simulators. Commercially available software packages provide access to technical support, ongoing software and model development, and access to more software and simulation features in general. Purchasing a license not only brings with it existing features, but also supports development and, typically, access to that development in the future.

Commercial simulators vary in the number and complexity of the processes that they can model; features included in each package also vary. In addition, the internal structure of software tools is not the same, complicating not only model output but also model input. Nevertheless, regardless of the software tool being used, the principle behind these dynamic simulators is the same. For the most part, these simulators are designed to be user friendly and help modelers dynamically simulate their model of interest. These software packages that provide users with ease of use and extraordinary computation capabilities should not be confused with models that are being simulated. Several software companies have developed proprietary (and often unpublished) models that are included with their package, but the distinction between the simulator (i.e., the tool) and the model should be clear. That is, simulators represent the user interface that helps with input, output, and organization of data and are separate from the models they are running.

3.3.4 Typical Dynamic Simulator Features

Because software developers are constantly changing or updating their products, it is important to keep in mind that not all products have all features. The features listed in the following subsections can be used as a guide to various capabilities that might be available, but readers are encouraged to investigate the individual features of specific products to ensure that the products have the capabilities they need. Available commercial software packages include a number of different unit process models and features that might be of interest such as those included in Section 3.3.4.1.

3.3.4.1 Basic Features

3.3.4.1.1 Process Inputs

Typical influent options include standard influent wastewater objects that can be configured to add steady or dynamic flows and loads typically characterized by a fractionated COD load with nitrogen and phosphorus components. Models for external carbon (acetate and methanol) and chemical addition (metals and pH control) are also possible.

3.3.4.1.2 Preliminary Treatment

Preliminary treatment object models for unit processes like grit removal, equalization tanks, and precipitation chambers are available.

3.3.4.1.3 Primary Treatment

Primary treatment typically involves various clarification models. Empirical and flux-based models are available and these models can be configured into the appropriate geometric shape (e.g., rectangular, circular, flat-bottom, conical-shaped hopper, etc.).

3.3.4.1.4 Suspended Growth

Available suspended growth unit processes include the ability to model completely mixed tanks, plug-flow tanks, oxidation ditches, membrane bioreactors, sequencing batch reactors, lagoons, and deep-shaft reactors. Some packages also advertise the ability to model specialty objects like high-purity-oxygen processes and powdered activated carbon.

3.3.4.1.5 Aeration Systems

In aerated unit processes, aeration is a critical part of the simulator setup and output. The simulators have different combinations of a number of aeration options including dissolved oxygen controllers, airflow inputs for diffused systems, and power inputs for mechanical systems. In addition, features like alphas, betas, air temperatures, humidity, surface infiltration, flows per diffuser, flow limits, and power availability can be inputted in some simulators to increase the realism of the aeration system model.

3.3.4.1.6 Biofilms (Attached Growth)

Biofilm-dependent processes are also available in many of the simulators including trickling filters, rotating and submerged biological contactors, aerated and unaerated biofilters, and integrated fixed-film reactors.

3.3.4.1.7   Secondary Settling

Like primary treatment, secondary settling can be modeled through various empirical or flux-based options and simulated in typical geometric configurations.

3.3.4.1.8   Digestion

Aerobic digestion and anaerobic digestion are available in most of simulators. Some simulators incorporate their available anaerobic digestion models through the use of state variable interfaces (Nopens et al., 2009), while others use an integrated model approach.

In the interface approach, different published models are connected through algorithms that equate the state variables between each model (i.e., ASM1→ADM1 or ADM1→ASM1). In the integrated model approach, all state variables are active in all the unit processes even if they are zero. That is, in the integrated approach, anaerobic state variables are integrated in aerobic unit processes and aerobic state variables are integrated in anaerobic unit processes. The approach used in each package is software-specific and each approach has both advantages and disadvantages. (This brief explanation is included to make readers aware of these differences.)

3.3.4.1.9   Other Processes

Other unit processes are also available in some simulators including filtration, disinfection, thickening, dewatering, operating costs, energy production, and hydraulic modeling.

3.3.4.2   Numerical Tools

In addition to simply running the various available models, many simulators offer some advanced numerical tools that might be of interest including the capability to set up Monte Carlo simulations, postsimulation data examination, sensitivity analyses (e.g., to analyze how sensitive the model is to a set of parameters), and parameter value search routines (e.g., to help calibrate a model output or tune a controller).

Several simulators also have the capability to model controllers and control strategies. Water resource recovery facilities often have control routines (e.g., for aeration, sludge wastage, flow control, and chemical dosing) and, to predict the control behavior, it is advantageous to be able to replicate the control logic as closely as possible inside the system model. Many simulators include various forms of control modeling. Some are built-in, some are add-on features, and some are completely customizable, making it possible to model various complex control strategies if necessary.

3.3.4.3   Customization

A final feature that is often useful is the ability to customize the simulation and/or models being run. However, the ability to customize software varies. Some simulators allow the user to add equations to the base models, where others allow the user to alter the existing equations and add new equations as required. Furthermore, in some instances, the ability to customize software extends to the ability to write custom code (e.g., for control strategy implementation or to calculate a site-specific variable not typically calculated). Customization is an advanced feature that is not always necessary, but facilities often have unique characteristics that are essential to duplicate for a predictive model and these unique issues often require some degree of customization.

3.3.4.4   Modeling Tools

Regardless of the tool chosen for modeling work, simulator support tools are needed. Some of these tools are built into simulators and some have been developed as stand-alone tools. The following subsections outline some useful support tools that are available.

3.3.4.4.1   Influent Fractionation, Data Handling, and Reporting

Data handling is one of the most time-consuming aspects of a modeling project. Many modelers have developed their own tools to help with this, but, as the simulators mature, more and more of these support tools are being incorporated into the software itself.

In any wastewater treatment model, dynamics in the influent for the most part drive the dynamics in the process and the influent COD fractionation is a critical part of that dynamic. To make the fractionation of the influent into its component parts easier, many of the simulators now include an influent fractionation tool. These tools are offered inside the simulator or as stand-alone tools, but the theory is the same. Because the models require a fractionation that cannot be explicitly measured, these tools help the user input measured data, estimate unknowns, and output a fractionation that can be used in the model.

In terms of data handling, there are two aspects that require significant time. Each of the simulators has devised different methods to input and output data, but, regardless of the simulator, input data needs to be formatted in a specific way.

Some simulators allow data to be prepared in external files and others require that data be pasted or entered in defined forms. This data could be operational data used by the model to vary the strength of the incoming wastewater, for instance, or they could be measured data being used to compare the model predicted result to the measured observation for calibration or validation exercises. Outputting data from the simulator is also important for postsimulation evaluation. Some simulators allow the user to output data to various formats after the simulation is complete, while others require that the output of data be defined prior to the start of the simulation. Data handling both internally and externally are important features and each simulator has developed different tools for that process.

Reporting is another advanced tool that can be used to get model information out of the simulator in a usable form. In the past, simulators might have output the data in a simulator–specific format. However, more recently, developers have included tools to output data in common formats such as Microsoft Word or Excel.

3.3.4.4.2 *Linking to Other Software*

For advanced modelers, some developers have increased the capability of the simulator by enabling real-time links to other software tools. For instance, Dynamic Data Exchange (DDE) has been used to exchange data between certain simulators and common Windows programs like Microsoft Excel. Although not widely used, this link can be used to input data directly to the simulator or to accept output data from a model run. In instances where multiple runs are set up, the DDE link could be used to output specific data to simplify the comparison of multiple outputs. Dynamic Data Exchange can be used to run the simulator in special situations or to perform supplementary calculations outside the simulator.

Matlab is another external software package that can be linked to modeling tools. The controller toolbox in Matlab, for instance, may provide functions and operations that exceed the capabilities of the simulators, and use of these Matlab features may be beneficial. Some of the simulators allow code to be set up in Matlab and data to be exchanged bidirectionally as the simulation is running. This feature allows the user to make use of Matlab capabilities without the need to code the functions directly into the simulator they are using.

3.3.4.4.3 *Integrated Modeling*

The progression of modeling in the water resource field has broadened the use of wastewater models to other fields and has started to put a greater emphasis on integrated modeling. Recent trends suggest that the ability to link or integrate waste-water simulators either directly or indirectly through a file sharing system to broader catchment models (i.e., sewers and rivers) will be important. Some of the simulators have this capability already; however, if this trend continues, then seamless integration will become important.

3.4 Simulator Companies

This section lists some of the common simulators used in the market at the time of publication of this manual. Company Web sites are provided so that the reader can research products to ensure they acquire the software that will best suit their needs. The following list is intended to serve as a starting point and not an all-inclusive list of available simulators:

- ASIM (Holinger, http://www.holinger.com)
- BioWin (EnviroSim Associates Inc., http://www.envirosim.com)
- CapdetWorks (Hydromantis Environmental Software Solutions, Inc., http://www.hydromantis.com)
- GPS-X (Hydromantis Environmental Software Solutions, Inc., http://www.hydromantis.com)
- JASS (Uppsala University, http://www.it.uu.se/research/project/jass/)
- Plan-It STOAT (Water Research Centre, http://www.wrc.co.uk)
- SIMBA (ifak System, http://www.ifak-system.com/products/simulation-software)
- STOAT (Water Research Centre, http://www.wrc.co.uk)
- WEST (Danish Hydraulic Institute, http://www.dhigroup.com)

3.5 Considerations

When choosing a simulator, the reader is reminded to consider the intended use of the simulator and intangible items such as the option for support, a user-friendly interface, and available features. The reader is also reminded that not all code in commercial products is viewable, which could hamper understanding and troubleshooting when problems arise. Similarly, by buying a commercial simulator, the user is relying on developers to not make mistakes in their implementation and interpretation of the current literature.

4.0 GENERAL-PURPOSE SIMULATORS

4.1 Introduction

General-purpose simulators provide a framework for building one's own model within a broad-based, commercially available package. General-purpose simulators can be classified as block diagram simulators and general-purpose simulation languages.

Block diagram simulators allow simulation of continuous-time (differential equations) and discrete-time (difference equation) blocks. Simulink by The Math-Works, Inc. (Natick, Massachusetts) (an extension of Matlab), SystemBuild by National Instruments (Austin, Texas), and EASY-5 by MSC Software Corporation (Santa Ana, California) are some examples of commonly used block diagram simulators on the market. General-purpose simulation languages typically are based on a textual definition of the models and were originally command-driven. As such, they are often considered "high-level" languages for simulation. Some general-purpose simulation languages on the market include acslX by AEgis Technologies Group, Inc. (Huntsville, Alabama) and Aspen Plus by Aspen Technology (Burlington, Massachusetts).

General-purpose simulator tools present the greatest degree of flexibility in terms of incorporating biological, chemical, or physical processes to a wastewater treatment process model. General-purpose simulators are not restricted to wastewater process modeling and can be used to model any dynamic process that can be specified with a set of process variables, kinetic reactions, and stoichiometric relationships. Because of this high degree of flexibility, these simulators typically are used in academic research and not in engineering practice.

While general-purpose simulators offer the greatest degree of flexibility, the model user needs to specify all the reactions and system parameters when building a process model for a WRRF. Often, the flexibility offered by general-purpose simulators does not justify the additional time required to maintain the model.

4.2 Considerations

General-purpose simulators service a specific market between commercial simulators and spreadsheets. Some commercially available simulators are based on a general-purpose simulator platform. This gives them some customizable and model input advantages (because the models are already coded in the software); however, because the commercial code is locked up, not all the code is available, hence, some of the inherent advantages of the general-purpose simulator are lost. By using a general-purpose simulator, users ensure that they have access to all the model code. A further advantage is that the code can be written in a language that the user is familiar with (e.g., VB, Fortran, or Matlab). These simulators are popular with academics and programmers with special goals because the user has full control of the code. However, as with spreadsheet modeling, the resulting code is prone to errors so a great deal of care has to be taken. In some instances, shareware model code exists that can help eliminate coding errors. However, not all models are available, which means that the development of a wastewater treatment simulator in one of these tools is only necessary for unique modeling projects or if control of the code is essential.

5.0 REFERENCES

Alex, J.; Beteau, J. F.; Copp, J. B.; Hellinga, C.; Jeppsson, U.; Marsili-Libelli, S.; Pons, M. N.; Spanjers, H.; Vanhooren, H. (1999) Benchmark for Evaluating Control Strategies in Wastewater Treatment Plants. *Proceedings of the European Control Conference*; Karlsruhe, Germany, Aug 31–Sept 3.

Copp, J. B. (Ed.) (2002) *The COST Simulation Benchmark—Description and Simulator Manual*; Office for Official Publications of the European Communities: Luxembourg.

Henze, M.; Gujer, W.; Mino, T.; van Loosdrecht, M. C. M. (2000) *Activated Sludge Models: ASM1, ASM2, ASM2d, and ASM3*; Scientific and Technical Report No. 9; IWA Publishing: London, U.K.

Mehrotra, A. S.; Sedran, M. S.; Pincince, A. B. (2005) Workshop 210 Guidelines for Use of Activated Sludge Models in Practice: Comparison of Calculations Using Simulation Packages. *Proceedings of the 78th Annual Water Environment Federation Technical Exhibition and Conference* [CD-ROM]; Washington, D.C., Oct 29–Nov 2; Water Environment Federation: Alexandria, Virginia.

Nopens, I.; Batstone, D. J.; Copp, J. B.; Jeppsson, U.; Volcke, E.; Alex, J.; Vanrolleghem, P. A. (2009) An ASM/ADM Model Interface for Dynamic Plant-Wide Simulation. *Water Res.*, **43** (7), 1913–1923.

Pons, M. N.; Spanjers, H.; Jeppsson, U. (1999) Towards a Benchmark for Evaluating Control Strategies in Wastewater Treatment Plants by Simulation. *Proceedings of 9th European Symposium on Computer Aided Process Engineering*; Budapest, Hungary, May 31–June 2.

Sedran, M. A.; Mehrotra, A. S.; Pincince, A. B. (2006) The Dangers of Uncalibrated Activated Sludge Simulation Packages. *Proceedings of the 79th Annual Water Environment Federation Technical Exhibition and Conference* [CD-ROM]; Dallas, Texas, Oct 21–25; Water Environment Federation: Alexandria, Virginia.

Takács, I.; Dudley, J.; Snowling, S. (2007) A Closer Look at the Dangers of Uncalibrated Simulators. *Proceedings of the 80th Annual Water Environment Federation Technical Exhibition and Conference* [CD-ROM]; San Diego, California, Oct 13–17; Water Environment Federation: Alexandria, Virginia.

Chapter 5

Dedicated Experiments and Tools

(continued)

137

1.0   INTRODUCTION

This chapter provides an introduction to dedicated experiments and modeling tools that may be used during certain phases of the modeling process, particularly the calibration step. The chapter discusses methods for wastewater and biokinetic characterization, approaches for analyzing the mixing behavior of unit processes in water resource recovery facilities (WRRFs), aeration testing, sludge settling characterization, and, finally, a range of tools that support the model calibration itself (i.e., parameter estimation methods to optimize for a range of criteria expressing the quality of a model with respect to acquired data, sensitivity analysis to select the parameters to be estimated, and uncertainty analysis to get a feeling for uncertainty in modeling results). The purpose of this chapter is to give enough background for modelers to decide which method to use and to provide background for the selection of protocols made in the procedures presented in Chapter 8. A wide range of references to literature, guidelines, and method descriptions are provided in this chapter for the reader to refer to for more details.

2.0   WASTEWATER CHARACTERIZATION METHODS

Influent chemical oxygen demand (COD), nitrogen (N), and phosphorus (P) fractions have to be determined according to the model used. This section focuses on measurement of specific characteristics.

Required model inputs (state variables) are typically calculated as fixed percentages (fractions) based on averages of the measured components. However, this is an assumption, and model inputs often vary over the course of a day or week or with weather conditions. Special care should be taken when intermittent industrial loads are discharged or with seasonal load variations because these atypical conditions may require specific investigations to properly characterize the influent loads.

Table 5.1 provides a list of methods that can be used to characterize wastewater streams in terms of COD fractions. Experience has shown that the proposed methods may lead to different fractions altogether (Fall et al., 2011; Gillot and Choubert, 2010). This also explains why the values obtained through measurements are often modified in the subsequent calibration step.

TABLE 5.1 Experimental methods for COD fractionation of STOWA.[a]

	STOWA	WERF
$COD_{TOT,B}$	Long-term BOD tests	-
S_U	Effluent filtration (0.45 μm)	
	$S_U = 0.9\ COD_{EFF,f0.45}$ (low loaded system) $S_U = 0.9\ COD_{EFF,f0.45}$-1.5 $BOD_{5,EFF}$ (high loaded system)	$S_U = COD_{EFF,f0.45}$
X_U	Deduced from previously determined fractions: $X_U = COD_{TOT} - COD_{TOT,B} - S_U$	SBR operation or full-scale data + calibration of an activated sludge model
S_B	Filtration (0.1μm) $S_B = COD_{f0.1} - S_U$	Flocculation + filtration (0.45μm) $S_B = COD_{ff0.45} - S_U$ OUR-based respirometric methods
XC_B	Deduced from $COD_{tot,B}$ and S_B $XC_B = COD_{TOT,B} - S_B$	Deduced from previously determined fractions: $XC_B = COD_{TOT} - X_U - S_U - S_B$
X_{OHO}, X_{ANO}	Neglected	OUR-based respirometry if required

[a]$COD_{TOT,B}$, total biodegradable COD; S_U, unbiodegradable soluble COD; X_U, unbiodegradable particulate COD; S_B, readily biodegradable (soluble) COD; XC_B, slowly biodegradable (particulate and colloidal) COD; X_{OHO}, ordinary heterotrophic organisms; X_{ANO}, autotrophic nitrifying organisms; $COD_{EFF,f0.45}$, COD of a filtered (0.45-μm) effluent sample; $COD_{f0.1}$, COD of a filtered (0.1-μm) sample; $COD_{ff0.45}$, COD of a sample that is flocculated first and then filtered (0.45 μm); COD_{TOT}, total COD; $BOD_{EFF,f0.45}$, BOD of a filtered (0.45-μm) effluent sample; and $BOD_{5,EFF}$, 5-day BOD of an effluent sample.

(Roeleveld and van Loosdrecht, 2002) and WERF (Melcer et al., 2003) (nomenclature according to Corominas et al. [2010]).

No generally accepted method has been established yet and, therefore, the following steps are recommended (nomenclature according to Corominas et al. [2010]):

- The readily biodegradable fraction (S_B) should be obtained either by respirometry or by physico-chemical methods that include a flocculation step to ensure that the colloidal matter becomes part of the slowly biodegradable fraction.

- The un-biodegradable fraction (S_U) is obtained by COD analysis of a filtered (0.45-μm) effluent sample ($COD_{EFF,f0.45}$).

- The total biodegradable COD fraction of an influent ($S_B + XC_B$) is critical to properly simulate the oxygen demand of the process and its total removal efficiency. To obtain good values of the total biodegradable COD fraction ($COD_{TOT,B}$), long-term biochemical oxygen demand (BOD) measurements or other types of respirometry can be used.

The choice between either a physico-chemical or respirometry method is often based on available equipment and experience. In general, one can say that the former method is easier to carry out (see Chapter 8, Section 2.3.1, for a description and further references). However, it does not measure the biologically relevant property of the wastewater that is used in activated sludge models. Rather, it fractionates COD on the basis of size and not on the basis of rate of biodegradation, which may be important (e.g., for denitrification).

Nitrogen and phosphorus species such as ammonium-ammonia (NH_x-N), nitrite-nitrate (NO_x-N), phosphate (PO_4-P), total phosphorus, total nitrogen, total Kjeldahl nitrogen (TKN), and total soluble fractions are typically obtained through standard analysis. Organic nitrogen and phosphorus fractions are calculated by difference. More information on this topic can be obtained in an early extensive review on characterization methods by Petersen et al. (2003), an International Water Association (IWA) scientific and technical report by Rieger et al. (2012), and a recent critical review of wastewater characterization methods by Choubert et al. (2012).

An important variable in practice is the concentration of total suspended solids (X_{TSS} in model notation). It consists of a volatile part (volatile suspended solids, or VSS) and an inorganic part (inorganic suspended solids; ISS = TSS – VSS). However, the approach for which X_{TSS} is introduced as a state variable in an activated sludge model is not completely described in model publications and needs to be carefully set up. Only then can it become a useful variable that can be linked to TSS measurements.

Respirometry is defined as the measurement and interpretation of the oxygen-uptake rate, r_O, of activated sludge (Spanjers et al., 1998). In general, r_O consists of the following two components: the exogenous oxygen-uptake rate ($r_{O,ex}$), which is the immediate oxygen uptake needed to degrade a substrate, and the endogenous oxygen-uptake rate ($r_{O,end}$). Different definitions of $r_{O,end}$ appear in the literature. The definition applied here is that the $r_{O,end}$ is the oxygen-uptake rate in absence of readily biodegradable substrate. The exogenous oxygen-uptake rate is calculated from the total uptake rate by subtracting $r_{O,end}$.

Figure 5.1 illustrates the conceptual idea of respirometry. The degradation of substrate S_1 and S_2 (Figure 5.1a) results in a total exogenous uptake rate $r_{O,ex}$

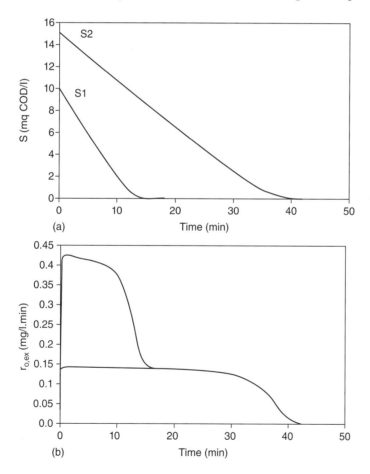

FIGURE 5.1 Degradation of two substrates S_1 and S_2 (a) leading to the conceptual $r_{O,ex}$ respirometric data set and (b) endogenous respiration rates are already subtracted from the total respiration rates measured. (Petersen et al., 2003).

(Figure 5.1b). Figure 5.1b illustrates a typical respirogram (i.e., a time course of respiration rates) with an initial peak in $r_{O,ex}$ caused by oxidation of the most readily biodegradable matter (S_1) followed by, in this instance, one "shoulder" in the $r_{O,ex}$ profile, where component S_2 continues to be degraded. Thus, in this example, the contribution of S_1 and S_2 to total $r_{O,ex}$ can easily be distinguished. It is important to realize that the oxygen demand required to oxidize the substrates can be calculated from the integral under the curve of respiration rates. Biodegradable COD is then calculated as $(1 - Y_H)$ times this integral, that is

$$S_1 = (1 - Y_H) \int r_{o,ex}^1 dt \qquad (5.1)$$

$$S_2 = (1 - Y_H) \int r_{o,ex}^2 dt \qquad (5.2)$$

where

Y_H = the heterotrophic yield coefficient

$r^1_{O,ex}$ and $r^2_{O,ex}$ = the oxygen-uptake rates corresponding to the first and second substrate, respectively

Figure 5.2 shows a typical respirometric data set obtained after injection of a sample of municipal wastewater in an activated sludge filled batch reactor. In this instance, the interpretation of the wastewater composition is based on Activated

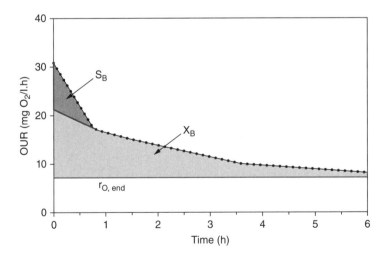

FIGURE 5.2   Typical batch respirometric data set after injection of a municipal wastewater sample in activated sludge. (Kappeler and Gujer [1992]).

Sludge Model No. 1 (ASM1) with S_B (readily biodegradable) and XC_B (slowly biodegradable) fractions differentiated on the basis of their kinetic profiles, XC_B being degraded more slowly. It is important to note that XC_B recorded in such relatively short respirometric experiments does not represent all XC_B, rather, only the faster, degradable part. Longer respirometric experiments are required such as the BOD method proposed by Roeleveld and van Loosdrecht (2002). This method provides the total biodegradable COD that allows calculating XC_B by subtracting S_B (Melcer et al., 2003). Melcer et al. (2003) also describe an alternative method using COD analyses and filtration.

3.0 BIOKINETIC CHARACTERIZATION

The following sources of information can be used to obtain values for the kinetic and stoichiometric parameters of activated sludge models (Petersen et al., 2002):

- Default parameter values from literature. It is important to note that default parameter values are often not the originally published values (called *original parameter values*), but were derived from a consensus-building process in which the profession has agreed that these values are a good starting point for a modeling study.

- Full-scale facility data.
  - Average or dynamic data from grab or time/flow proportional samples
  - Conventional mass balances of the full-scale data
  - Online data
  - Measurements in reactors to characterize process dynamics

  Parameter values are obtained through fitting the model until simulation results agree sufficiently with the data (i.e., a calibrated value is obtained).

- Bioassays, that is, different kinds of laboratory-scale experiments with wastewater and activated sludge from the full-scale facility under study. There are a number of experimental setups that have been created that allow the direct calculation of the parameter value from the data. Such value is called a *measured parameter value*. In a number of other methods, a (simplified) model is fitted to the data and one or more parameter values are estimated. In those instances, calibrated values are obtained.

Default kinetic parameters included in activated sludge models generally can describe the performance of most systems treating typical domestic wastewater. However, in some instances, wastewater characteristics and the resulting bacterial community in the biological treatment process are so unusual compared to typical domestic wastewater that the system performance cannot be explained using typical kinetic parameters. Examples of such an instance include a WRRF receiving significant contributions of industrial wastewater or the use of external carbon sources with unusual degradation kinetics (e.g., methanol and glycerol). In these instances, it is beneficial to conduct kinetic studies to better understand the reason for deviations from model predictions and to adjust the relevant model parameters. Kinetic studies can reveal possible causes for unusually slow (or fast) nitrification or denitrification rates.

As mentioned previously, there are two approaches for determining biokinetic parameters (Vanrolleghem et al., 1999). Direct methods focus on specific parameters that can directly be evaluated from the measured data (e.g., nitrification rate deduced from the nitrate data). The values obtained are known as *measured parameter values*. The second approach uses manual or automated optimization methods that use a more or less simplified model that is fitted to the measured data (i.e., calibration). The latter methods use numerical techniques to find parameter values that lead to the smallest deviation between model predictions and measurements. The values obtained are known as *calibrated parameter values*.

Care should be taken when transferring model parameters obtained from laboratory-scale experiments with activated sludge to the full-scale installation (Gernaey et al., 2004). A batch experiment with activated sludge provides much more detailed information about the reaction kinetics compared to full-scale WRRF data, but it may be that the information is reflecting a different behavior than what occurs in the full-scale data. This behavior may be altered because of differences in feeding pattern, environmental conditions such as pH, temperature, mixing intensity or surface-to-volume ratio, or sludge history. Petersen et al. (2003) provide an extensive discussion on this transferability issue.

Two kinetic characteristics that have been the subject of comprehensive method development relate to the nitrification and denitrification rate, as outlined in the following sections. Many dedicated methods have been developed for biokinetic characterization; however, these go beyond the scope of this chapter. The reader is referred to reviews by Melcer et al. (2003) and Petersen et al. (2003) for more information.

3.1 Nitrification Rate

The nitrification rate has often been highlighted as one of the most important factors influencing the required tank volume/aeration time of a WRRF. Even though parameters for nitrification are well documented in the literature, the nitrification rate may be affected by inhibiting or even toxic influent components. Melcer et al. (2003) mentioned two main approaches to determine the maximum nitrifier growth rate using, as mentioned previously, bioassays (a direct method) and full-scale data (interpreted with the optimization method).

3.1.1 Bioassay Methods

Melcer et al. (2003) describes several bioassay methods. The reader is referred to this Water Environment Research Federation (WERF) report for further details on procedures and data interpretation. The following are summaries of the tests:

- Low food-to-microorganism ratio (F/M) tests—nitrifying mixed liquor is combined with influent wastewater containing ammonia. Because of the low F/M, the nitrate production response over time is linear. The main disadvantage of this method is that the nitrifier concentration must be estimated or a sequencing batch reactor (SBR) must be operated for approximately three sludge ages.

- High F/M batch tests—in this test, a relatively small concentration of nitrifying biomass is spiked with ammonia, and the nitrate and nitrite production response is monitored for a period of approximately 4 days (Figure 5.3). Determination of the nitrifiers' maximum growth rate using this test does not require knowledge of the nitrifier concentration, and it does not require the operation of an SBR for an extended period of time.

- Washout test—this test consists of operating a flow-through reactor with nitrifying biomass at a solids retention time (SRT) shorter than the required retention time for nitrification. As with the high F/M tests, the determination of the nitrifier maximum growth rate using the washout test does not require knowledge of the nitrifier concentration or operation of an SBR for an extended period of time.

3.1.2 Simulation of Full-Scale Dynamic Behavior

Determination of the maximum nitrifier growth rate through model simulation of dynamic response behavior can be accomplished by fitting a model prediction to an observed dynamic response in ammonia and/or nitrate and nitrite concentrations

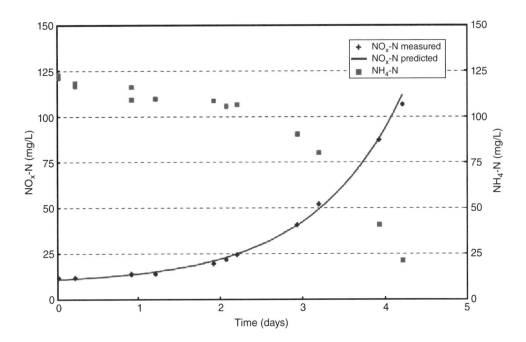

FIGURE 5.3 High F/M batch test results for direct assessment of the nitrifier growth rate. (Melcer et al., 2003).

(Kristensen et al., 1992, 1998). To ensure that the maximum nitrifier growth rate is observed, one must use only those data collected in the absence of limitation because of ammonia, dissolved oxygen, or other substrates. More sophisticated parameter estimation methods (Kristensen et al., 1998; Van Hulle et al., 2007) that simultaneously estimate half-saturation constants for oxygen and ammonia can be applied to deal with these situations.

To maximize the reliability with which parameters can be estimated from experimental data, the principles of experimental design could be applied (Dochain and Vanrolleghem, 2001). While an experiment can be designed using a rigorous model-based procedure, in practical terms it often boils down to selecting those experimental conditions in which extensive dynamics are visible in the variables. The more variation can be seen in the data (either induced by manipulating the process or by external disturbances) the better the estimation will be of nitrification parameters. An example of such simulation-based estimation of nitrification (and denitrification) parameters is given for an SBR study by Corominas et al. (2011).

3.2 Denitrification Rate

Kinetic studies also are common when evaluating specific process changes to achieve or improve denitrification, particularly when alternative carbon sources are to be used. Indeed, unusual carbon sources may result in specific kinetics parameter values that were not considered when the default kinetic values for activated sludge models were established. More importantly, a specialized microbial population may develop that will use the added carbon source with different kinetics than the wastewater COD that is degraded by traditional heterotrophs (Torres et al., 2011). This may require extension of the model structure. A critical evaluation of the applicability of default model parameters is always required, and such evaluation may sometimes suggest kinetic testing. For characterization of denitrification kinetics again, bioassays and simulation of full-scale data can be used.

3.2.1 Bioassay Methods

The parameter eta (η) that characterizes the reduction in rate under anoxic vs aerobic conditions can be estimated by comparing the oxygen utilization rate (OUR) (r_O) and the nitrate utilization rate (NUR) (r_{NO3}) in aerobic and anoxic batch tests using the same mixed liquor and organic substrate (Kristensen et al., 1992), as follows:

$$\eta = 2.86 \frac{r_{NO3,ex}}{r_{O,ex}} \qquad (5.3)$$

The subscript "ex" refers to the exogenous rate, that is, the substrate-induced rate that is calculated by subtracting the endogenous rate from the overall r_O and r_{NO3} data (see Section 2.0 in this chapter).

The OUR is determined by monitoring the decrease of dissolved oxygen following the addition of a readily biodegradable substrate, such as acetate. The NUR is determined by measuring the nitrate decrease when adding the same organic substrate (e.g., acetate) to mixed liquor under anoxic conditions (no oxygen, only nitrate present). Details for these tests can be found in work by Melcer et al. (2003). An example of high-frequency NUR and OUR data collected using nitrate and dissolved oxygen sensors in a batch setup is given in Figure 5.4.

3.2.2 Simulation of Full-Scale Dynamic Behavior

Effluent or in-process nitrate concentration data can be used as a target variable to calibrate the denitrification kinetics parameters, the latter of which typically are more information-rich and leads to more reliable estimates. It is important to ensure that the correct amount of organic substrate is being made available to the denitrifiers

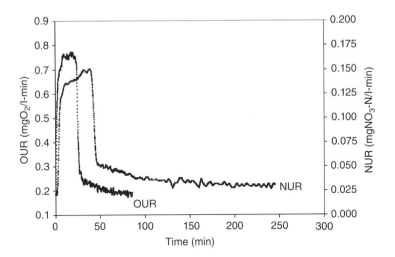

FIGURE 5.4   Comparison of acetate-based OUR (47 mg COD/L acetate added) and NUR (39 mg COD/L acetate added) determinations using dissolved oxygen and nitrate sensors in a batch setup. Note the endogenous respiration rate (0.2 mg O_2/L·min; 0.03 mg NO_3-N/L·min) that needs to be subtracted to obtain the exogenous rates required for the kinetic parameter. (Sin and Vanrolleghem, 2004).

during the simulation. This means that the influent composition (see Section 2.0) and hydrolysis/fermentation parameters (e.g., with methods proposed by Kappeler and Gujer [1992]) must be determined beforehand.

4.0   HYDRAULIC CHARACTERIZATION

As explained in Section 3.1.4 of Chapter 2, practical mixing behavior in reactors can be modeled in two ways. Either the advection-dispersion equation is used, leading to the use of a partial differential equation that requires specific solvers that are computationally slow. The alternative approach that has been adopted in the wastewater industry is based on a discretization of the spatial dimension. This results in the tanks-in-series model (Gujer, 2009).

The following approaches can be used to determine the number of tanks-in-series to be used: the empirical equation approach and an experimental approach based on tracer testing. These are briefly introduced in the following section.

4.1 Empirical Equation Approaches

Chambers and Jones (1988) at WRc plc (Swindon, U.K.) developed the following simple empirical relationship between the number of tanks-in-series (N) and the geometry and flow-through rate of a reactor:

$$N = 7.4 \frac{L}{WH} Q_{in} \tag{5.4}$$

where

N = the number of tanks-in-series describing the mixing regime of the reactor (-)

W, H, and L = the dimensions of the tank (i.e., width, height, length [m])

Q_{in} = the flowrate through the tank (m³/s)

For large facility-wide models or complex submodels, this formula may result in an overly complex model layout. The modeler should always use common sense when balancing model complexity with the model objective. If the formula recommends 10 units in series to simulate an aeration basin but the modeler knows that there are only five different diffuser grid types, it may be sufficient to include only five reactors in series. On the other hand, if a square anoxic zone is partitioned with three baffle walls, it may be important to include three reactors in series to simulate the plug flow pattern. Model calibration and validation efforts will confirm whether what the modeler has developed included the appropriate level of complexity.

An alternative, more accurate (but also more complex) empirical equation was proposed by Fujie et al. (1983). It analyzes the number of tanks describing a reactor zone, using the Peclet number, Pe, as follows:

$$N = \frac{Pe^2}{2}(Pe - 1 + e^{-Pe}) \tag{5.5}$$

With $Pe = \dfrac{\mu L}{E_L}$ (-) with u the average velocity (cm/s), L the zone length of the reactor (cm), and E_L the longitudinal dispersion coefficient (cm²/s) that is calculated as follows:

$$E_L = 0{,}0115\left(1 + \frac{H}{L}\right)^{-3} \mu_g^{-0.34} a_d \Phi^{m_d}(H + W) \tag{5.6}$$

$$\Phi = h u_g \left(\frac{h}{H}\right)^{1/2}\left(\frac{H}{W}\right)^{1/3} \tag{5.7}$$

TABLE 5.2   Values of Fujie empirical constants a_d and m_d as a function of diffuser type and Φ as specified in Eq. 5.7.

Type of air diffuser	$\Phi(cm^2/s)$	m_d	a_d
Fine bubble types[a]	$\Phi \leq 20$	0.64	7.0
	$\Phi > 20$	0.46	12.0
Coarse bubble types[b]	$\Phi \leq 20$	0.78	3.5
	$\Phi > 20$	0.56	4.9

[a]Porous plates and tubes.
[b]Perforated plates and tubes, single nozzles, and others.

where

H and W = the height and width (cm) of the reactor zone, respectively

u_g = superficial gas velocity (cm/s) calculated from the airflow rate and the zone surface area

h = diffuser depth (cm)

a_d and m_d = Fujie empirical constants that can be read from Table 5.2

Makinia and Wells (2005) reported that Fujie's equation gives satisfactory results for most standard cases. A spreadsheet with these two estimation formulae is provided by Rieger et al. (2012).

4.2   Tracer Testing

An experimental approach to determine the number of tanks describing the mixing regime of a reactor is to perform an experimental tracer study by injecting a pulse of an inert tracer or to increase its concentration for a prolonged period at the reactor inlet. The inert nature of the tracer is important because the substance should not adsorb to the sludge and should not be degraded. To verify this, a mass balance should be made to ensure that all tracers have been recovered by the end of the tracer experiment. The time series of recovered tracers is then measured at the outlet of the reactor under study. Although this experimental approach is more time-consuming and expensive, it is more accurate. It is important to note that this does not account for any wet weather flow, unless the tracer test is repeated with different inflow rate conditions. Another significant effect on mixing behavior is caused by aeration. High aeration intensity typically leads to increased mixing and, therefore, to less plug flow behavior (modeled as fewer tanks in series).

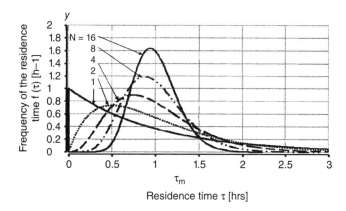

FIGURE 5.5   Ilustration of the effect of a different number (*N*) of continuously stirred tank reactors in series to describe mixing behavior. The *y* axis represents RTD and τ_m represents the dimensionless time based on the mean residence time in all *N* tanks.

Typical model-based responses to tracer injections are given in Levenspiel (1999). The pulse injection case is reproduced in Figure 5.5, and similar figures can be found for the step change experiment (the reader should note that a step is the integral of a pulse and that, because of linear system equations, step responses are the integrals of the pulse responses). When performing experiments, recycle flows typically present in WRRFs can complicate the interpretation. To allow for the simple interpretation of tracer test data, recycling could be switched off. In such instances, simple calculations allow finding the number of tanks-in-series from the data (Gujer, 2009).

In instances where more complicated flow patterns are observed (short-circuiting, recycles), an approach using simulation models with different configurations of tanks and recycles can be simulated in a facility simulator (using an inert soluble component as a surrogate for the tracer). The best-fitting model determines the configuration of tanks to be used subsequently. Figure 5.5 shows that using one tank gives rise to the typical exponential decay model response of an ideally mixed tank. Increasing the number of tanks results in the buildup of a peak, which becomes higher in absolute value and whose occurrence is shifted further in time. This is the typical behavior of a complete plug flow system. When the number of tanks reaches infinity, a perfect plug flow response is approximated, that is, all tracer appears at once at the outlet after a time corresponding to hydraulic residence time (HRT).

The following procedures are recommended when setting up a tracer experiment:

- Choice of tracer—the tracer should be a non-biodegradable, non-adsorbing compound. Tracer materials that are often used are a lithium salt, rhodamine, or bromide. The choice depends on measuring accuracy, reachable measuring frequency (e.g., through online sensors or only possible with laboratory analysis), toxicity of the tracer, or other negative effects on the environment. The measuring accuracy/limit of detection together with the flowrate determines the required tracer mass load.

- Injection of the tracer—the tracer should be injected as close as possible to the entrance of the reactor/facility section under study.

- Data collection—samples should be taken as close as possible to the exit of the reactor under study. Samples should be collected for a period that covers at least three to five HRTs. During this period, about 20 to 50 samples should be taken. If possible, a first screening of tracer dynamics should be conducted and only samples taken at times where the highest variations are visible should be analyzed further if the budget is limited. Considering Figure 5.5 and the typical mixing regime in bioreactors (three to 10 tanks-in-series), most samples should be taken before HRT is reached, with some samples taken to monitor the tailing. It may also be useful to wait until the recycle kicks in (i.e., until the tracer comes back with the recycle flows). While these additional dynamics complicate interpretation, they give valuable information.

An example of a tracer test performed at full scale is shown in Figure 5.6 (De Clercq et al., 1999). In this example, it was observed that using one tank is not sufficient to model the mixing behavior. However, when using two tanks-in-series, the performance of the model is satisfactory.

Flow splitting also deserves attention when modeling WRRFs. Splitting chambers are typically modeled as ideal flow splitters. However, in reality, this is often not the case, and this can have a large effect on modeling results. One way to investigate an influent splitting work is to investigate the sludge concentrations that occur in the different lanes. Theoretically, these should be the same when an equal loading is provided. The respective sludge concentrations in the different lanes for the aforementioned example are given in Figure 5.7. In terms of equal flow distribution, a line through the origin with a 45-deg slope should be found. Figure 5.7 shows that this is the case for lanes AS1 and AS3. However, lane AS2 exhibits a much larger sludge

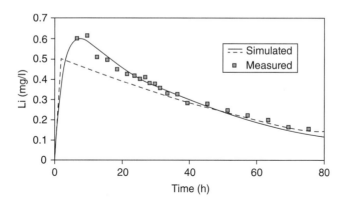

FIGURE 5.6 Illustration of model fits to RTDs in train 1 recorded using Lithium as tracer, with one tank (dashed line) and two tanks in series (solid line). Note the bump on the graph at approximately 35 hours caused by recycle of tracer.

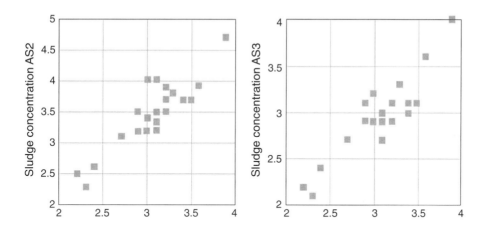

FIGURE 5.7 Illustration of differences in sludge concentration in different lanes caused by an improperly functioning flow-splitting works.

concentration (+20%), suggesting that more sludge and less wastewater is going to that lane, leading to a loss in treatment performance. Closer investigation of the influent splitting works revealed shortcut flows of both the influent and return activated sludge (RAS) from the secondary settlers.

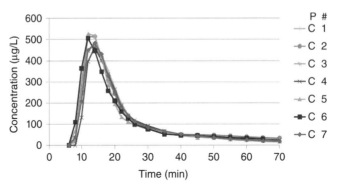

FIGURE 5.8   Illustration of rhodamine wastewater treatment tracer test results for detection of unequal flow splitting in a multilane primary clarifier system. (Tik and Vanrolleghem, 2012).

A tracer study can also be used to identify unequal flow splitting. The residence time will be different in instances where unequal flow loading occurs. A recently studied multilane primary clarifier system (Tik and Vanrolleghem, 2012) exhibited excellent splitting as could be deduced from a tracer test conducted by injecting a pulse of rhodamine WT and collecting samples at each of the seven primary clarifier overflows. The almost exact match of the seven tracer curves in Figure 5.8 illustrates good flow distribution, with the exception of PC#6, which appears to get some more load.

5.0 AERATION TESTING UNDER PROCESS CONDITIONS

The performances of aeration systems in WRRFs are typically guaranteed and measured in clean water according to a standardized method, the unsteady-state clean water test (ASCE, 2007). Translation from clean water conditions to process conditions requires determination of the alpha factor, which depends on many parameters (see Chapter 8). Oxygen transfer rates can also be characterized under process conditions using several methods (ASCE, 1997; Capela et al., 2004). The following two methods, in particular, can be applied: the offgas analysis method developed by Redmon et al. (1983) and the hydrogen peroxide method originally proposed by Kayser (1979).

FIGURE 5.9   Sampling hood. (Credit: Y. Fayolle, Irstea).

5.1   Offgas Method

The offgas method is based on a gas-phase mass balance. The gas bubbles at the liquid surface are collected using floating hoods (Figure 5.9). At each hood location i, the offgas flowrate ($q_{e,i}$) is measured and the molar oxygen concentration is determined to compute the standard oxygen-transfer efficiency under process water (field SOTE, $SOTE_{f,i}$), as follows:

$$SOTE_{f,i} = 1 - \frac{y_s(1 - y_e)}{y_e(1 - y_s)} \qquad (5.8)$$

where y_e, y_s is the molar oxygen concentration in the insufflated air and in the offgas (-).

The overall $SOTE_f$ of the aeration system is obtained by weighting the $SOTE_f$ values by the offgas flowrates collected at each test location, as follows:

$$SOTE_f = \frac{\Sigma\, q_{e,i} \times SOTE_{f,i}}{\Sigma\, q_{e,i}} \qquad (5.9)$$

Full-scale aeration testing using the offgas method requires prior definition of a representative gas-sampling plan such as that exemplified in Figure 5.10.

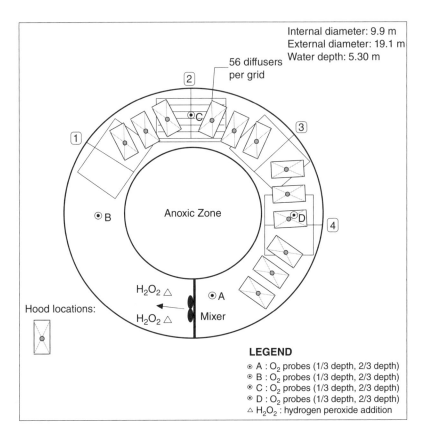

Internal diameter: 9.9 m
External diameter: 19.1 m
Water depth: 5.30 m

56 diffusers per grid

Anoxic Zone

Hood locations:

H_2O_2 △

H_2O_2 △   Mixer   ⊙ A

LEGEND
⊙ A : O_2 probes (1/3 depth, 2/3 depth)
⊙ B : O_2 probes (1/3 depth, 2/3 depth)
⊙ C : O_2 probes (1/3 depth, 2/3 depth)
⊙ D : O_2 probes (1/3 depth, 2/3 depth)
△ H_2O_2 : hydrogen peroxide addition

Figure 5.10   Illustration of an offgas sampling plan. (Credit: Capela et al. [2004]).

Gas-sampling plans vary from one facility to another depending on the geometry of the tank and the distribution pattern of the aeration system.

5.2   Hydrogen Peroxide Method

The hydrogen peroxide method allows the determination of $k_L a_f$ by monitoring the dissolved oxygen concentration over time after adding hydrogen peroxide (H_2O_2) to create a perturbation from steady-state conditions. The dissolved oxygen concentration during the deaeration period can be written as follows (ASCE, 1997):

$$C = C_R - (C_R - C_0) \times e^{-(k_L a_f + Q/V)xt} \tag{5.10}$$

where

    C = actual dissolved oxygen concentration (mg/L)

    C_R = the concentration at steady state in process water (mg/L)

    C_0 = initial dissolved oxygen concentration (mg/L)

    $k_L a_f$ = volumetric mass-transfer coefficient in process water ($1/T$)

    Q = volumetric wastewater flowrate (m³/h)

    V = tank volume (m³)

The addition of H_2O_2 is performed maintaining a constant power level. The goal is an increase in the dissolved oxygen concentration above the steady-state dissolved oxygen concentration (C_R) higher than 10 mg/L. For a 35% H_2O_2 solution, the volume to inject is determined as follows:

$$V_{H_2O_2,35\%} = 5,37\Delta C V 10^{-6} \tag{5.11}$$

where

    $V_{H_2O_2,35\%}$ = volume of peroxide solution (35%) to inject (L)

    ΔC = increase in dissolved oxygen concentration (mg/L)

    V = tank volume (L)

An example of a deaeration curve is presented in Figure 5.11.

6.0 SLUDGE SETTLING CHARACTERIZATION

A detailed settling characterization is necessary if the settling performance and the reactions during settling influence overall system behavior (e.g., effluent COD, nitrogen, and phosphorus). In addition, the target(s) might be the optimization of settling or may be the amendment of effluent suspended solids (ESS) removal. Hindered and compression settling are the main phenomena taking place in secondary settlers. Primary settlers that are characterized by discrete and flocculation-type settling are discussed at the end of this section.

In addition to the sedimentation process, reactions such as denitrification (leading to gas-bubble formation that disturbs settling) and phosphorus release in the settler may have to be considered (Henze et al., 1993; Koch et al., 1999; Wouters-Wasiak et al., 1996). This can be checked by calculating the mass balance over nitrate-phosphate or by site observations such as the occurrence of nitrogen gas bubbles trapped in the flocs (Henze et al., 1993). Information on flow velocity measurements is given at the end. They are of special interest when carrying out a detailed evaluation of clarifiers using computational fluid dynamics (CFD) models.

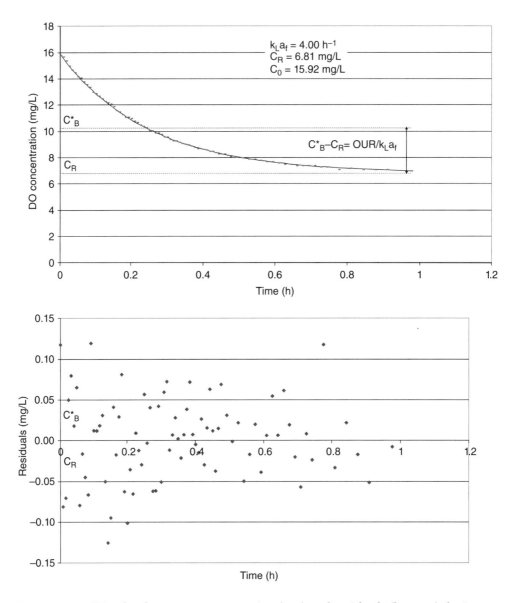

FIGURE 5.11   Dissolved oxygen concentration (top) and residuals (bottom) during a deaeration test after the addition of hydrogen peroxide.

6.1 Full-Scale Tests

Estimation of a settler model's parameters can be performed by fitting the model outputs to time series of flowrates and total suspended solids (TSS) concentrations measured at the clarifier's inlet, underflow, and effluent. The latter data can either be obtained by offline TSS measurements or by turbidity sensors. Provided they are properly handled, the latter have reached a level of reliability and precision that allows their use for model calibration (Vanrolleghem and Lee, 2003). It is important to realize that best calibration performance will be obtained under dynamic conditions because these better expose the dynamics of the settler.

For more advanced models, it may be necessary to supply data taken within the clarifier. Turbidity sensors that move up and down in the clarifier or ultrasonic profilers may provide time series of the sludge blanket height and even sludge profiles (Figure 5.12).

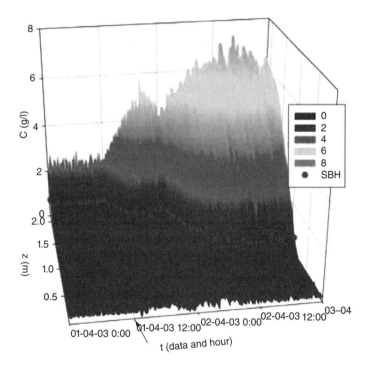

FIGURE 5.12 Sludge concentration profiles and sludge blanket height (SBH) during stress testing a secondary clarifier at the Heist water resource recovery facility. (De Clercq, 2006).

Highly informative data sets, in particular, can be obtained from clarifier stress tests (Wahlberg, 2002) that involve hydraulically stressing an existing final clarifier such that the dynamic clarifier performance may be monitored in effluent and underflow. This is achieved by taking units offline until the targeted surface overflow rate (SOR) and solids loading rates (SLRs) are reached. Operating performance is continuously monitored until performance reaches a determined "failure" point, such as high effluent TSS or high blanket level, at which point the stress testing is ended and the offline units are returned to service. Parameters monitored for each operating clarifier include sludge blanket levels, mixed liquor suspended solids, ESS, RAS, TSS, sludge volume index (SVI), flowrates, dispersed suspended solids (DSS), and flocculated suspended solids (FSS).

6.2   Laboratory Testing

6.2.1   Hindered Settling Parameters

Hindered zone settling of sludge is typically modeled using the Vesilind equation or alike that relates settling velocity to sludge concentration. In simulation studies, the two kinetic parameters of the following equation, Vo and k, are often obtained from correlations with SVI data (e.g., Daigger and Roper [1985]):

$$Vo = 7.8 \text{ m/h;} \qquad k = 0.148 + 0.0021 \text{ SVI}$$

Critical reviews of such correlations can be found in literature by Bye and Dold (1999) and Giokas et al. (2003).

Settling parameters can be determined directly by using a series of batch column settling tests, measuring settling velocities at different sludge concentrations (Bye and Dold, 1999). Different concentrations are obtained by dilution of RAS with effluent. At least six different concentrations should be tested (Ekama et al., 1997). The individual settling velocities are measured following the procedure described in *Standard Method* 2710 E for the evaluation of the zone settling rate (APHA et al., 2005), that is, the descent of the sludge–water interface is monitored at regular intervals for 30 to 60 minutes (Figure 5.13). The slope of the linear part of the curve of the interface displacement is defined as the zone settling velocity.

Batch settling tests are typically performed in a 1-m tall by 15-cm diameter settling column provided with a stirring mechanism to minimize wall effects. Automated systems that can be installed in the field for online characterization of settling performance have been developed (Vanrolleghem et al., 2006).

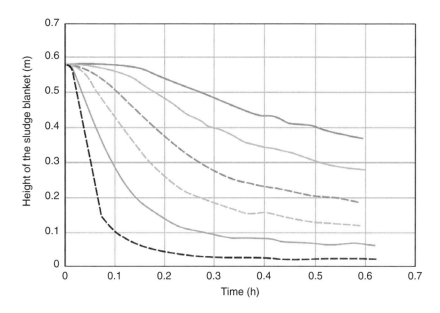

FIGURE 5.13   Batch settling curves for different initial sludge concentrations (from bottom to top: 3, 5, 7.7, 9.7, 12.7, and 15.6 g/L).

6.2.2   Nonsettleable Fraction

The nonsettleable fraction of the suspended solids, f_{ns}, can readily be measured in a settling column analysis (Takács et al., 1991) or, as mentioned previously, from a time series of effluent TSS measurements.

To gain more insight to the ultimate settleability of suspended solids, the DSS/FSS test can be conducted (Ekama et al., 1997; Wahlberg, 2002). Conducting the DSS/FSS test requires measurement of TSS in three samples, two of which must be collected in a special manner, as follows:

- The FSS sample consists of supernatant from a settled mixed liquor sample following flocculation in a standard 2.0-L rectangular beaker flocculation apparatus. The FSS sample represents the best performance (lowest ESS) expected from the secondary clarifier because, theoretically, ideal flocculation and settling occurred prior to sampling the supernatant.

- The DSS sample consists of secondary clarifier effluent collected using a modified Kemmerer sampler at the clarifier effluent weir (Ekama et al., 1997;

Parker et al., 1970). Immediately following sample collection, a portion of the initial Kemmerer sampler contents becomes the ESS sample. Following 30 minutes settling in the Kemmerer sampler of solids that did not settle in the clarifier, the supernatant becomes the DSS sample.

Repeating DSS/FSS tests several times provides improved confidence in the results because of the high variability in clarifier ESS. The DSS/FSS test presently offers the most practical method to analyze secondary clarifier performance.

6.3   Liquid Velocity Measurements in Clarifiers

Velocity measurements are important to validate a CFD model. Although the residence time distribution (RTD) describes the hydraulic behavior of the reactor, it only allows indirect knowledge about the flow pattern. Hence, to validate the flow field, velocity data are essential. Many possibilities exist to measure liquid velocities, which vary between zero and 80 mm/s in secondary settling tanks (Anderson, 1945; Bretscher et al., 1992; Kinnear, 2000; Ueberl and Hager, 1997).

The basic method proposed by Anderson (1945) is still in use, although other measurement techniques are available. These alternative flow velocity devices can be subdivided into the following three groups: mechanical (Lindeborg et al., 1996; STOWa, 2002a,b), electromagnetic, and acoustic Doppler velocity meters (Deininger et al., 1998; Kinnear and Deines, 2001). Whereas the former two techniques measure true liquid velocity, the acoustic Doppler velocity meter measures the velocity of small particles suspended in the liquid.

6.4   Primary Clarifiers

Attention to primary clarifier modeling has been lacking in wastewater treatment modeling for a long time. Often, modeling only started with primary effluent. However, increased attention on energy-neutral WRRFs and the role primary clarifier performance plays in this has put this unit process back in the spotlight (Crawford et al., 2010).

Current primary clarifier models describe settling according to Stokes' law, even though flocculent settling conditions may occur. The latter process warrants analysis and inclusion in the model description, especially in chemically enhanced primary clarifiers. While analysis using coagulation/flocculation tests is well established (Melcer et al., 2010), the models are not really applied and the effect of chemical addition is simply included as an improved separation efficiency or higher settling velocity.

The parameters of primary clarifier models are typically deduced from inlet–outlet TSS-removal performance analysis (e.g., Lessard and Beck, 1988) or batch settling experiments conducted over different settling times (Otterpohl, 1995). The settleable solids analysis after 2 hours of sedimentation gives a good indication of the maximal removal efficiency a primary settler can achieve. Supernatant TSS concentrations indicate the nonsettleable solids.

Recently, more detailed experimental methods are developed to characterize settling in primary clarifiers. Using the ViCAs protocol originally developed for stormwater characterization (Chebbo and Gromaire, 2009), the distribution of settling velocities of raw and clarified wastewater can be obtained (Figure 5.14) (Maruejouls et al., 2011). From such graphs, it is easy to deduce which mass fraction will be removed by a primary clarifier, and models that use settling velocity classes of particles are currently being developed to better describe the dynamics of primary clarifiers (Bachis et al., 2012).

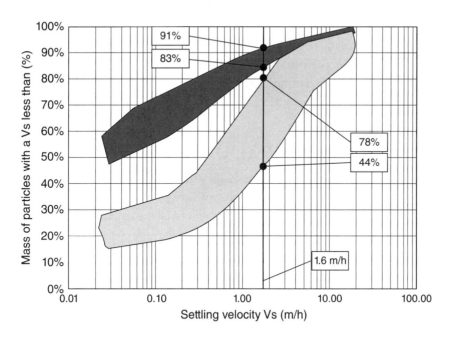

FIGURE 5.14 Settling velocity (Vs) distribution curves for dry weather wastewater. "Dark" is the Vs distribution range of wastewaters from the effluent of primary settling. "Pale" is the Vs distribution range of wastewaters from the influent of primary settling. (Maruejouls et al., 2011).

A ViCAs curve must be interpreted as follows: the lower the curve the larger the fraction of rapidly settling particles. Considering a sedimentation velocity of 1.6 m/h to be the typical design overflow rate for primary sedimentation units (Metcalf and Eddy, 2003), Figure 5.14 shows that between 83% and 91% of the particle masses at the outlet of the primary settler have a settling velocity lower than their design value (1.6 m/h). Furthermore, between 44% and 78% of the influent particle masses have settling velocities lower than 1.6 m/h, resulting in 56% to 22% of particle masses that can be intercepted by a primary settler.

The primary clarifiers ahead of the biological reactors also have a vital importance because there might be activation and/or interactions between physico-chemical and biological reactions. For instance, in addition to biological reactions, COD removal can be expressed as a physico-chemical reaction such as flocculation and solubilization of the particulate COD fractions into soluble fractions. The biomass in wastewater also induces biological processes such as fermentation, acidification, and ammonification. The settling and biological reactions should, therefore, be combined in the model if the effect of the primary clarifier on the overall process performance becomes evident (Lessard and Beck, 1988). The effect of the primary clarifier on wastewater composition can be evaluated by performing some relevant on-site experimental analyses such as filtered/unfiltered COD, ammonium, phosphorus, pH, VFA analysis, and so on in the influent and effluent streams of a primary clarifier.

7.0 MODEL ANALYSIS TOOLS

After data are collected and analyzed for quality, the model is adjusted to them with the goal of getting a model that describes the observed reality in an acceptable manner. This activity is called *calibration*, and entails adjustment of (some) model parameters (within reasonable ranges) in such a way that the model describes the collected data well and fulfills the requirements set out in the objectives of the modeling study. Both the statements, "describes … in an acceptable manner" and "describes the data well", have been used in a subjective way in many modeling projects, but there is increased consensus that the quality of a model should be quantified by objective criteria (Hauduc et al., 2011).

Not all parameters of a facility model (of which there are many) should be adjusted during calibration. Some authors even state that, for municipal WRRFs, no adjustment of kinetic and stoichiometric parameters is needed as the model with default parameters provides a prediction performance that is sufficient for the model purpose at hand. In most model applications, however, some model adjustment is required.

Some model parameters may obtain their value directly from some of the aforementioned dedicated experiments (e.g., settling properties, growth rates, and wastewater fractions), although these values could best be considered good initial estimates for further calibration steps given the issues with transferability of dedicated experiments to full-scale behavior (Gernaey et al., 2004; Petersen et al., 2003). For the remaining, often large number of parameters, there is often insufficient data available to obtain good estimates for all of them (i.e., the information content of the data set is insufficient to practically identify all parameters). Optimal design of experiments is one solution to this identifiability problem, yielding data that lead to better estimates of parameters (Vanrolleghem et al., 1995).

In general, however, resources are not available to estimate all parameters and, therefore, one is content with estimating those parameters that make a difference with respect to the quality of the model's predictions. Sensitivity and uncertainty analysis methods allow dealing with this and are discussed in Sections 7.3 and 7.4.

7.1 Goodness-of-Fit

In wastewater treatment modeling, evaluation of model quality is often based on qualitative comparisons between simulation results and observed data. Although such visual evaluation is useful, it does not provide an objective assessment of the quality of a calibration parameter set. Moreover, it cannot be used in an automatic calibration procedure.

Environmental sciences (in particular, hydrology) commonly use mathematical comparisons of predicted and observed values (Dawson et al., 2007). In wastewater treatment, several target constituents are typically considered simultaneously during model calibration (sludge production, TSS, COD, nitrogen, and phosphorus in the effluent). Although a review of quality criteria is presented in Dochain and Vanrolleghem (2001), quantitative criteria are rarely determined in wastewater treatment modeling (Ahnert et al., 2007; Petersen et al., 2002; Sin et al., 2008).

Depending on modeling objectives, the goodness-of-fit of a model can be defined as the capability of the model to capture one or several of the following characteristics of observed data: mean, timing, and magnitude of peaks or typical periodical variations (diurnal, weekly, seasonal, etc.). For example, if a specific effluent limit of a facility is based on a monthly average, there is little sense in evaluating the accuracy of the fit of each single peak. However, if peak effluent limits have to be met, a criterion evaluating the fit of peaks should be used. Thus, to characterize the goodness-of-fit of the model, different quality criteria may be needed.

Hauduc et al. (2011) selected 31 quantitative goodness-of-fit criteria in a comprehensive literature review. They were grouped according to two classification systems. The first classification scheme is inspired by Dawson et al. (2007) and groups the criteria into the following six main classes:

1. Single event statistics—in instances where modeling objectives require accurate simulation of events (e.g., the ability of the WRRF to handle storm flows or toxic peaks), criteria are needed to characterize the goodness-of-fit of the model for this event. The goal of the single-event statistics peak difference (Gupta et al., 1998) and percent error in peak (Dawson et al., 2007) is to characterize the difference between the maximum observed and maximum modeled value.

2. Absolute criteria from residuals—absolute criteria are based on the sum of residuals (difference between observed O_i and predicted P_i values, respectively, at time step i), generally averaged by the number of data, n. A low value of this criterion means good agreement between observation and simulation (with γ an exponent).

$$E_\gamma = \frac{1}{n}\sum_{i=1}^{n}(O_i - P_i)^\gamma \qquad (5.12)$$

3. Residuals relative to observed values—at each time step, error is related to the corresponding observed or modeled value. A low value of this criterion means good agreement between observation and simulation.

$$RE_\gamma = \frac{1}{n}\sum_{i=1}^{n}\left(\frac{O_i - P_i}{O_i}\right)^\gamma \qquad (5.13)$$

4. Total residuals relative to total observed values—for these criteria, the sum of errors is related to the sum of observed values, without any correspondence to the time step. A low value of this criterion means good agreement between observation and simulation.

$$TRE_\gamma = \frac{\sum_{i=1}^{n}(O_i - P_i)^\gamma}{\sum_{i=1}^{n}O_i^\gamma} \qquad (5.14)$$

5. Agreement between distributional statistics of observed and modeled data—these criteria are not based on error comparison, but on a comparison between cumulative distributions of modeled and observed data. In the

wastewater field, these criteria can be relevant for influent and effluent pollutant loads by summing the fluxes.

6. Comparison of residuals with reference values and with other models—these criteria compare the residuals with residuals obtained with a reference model $\tilde{P}$, such as a model describing the mean value ($\tilde{P}_i = \bar{O}$) or the previous value ($\tilde{P}_i = O_{i-1}$) (with α an exponent).

$$CE_{\alpha,\gamma} = 1 - \frac{\sum\limits_{i=1}^{n}(O_i^\alpha - P_i^\alpha)^\gamma}{\sum\limits_{i=1}^{n}(O_i^\alpha - \tilde{P}_i^\alpha)^\gamma} \tag{5.15}$$

In a second classification system, Hauduc et al. (2011) classified the 31 quality criteria as six main characteristics of the adjustment of the predicted values to the observed data set. Indeed, the study showed that the criteria clustered in only six different types, each focusing on one of the following objectives:

1. Criteria evaluating the mean error
2. Criteria evaluating the bias
3. Criteria that emphasize large errors
4. Criteria that emphasize small errors
5. Criteria evaluating peak magnitudes
6. Criteria evaluating event dynamics

Strong correlations exist between the values obtained with members of the same cluster, indicating that there are redundant criteria that do not add anything to the model quality evaluation. On the contrary, the existence of redundant criteria confuses communication because two groups of modelers may use different criteria without realizing that they are pursuing the same modeling objective. In addition, experience with criteria is divided among members of the cluster, meaning that less is known about the criteria and the interpretation of the values they take.

7.2 Parameter Estimation

Parameter estimation consists of determining the "optimal" values of the parameters of a given model with the aid of measured data. Figure 5.15 presents a schematic of the basic idea behind parameter estimation.

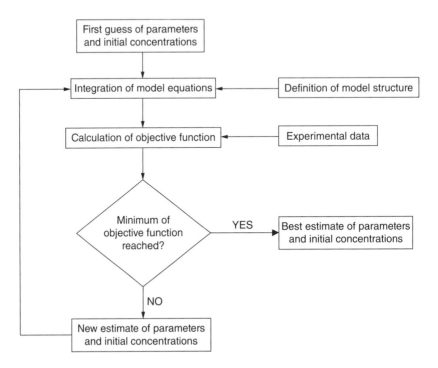

FIGURE 5.15	Illustration of parameter estimation routine. (Modified from Wanner et al. [1992]).

Initially, model structures, of which selected parameters need to be estimated, and experimental data need to be defined. Moreover, first guesses of the parameters to be estimated have to be given. The parameter estimation procedure then basically consists of minimizing an objective function *J*, which, for example, can be defined as the weighted sum of squared errors between the model output and data. When the objective function reaches a minimum with a certain given accuracy, optimal parameter values are obtained, as illustrated in Figure 5.16.

The modeler can conduct this search for the best parameter values in a trial-and-error manner until he or she finds a satisfying result for the quality criterion (it is important to keep track of the parameter sets already evaluated so that searches are not repeated). Other modelers use automated optimization algorithms. Although numerical techniques for automatic estimation will not be discussed here, the reader is referred to literature by Dochain and Vanrolleghem (2001).

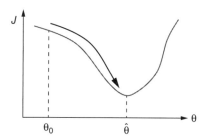

FIGURE 5.16 Minimization of a model quality criterion value J starting from an initial parameter estimate θ_0 to converge gradually (arrow) to the final, best estimate $\hat{\theta}$.

It is important to note, however, that because of the high complexity of the optimization problem caused by numerous parameters and the nonlinear nature of wastewater treatment models (Sin et al., 2008), it is cumbersome to apply automated calibration techniques. Rather, a combination of trial and error with intermittent use of an optimization algorithm is advised.

Indeed, a significant problem encountered in calibration of treatment models is the lack of identifiability of model parameters. *Identifiability* is the ability to obtain a unique combination of parameter values describing a system's behavior. This subject is dealt with in great detail in literature by Dochain and Vanrolleghem (2001). Here, it should only be stressed that a typical problem related to the calibration of wastewater models is that more than one combination of influent characteristics and model parameters can give the same good description of collected data (Dupont and Sinkjær, 1994; Kristensen et al., 1998). While this is acceptable for some model objectives (e.g., description of data), it is not for others (e.g., prediction for new situations as typically found in upgrade studies). In the latter instance, one must either reduce the number of parameters one can estimate (using expert knowledge or sensitivity analysis) or collect additional data to provide the information with which the parameter can be assessed.

7.3 Sensitivity Analysis

A sensitivity analysis studies the "sensitivity" of the outputs of a system (variables of interest) to changes in parameters. It also allows for ranking the model parameters according to how much they influence model outputs. Finally, a sensitivity analysis can be used to identify which model parameters can be estimated based on a given set of measurements. Only the most influential parameters will be retained for calibration.

There are two types of sensitivity analysis, local sensitivity analysis (LSA) and global sensitivity analysis (GSA). While GSA provides an "average" sensitivity of the nonlinear model outputs within a defined range of parameter values, LSA evaluates how much model outputs will change if only a small change is applied to the parameter values in the neighborhood of the assumed parameter values. Typical LSAs are carried out by altering the parameters by 1% of their assumed value, although De Pauw and Vanrolleghem (2006) warned of errors that occur when changes that are too large are applied.

A state-of-the-art LSA method for selecting parameters is one proposed by Brun et al. (2002). In this method, an overall sensitivity measure (δ) is calculated on the basis of scaled sensitivity values $s_{i,j}$ for each of the n model outputs i with respect to the parameters j:

$$\delta_j = \sqrt{\frac{1}{n}\sum_{i=1}^{n} s_{i,j}^2} \qquad (5.16)$$

$$s_{i,j} = \frac{\Delta p_j}{sc_i} \cdot \frac{\partial y_i}{\partial p_j} \qquad (5.17)$$

where
  Δp_j = the uncertainty range of the parameter p_j
  sc_i = a scale factor
  n = the number of model outputs considered

A large δ means that a change of Δp_j in parameter p_j has a substantial effect on the considered model output(s). Model parameters are typically assigned to three uncertainty classes according to Brun et al. (2002). Parameters from uncertainty class 1 (e.g., stoichiometry) have a low uncertainty (5% of the default parameter value); class 2 (growth rates) has an uncertainty range of 20% of the default parameter value; and 50% is recommended for parameters from uncertainty class 3 (e.g., half-saturation constants). Often-used scaling factors are (1) the model output value (or its average if a dynamic simulation is performed) or (2) the model output measurement error standard deviation. Ranking parameters on the basis of the δ_j values is the basis for selection of parameters to be estimated. Boltz et al. (2011) calculated δ_j for three bulk concentration predictions and four biofilm fluxes, respectively, at two different temperatures. It is important to note that the choice of output variables (i.e., expressing the interest of the modeler) determines which parameters are most influential and that temperature has a significant effect on δ_j values and ranking parameters in terms of importance for the model calibration.

With increasing available computing power, GSA methods are gaining in popularity as they can overcome the main problem of LSA, which is that LSA results are only valid for the parameter and input values used during the analysis. Because of the strongly nonlinear nature of the WRRF models, LSA results (e.g., the importance ranking of the parameters) are dependant on these factor values. Conversely, GSA methods assess how model outputs are influenced by the variation of the model input factors over their entire range of uncertainty (Saltelli et al., 2004). The GSA may help modelers select important factors ("factors prioritization") and noninfluential factors ("factors fixing") and identify interactions among factors. More specifically, by means of factors prioritization, model input factors that have the greatest effect on model outputs are identified. Conversely, the factors-fixing setting leads to the identification of factors that may be fixed at any given value over their uncertainty range without reducing the output variance (Saltelli et al., 2004).

The GSA is based on extensive simulation, typically using Monte Carlo simulation. Depending on the number of parameters to be considered and the GSA method used, hundreds to tens of thousands of simulations are required to adequately cover the entire range of uncertainty of the factors.

In Saltelli (2000), GSA methods are classified as (1) global screening methods (e.g., Morris screening method [Campolongo et al., 2007; Morris, 1991]); (2) variance decomposition methods such as Fourier amplitude sensitivity testing (FAST), extended-FAST (E-FAST), and the Sobol indices method (Saltelli et al., 1999; Sobol, 2001); and (3) regression/correlation-based methods such as the standardized regression coefficients (SRCs) method (Saltelli et al., 2008). Although it is beyond the scope of this chapter to explain in detail how these methods actually work, their interpretation and how they can be used in model parameter selection are relevant.

Mannina et al. (2012) suggest a common terminology on the basis of definitions drawn from literature by Saltelli (2000), Campolongo et al. (2007), and Pujol (2009). The first definition comes with the SRC method which, by defining a cutoff threshold (CFT), distinguishes between the following two factors (Figure 5.17a):

- Important factors—if sensitivity > CFT

- Nonimportant factors—if sensitivity < CFT

Important factors represent those model factors that have a high sensitivity coefficient and where, therefore, the modeler should pay more attention. Conversely, nonimportant factors are those model factors characterized by a low sensitivity coefficient. Linear models can be fixed anywhere in their variation ranges. For nonlinear

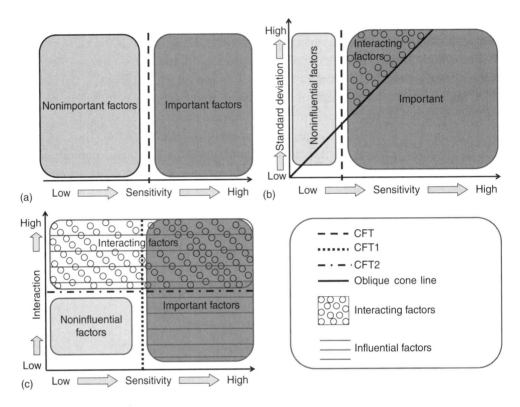

FIGURE 5.17 Schematic overview of the suggested terminology for differentiating input factors according to different GSA methods—(a) SRC, (b) Morris screening, and (c) E-FAST. (Mannina et al., 2012).

models (e.g., WRRF models), however, some nonimportant factors cannot be fixed because of interactions with other input factors (as discussed later in this section).

Morris screening provides a second type of classification of input factors. It allows for implicitly distinguishing between three different types of factors with respect to the mean and the standard deviation of the sensitivity, as follows (Figure 5.17b):

- Important factors—if mean sensitivity > CFT
- Interacting factors—if mean sensitivity > CFT and the standard deviation of the sensitivity is above a specified cone line
- Noninfluential factors—if mean sensitivity < CFT

In particular, the method as modified by Campolongo et al. (2007) basically defines a cone whose edges are set by a CFT and an oblique line that is a statistical function of the mean and standard deviation of the sensitivity (Figure 5.17b) (quantitative characteristics are given here).

The E-FAST method distinguishes four classes of factors on the basis of two CFTs (CFT1 and CFT2) as follows (Figure 5.17c):

- Important factors—if sensitivity > CFT1
- Interacting factors—if interaction > CFT2
- Influential factors—if sensitivity > CFT1 or interaction > CFT2
- Noninfluential factors—if sensitivity < CFT1 and interaction < CFT2

Noninfluential factors that can be identified by both Morris screening and the E-FAST method can be fixed anywhere within their range of uncertainty without changing the model output variance.

In terms of computational load, there is nonconclusive evidence that the SRC method is the least computationally expensive, followed by the Morris screening method and, finally, the E-FAST and Sobol methods. However, issues with convergence of the methods need to be studied further (Benedetti et al., 2011; Yang, 2011)

7.4   Uncertainty Analysis

In the field of wastewater treatment modeling, uncertainty analysis is increasingly recognized as an essential tool that, next to simulation results, also provides a quantitative expression of the reliability of those results (Belia et al., 2009). Next to the expression of uncertainty bounds on results, uncertainty studies can also be used to provide insight to the role of parameter and input uncertainty on model output uncertainty. Finally, uncertainty analysis can also be a means to prioritize dealing with uncertainties and to focus research efforts on the most problematic points of a model. As such, it helps to prepare future measurement campaigns.

The following steps can be taken to conduct a simple uncertainty analysis: (1) identification of the main uncertainty sources, (2) characterization of parameter and input uncertainty, and (3) propagation of the uncertainties into the model outputs.

7.4.1   Step 1—Identification of the Main Uncertainty Sources

In literature, sources of uncertainty have been considered from the perspective of where they are located in a generic model (Refsgaard et al., 2007; Walker et al., 2003). Thus, these authors identified three or four main areas that introduce uncertainties to model predictions. These are model inputs (i.e., any type of data needed to perform a simulation; e.g., influent flow and wastewater characteristics); model structure (e.g., activated sludge model and clarifier model); and model parameters. Uncertainty in the inputs is caused by random variations of the system (e.g., weather) and to errors in measurements (e.g., imprecise sampling and measurement techniques). Uncertainty in the model is caused by incomplete understanding of the modeled processes and/or the simplified descriptions of the processes chosen to be included in models. A fourth source of uncertainty results from implementation of the models in software packages (e.g., numerical integration, bugs, and solver settings) (Claeys et al., 2010; Copp et al., 2008).

To provide a more intuitive method of identifying sources of uncertainty, Belia et al. (2009) proposed that the focus be shifted from the location of uncertainty within the model to when this uncertainty is introduced during a typical modeling project. To aid in this analysis, typical steps of a standard modeling project can be used (Refsgaard et al., 2005). The five steps, shown in the first row of Figure 5.18, are an intuitive sequence of tasks as suggested by the IWA Task Group on Good Modeling Practice (Rieger et al., 2012).

Uncertainty can be identified and evaluated at key times during a project as suggested by Refsgaard et al. (2007) and shown in Figure 5.18. This figure also includes a list of items for each project step that need to be selected or decided on and that identify a location of uncertainty. The figure, therefore, combines the traditional location of uncertainty within the model with a project-step-oriented or sequential approach.

Belia et al. (2009) provide an extensive list of the sources of uncertainty introduced during a typical modeling project.

7.4.2   Step 2—Estimation or Calculation of Uncertainty

Parameter uncertainty can be calculated from the covariance matrix. The latter is obtained during local sensitivity analysis or the calibration process if optimization methods are derivative-based so that the covariance matrix is calculated during optimization (Beck, 1987).

If no direct calculations are possible (e.g., for uncertainty on inputs, the uncertainties need to be estimated), one can divide the parameters and data in uncertainty classes (i.e., accurately known, very poorly known, and an intermediate class) and assign a percentage uncertainty to them. Reichert and Vanrolleghem (2001) adopted a similar approach to this. If direct calculation is impossible, other options are expert knowledge, questionnaires, or statistical calculation of uncertainties with historic data.

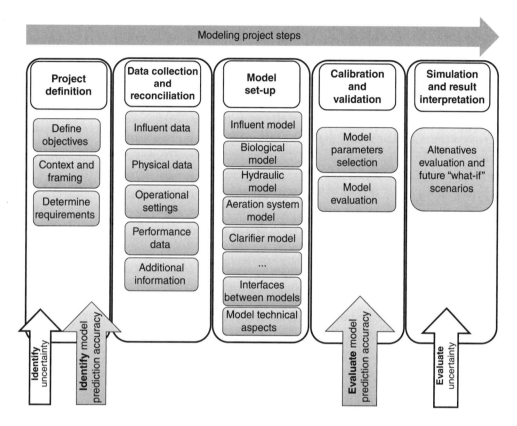

FIGURE 5.18 Typical modeling project steps including instances where model uncertainty and prediction accuracy should be identified and evaluated. (Adapted from Refsgaard et al. [2007] and Rieger et al. [2012]).

7.4.3 Step 3—Propagate Uncertainty through the Model

To propagate uncertainty in model outputs, different approaches are available that can be divided into the following two main groups: (analytical) uncertainty propagation equations or (probabilistic) Monte Carlo sampling-based methods.

7.4.3.1 Error Propagation Equations

Uncertainty can be propagated analytically through simple, linear, or nearly linear models. In the simple form of uncertain parameters and inputs X whose uncertainty can be characterized by the variance σ_x^2 of the uncertainty distribution, the variance of model output Y can be calculated from

$$\sigma_y^2 = \sum_{i=1}^{N}\left(\frac{\partial f}{\partial x_i}\right)^2 \cdot \sigma_i^2 \qquad (5.18)$$

This expression, and its generalized form that also considers correlation between the X, is known as the law of propagation of uncertainty in the guideline titled *Uncertainty of Measurement—Part 3: Guide to the Expression of Uncertainty in Measurement* (ISO, 2008).

This technique has been frequently used in wastewater modeling (Gujer, 2009), not only for linear models but also for nonlinear ones. This means that this method becomes an approximation in which "even for mildly nonlinear models, the results may be rather inaccurate" (Beven, 2009).

7.4.3.2   Monte Carlo Methods

With increasing computer power, Monte Carlo sampling-based methods have become more important in overcoming the limitations of analytical methods. Additionally, the uncertainty of model variables can be expressed by any distribution (e.g., uniform, normal, lognormal, and empirical distributions). In Monte Carlo methods, uncertain model components (e.g., model parameters and inputs) are sampled from prior probability distributions and the corresponding model results are calculated. Subsequently, from the distribution of the obtained model outputs, the uncertainty bounds (e.g., 5% and 95% percentiles) can be calculated.

Different methods for sampling the parameter space that are used range from random (brute-force) sampling to optimized Monte Carlo methods as Markov Chain Monte Carlo simulations (Kuczera and Parent, 1998). In addition, structured sampling of the parameter space is possible through use of a variety of methods (e.g., Latin Hypercube sampling).

Despite significant progress in computational power during the last few years, the problem still persists that when dealing with a large number of parameters and inputs it is difficult to draw enough samples for an adequate representation of the models' output uncertainty distribution, especially when computational requirements and model run times are high. Benedetti et al. (2011) reviewed the number of simulations that are typically required to obtain an adequate representation and concluded that the number depends on the model used and even the output considered.

Overall, the necessary number of simulations varied between 15 and 150 times the number of uncertain parameters and inputs considered.

8.0   REFERENCES

Ahnert, M.; Blumensaat, F.; Langergraber, G.; Alex, J.; Woerner, D.; Frehmann, T.; Halft, N.; Hobus, I.; Plattes, M.; Spering, V.; Winkler, S. (2007) Goodness-of-Fit Measures for Numerical Modeling in Urban Water Management—A Summary to Support Practical Applications. *Proceedings of the 10th Large Wastewater Treatment Plants Conference*; Vienna, Austria, Sept 9–13.

American Public Health Association; American Water Works Association; Water Environment Federation (2005) *Standard Methods for the Examination of Water and Wastewater*, 21st ed.; American Public Health Association: Washington, D.C.

American Society of Civil Engineers (1997) *Standard Guidelines for In-Process Oxygen Transfer Testing, (18- 96)*; American Society of Civil Engineers: New York.

American Society of Civil Engineers (2007) *ASCE/EWRI Standards: Measurement of Oxygen Transfer in Clean Water (2-06)*; American Society of Civil Engineers: New York.

Anderson, N. E. (1945) Design of Final Settling Tanks for Activated Sludge. *Sew. Works J.*, **17** (1), 50–65.

Bachis, G.; Vallet, B.; Maruejouls, T.; Clouzot, L.; Lessard, P.; Vanrolleghem, P. A. (2012) Particle Classes-Based Model for Sedimentation in Urban Wastewater Systems. *Proceedings of the International Water Association Particle Separation Conference*; Berlin, Germany, June 18–20.

Belia, E.; Amerlinck, Y.; Benedetti, L.; Johnson, B.; Sin, G.; Vanrolleghem, P. A.; Gernaey, K. V.; Gillot, S.; Neumann, M. B.; Rieger, L.; Shaw, A.; Villez, K. (2009) Wastewater Treatment Modeling: Dealing with Uncertainties. *Water Sci. Technol.*, **60**, 1929–1941.

Benedetti, L.; Claeys, F.; Nopens, I.; Vanrolleghem, P. A. (2011) Assessing the Convergence of LHS Monte Carlo Simulations of Wastewater Treatment Models. *Water Sci. Technol.*, **63**, 2219–2224.

Beven, K. (2009) *Environmental Modeling: An Uncertain Future?* Taylor & Francis: Abingdon, U.K.

Boltz, J. P.; Morgenroth, E.; Brockmann, D.; Bott, C.; Gellner, W. J.; Vanrolleghem, P. A. (2011) Systematic Evaluation of Biofilm Models for Engineering Practice: Components and Critical Assumptions. *Water Sci. Technol.*, **64**, 930–944.

Bretscher, U.; Krebs, P.; Hager, W. H. (1992) Improvement of Flow in Final Settling Tanks. *J. Environ. Eng.*, **118**, 307–321.

Brun, R.; Kühni, M.; Siegrist, H.; Gujer, W.; Reichert, P. (2002) Practical Identifiability of ASM2d Parameters—Systematic Selection and Tuning of Parameter Subsets. *Water Res.*, **36**, 4113–4127.

Bye, C. M.; Dold, P. L. (1999) Evaluation of Correlations for Zone Settling Velocity Parameters Based on Sludge Volume Index Type Measures and Consequences in Settling Tank Design. *Water Environ. Res.*, **71**, 1333–1344.

Campolongo, F.; Cariboni, J.; Saltelli, A. (2007) An Effective Screening Design for Sensitivity Analysis of Large Models. *Environ. Modeling Software*, **22** (10), 1509–1518.

Capela, S.; Gillot, S.; Héduit, A. (2004) Oxygen Transfer under Process Conditions: Comparison of Measurement Methods. *Water Environ. Res.*, **76**, 183–188.

Chambers, B.; Jones, G. L. (1988) Optimisation and Uprating of Activated Sludge Plants by Efficient Process Design. *Water Sci. Technol.*, **20** (4-5), 121–132.

Chebbo, G.; Gromaire, M.-C. (2009) VICAS—An Operating Protocol to Measure the Distributions of Suspended Solid Settling Velocities within Urban Drainage Samples. *J. Environ. Eng.*, **135** (9), 768–775.

Choubert, J.-M.; Rieger, L.; Shaw, A.; Copp, J.; Spérandio, M.; Soerensen, K.; Rönner-Holm, S.; Morgenroth, E.; Melcer, H.; Gillot, S. (2012) Rethinking Wastewater Characterization Methods—A Position Paper. *Proceedings of the 3rd International Water Association/Water Environment Federation Wastewater Treatment Modeling Seminar*; Mont-Sainte-Anne, Quebec, Canada, Feb 26–28.

Claeys, P.; van Griensven, A.; Benedetti, L.; De Baets, B.; Vanrolleghem, P. A. (2010) On Numerical Solver Selection and Related Uncertainty Terminology. *J. Hydroinformatics*, **12**, 241–250.

Copp, J. B.; Jeppsson, U.; Vanrolleghem, P. A. (2008) The Benchmark Simulation Models—A Valuable Collection of Modeling Tools. *Proceedings of the International Congress on Environmental Modeling and Software*; Barcelona, Spain, July 7–10; Vol. 2, pp 1314–1321.

Corominas, L.; Rieger, L.; Takács, I.; Ekama, G.; Hauduc, H.; Vanrolleghem, P. A.; Oehmen, A.; Gernaey, K. V.; van Loosdrecht, M. C. M.; Comeau, Y. (2010) New Framework for Standardized Notation in Wastewater Treatment Modeling. *Water Sci. Technol.*, **61**, 841–857.

Corominas, L.; Sin, G.; Puig, S.; Balaguer, M. D.; Vanrolleghem, P. A.; Colprim, J. (2011) Modified Calibration Protocol Evaluated in a Model-Based Testing of SBR Flexibility. *Bioprocess Biosyst. Eng.*, **34**, 205–214.

Crawford, G. V.; Fillmore, L.; Katehis, D.; Sandino, J. (2010) From Best Practices to the Plant of the Future: The WERF Optimization Program. *Proceedings of the 83rd Annual Water Environment Federation Technical Exhibition and Conference*; New Orleans, Louisiana, Oct 2–6; Water Environment Federation: Alexandria, Virginia; pp 6685–6694.

Daigger, G. T.; Roper, R. E. (1985) The Relationship Between SVI and Activated Sludge Settling Characteristics. *J.—Water Pollut. Control Fed.*, **57**, 859–866.

Dawson, C. W.; Abrahart, R. J.; See, L. M. (2007) HydroTest: A Web-Based Toolbox of Evaluation Metrics for the Standardized Assessment of Hydrological Forecasts. *Environ. Modeling Software*, **22**, 1034–1052.

De Clercq, B.; Coen, F.; Vanderhaegen, B.; Vanrolleghem, P. A. (1999) Calibrating Simple Models for Mixing and Flow Propagation in Waste Water Treatment Plants. *Water Sci. Technol.*, **39** (4), 61–69.

De Clercq, J. (2006) Batch and Continuous Settling of Activated Sludge: In-Depth Monitoring and 1D Compressive Modeling. Ph.D. Thesis; Faculty of Engineering; Ghent University, Belgium; pp 217.

De Clercq, J.; Devisscher, M.; Boonen, I.; Defrancq, J.; Vanrolleghem, P. A. (2005) Analysis and Simulation of the Sludge Profile Dynamics in a Full-Scale Clarifier. *J. Chem. Technol. Biotechnol.*, **80**, 523–530.

De Pauw, D. J. W.; Vanrolleghem, P. A. (2006) Practical Aspects of Sensitivity Function Approximation for Dynamic Models. *Mathematical and Computer Modeling of Dynamical Systems*, **12**, 395–414.

Deininger, A.; Holthausen, E.; Wilderer, P. A. (1998) Velocity and Solids Distribution in Circular Secondary Clarfiers: Full Scale Measurements and Numerical Modeling. *Water Res.*, **32**, 2951–2958.

Dochain, D.; Vanrolleghem, P. A. (2001) *Dynamical Modeling and Estimation in Wastewater Treatment Processes*; IWA Publishing: London; p 342.

Dupont, R.; Sinkjær, O. (1994) Optimization of Wastewater Treatment Plants by Means of Computer Models. *Water Sci. Technol.*, **30** (4), 181–190.

Ekama, G. A.; Barnard, J. L.; Günthert, F. W.; Krebs, P.; McCorquodale, J. A.; Parker, D. S.; Wahlberg, E. J. (1997) *Secondary Settling Tanks: Theory, Modeling and Operation*; IWA Scientific and Technical Report No. 6; IWA Publishing: London.

Fall, C.; Flores, N. A.; Espinoza, M. A.; Vazquez, G.; Loaiza-Návia, J.; van Loosdrecht, M. C. M.; Hooijmans, C. M. (2011) Divergence between Respirometry and Physicochemical Methods in the Fractionation of the Chemical Oxygen Demand in Municipal Wastewater. *Water Environ. Res.*, **83**, 162–172.

Fujie, K.; Sekizawa, T.; Kubota, H. (1983) Liquid Mixing in Activated Sludge Aeration Tank. *J. Fermentation Technol.*, **61**, 295–304.

Gernaey, K. V.; van Loosdrecht, M. C. M.; Henze, M.; Lind, M.; Jørgensen, S. B. (2004) Activated Sludge Wastewater Treatment Plant Modeling and Simulation: State of the Art. *Environ. Modeling Software*, **19**, 763–783.

Gillot, S.; Choubert, J. M. (2010) Biodegradable Organic Matter in Domestic Wastewaters—Comparison of Selected Fractionation Techniques. *Water Sci. Technol.*, **62**, 630–639.

Giokas, D. L.; Daigger, G. T.; von Sperling, M.; Kim, Y.; Paraskevas, P. A. (2003) Comparison and Evaluation of Empirical Zone Settling Velocity Parameters Based on Sludge Volume Index Using a Unified Settling Characteristics Database. *Water Res.*, **37**, 3821–3836.

Gujer, W. (2009) *Systems Analysis for Water Systems*; Springer-Verlag: Heidelberg, Germany.

Gupta, H. V.; Sorooshian, S.; Yapo, P. O. (1998) Toward Improved Calibration of Hydrologic Models: Multiple and Noncommensurable Measures of Information. *Water Resour. Res.*, **34**, 751–763.

Hauduc, H.; Neumann, M. B.; Muschalla, D.; Gamerith, V.; Gillot, S.; Vanrolleghem, P. A. (2011) Towards Quantitative Quality Criteria to Evaluate

Simulation Results in Wastewater Treatment—A Critical Review. *Proceedings of the 8th International Water Association Symposium on Systems Analysis and Integrated Assessment in Water Management*; San Sebastian, Spain, June 20–22; pp 36–46.

Henze, M.; Dupont, R.; Grau, P.; De la Sota, A. (1993) Rising Sludge in Secondary Settlers Due to Denitrification. *Water Res.*, **27**, 231–236.

International Organization for Standardization (2008) *ISO/IEC Guide 98-3:2008: Uncertainty of Measurement—Part 3: Guide to the Expression of Uncertainty in Measurement*, (GUM:1995); ISO/IEC Guide 98-3:2008; International Organization for Standardization: Geneva, Switzerland.

Kappeler, J.; Gujer, W. (1992) Estimation of Kinetic Parameters of Heterotrophic Biomass under Aerobic Conditions and Characterization of Wastewater for Activated Sludge Modeling. *Water Sci. Technol.*, **25** (6), 125–139

Kayser, R. (1979) Measurement of Oxygen Transfer in Clean Water and Under Process Conditions. *Prog. Water Technol.*, **11** (3), 23–36.

Kinnear, D. J. (2000) Evaluating Secondary Clarifier Performance and Capacity. *Proceedings of the Florida Water Resources Conference*; Tampa, Florida, April 16–19.

Kinnear, D. J.; Deines, K. (2001) Acoustic Doppler Current Profiler: Clarifier Velocity Measurement. *Proceedings of the 74th Annual Water Environment Federation Technical Exposition and Conference* [CD-ROM]; Atlanta, Georgia, Oct 13–17; Water Environment Federation: Alexandria, Virginia.

Koch, G.; Pianta, R.; Krebs, P.; Siegrist, H. (1999) Potential of Denitrification and Solids Removal in the Rectangular Clarifier. *Water Res.*, **33**, 309–318.

Kristensen, H. G.; Elberg Jørgensen P.; Henze, M. (1992) Characterization of Functional Micro-Organism Groups and Substrate in Activated Sludge and Wastewater by AUR, NUR, and OUR. *Water Sci. Technol.*, **25** (6), 43–57.

Kristensen, H. G.; la Cour Janssen, J.; Elberg Jørgensen, P. (1998) Batch Test Procedures as Tools for Calibration of the Activated Sludge Model—A Pilot Scale Demonstration. *Water Sci. Technol.*, **37** (4-5), 235–242.

Kuczera, G.; Parent, E. (1998) Monte Carlo Assessment of Parameter Uncertainty in Conceptual Catchment Models: The Metropolis Algorithm. *J. Hydrol.* (Neth.), **211**, 69–85.

Lessard, P.; Beck, B. (1988) Dynamic Modeling of Primary Sedimentation. *J. Environ. Eng.*, **114**, 753–769.

Levenspiel, O. (1999) *Chemical Reaction Engineering*, 3rd ed.; Wiley & Sons: Hoboken, New Jersey.

Lindeborg, C.; Wiberg, N.; Seyf, A. (1996) Studies of the Dynamic Behavior of a Primary Sedimentation Tank. *Water Sci. Technol.*, **34** (3-4), 213–222.

Makinia, J.; Wells, S. A. (2005) Evaluation of Empirical Formulae for Estimation of the Longitudinal Dispersion in Activated Sludge Reactors. *Water Res.*, **39**, 1533–1542.

Mannina, G.; Cosenza, A.; Viviani, G.; Vanrolleghem, P. A.; Neumann, M. B. (2012) Global Sensitivity Analysis for Urban Water Quality Modeling: Comparison of Different Methods. *Proceedings of the 9th International Conference on Urban Drainage Modeling*; Belgrade, Serbia, Sept 4–7.

Maruejouls, T.; Lessard, P.; Wipliez, B.; Pelletier, G.; Vanrolleghem, P. A. (2011) Characterization of the Potential Impact of Retention Tank Emptying on Wastewater Primary Treatment: A New Element for CSO Management. *Water Sci. Technol.*, **64** (9), 1898–1905.

Melcer, H.; Ciolli, M.; Lilienthal, R.; Ott, G.; Land, G.; Dawson, D.; Klein, A.; Wightman, D. (2010) Bringing CEPT Technology into the 21st Century. *Proceedings of the 83rd Annual Water Environment Federation Technical Exhibition and Conference* [CD-ROM]; New Orleans, Louisiana, Oct 2–6; Water Environment Federation: Alexandria, Virginia; pp 6685–6694.

Melcer, H.; Dold, P. L.; Jones, R. M.; Bye, C. M.; Takacs, I.; Stensel, H. D.; Wilson, A. W.; Sun, P.; Bury, S. (2003) *Methods for Wastewater Characterization in Activated Sludge Modeling*; Water Environment Research Foundation: Alexandria, Virginia.

Metcalf and Eddy, Inc. (2003) *Wastewater Engineering Treatment and Reuse*, 4th ed.; McGraw-Hill: New York.

Morris, M. D. (1991) Factorial Sampling Plans for Preliminary Computational Experiments. *Technometrics,* **33** (2), 161–174.

Otterpohl, R. (1995) Entwicklung eines mathematischen Modells für Vorklärbecken. *GWA,* **151**, 28–36 (in German).

Parker, D. S.; Kaufman, W. J.; Jenkins, D. (1970) *Characteristics of Biological Flocs in Turbulent Flows*; SERL Report No. 70-5; University of California, Berkeley: Berkeley, California.

Petersen, B.; Gernaey, K.; Henze, M.; Vanrolleghem, P. A. (2002) Evaluation of an ASM1 Model Calibration Procedure on a Municipal-Industrial Wastewater Treatment Plant. *J. Hydroinformatics*, **4**, 15–38.

Petersen, B.; Gernaey, K.; Henze, M.; Vanrolleghem, P. A. (2003) Calibration of Activated Sludge Models: A Critical Review of Experimental Designs. In *Biotechnology for the Environment: Wastewater Treatment and Modeling, Waste Gas Handling*; Agathos, S. N., Reineke, W., Eds.; Kluwer Academic Publishers: Dordrecht, The Netherlands; pp 101–186.

Pujol, G. (2009) Simplex-Based Screening Designs for Estimating Metamodels. *Reliability Eng. System Safety*, **94**, 1156–1160.

Redmon, D.; Boyle, W. C.; Ewing, L. (1983) Oxygen Transfer Efficiency Measurements in Mixed Liquor Using Off-Gas techniques. *J.—Water Pollut. Control Fed.*, **55**, 1338–1347.

Refsgaard, J. C.; Henriksen, H. J.; Harrar, W. G.; Scholten, H.; Kassahun, A. (2005) Quality Assurance in Model Based Water Management—Review of Existing Practise and Outline of New Approaches. *Environ. Modeling Software*, **20**, 1201–1215.

Refsgaard, J. C.; van der Sluijs, J. P.; Højberg, A. L.; Vanrolleghem, P. A. (2007) Uncertainty in the Environmental Modeling Process—A Framework and Guidance. *Environ. Modeling Software*, **22**, 1543–1556.

Rieger, L.; Gillot, S.; Langergraber, G.; Ohtsuki, T.; Shaw, A.; Takács, I.; Winkler, S. (2012) *Guidelines for Using Activated Sludge Models*; IWA Scientific and Technical Report; IWA Publishing: London.

Roeleveld, P. J.; van Loosdrecht, M. C. M. (2002) Experience with Guidelines for Wastewater Characterization in The Netherlands. *Water Sci. Technol.*, **45** (6), 77–87.

Saltelli, A. (2000) *Sensitivity Analysis*; Wiley & Sons: Chichester, U.K.

Saltelli, A.; Ratto, M.; Andres, T.; Campolongo, F.; Cariboni, J.; Gatelli, D.; Saisana, M.; Tarantola, S. (2008) *Global Sensitivity Analysis. The Primer*; Wiley & Sons: Chichester, U.K.

Saltelli, A.; Tarantola, S.; Campolongo, F.; Ratto, M. (2004) Sensitivity Analysis in Practice. A Guide to Assessing Scientific Models. In *Probability and Statistics Series*; Wiley & Sons: Chichester, U.K.

Saltelli, A.; Tarantola, S.; Chan, K. P. S. (1999) A Quantitative Model-Independent Method for Global Sensitivity Analysis of Model Output. *Technometrics*, **41** (1), 39–56.

Sin, G.; De Pauw, D. J. W.; Weijers, S.; Vanrolleghem, P. A. (2008) An Efficient Approach to Automate the Manual Trial and Error Calibration of Activated Sludge Models. *Biotechnol. Bioeng.*, **100**, 516–528.

Sin, G.; Vanrolleghem, P. A. (2004) A Nitrate Biosensor Based Methodology for Monitoring Anoxic Activated Sludge Activity. *Water Sci. Technol.*, **50** (11), 125–133.

Sobol, I. M. (2001) Global Sensitivity Indices for Nonlinear Mathematical Models and their Monte Carlo Estimates. *Mathematics and Computers in Simulation*, **55** (1-3), 271–280.

Spanjers, H.; Vanrolleghem, P. A.; Olsson, G.; Dold, P. (1998) *Respirometry in Control of the Activated Sludge Process*; International Association on Water Quality: London, U.K.

STOWa (2002a) *Optimalisatie van ronde nabezinktanks: Modelproeven*; Report Physical Secondary Settling Tank Model; Stichting Toegepast Onderzoek Waterbeheer (STOWa): Utrecht, The Netherlands (in Dutch).

STOWa (2002b) *Optimalisatie van ronde nabezinktanks: Ontwikkeling nabezinktank-model en evaluatie STORA-ontwerprichtlijn*; Report Physical Secondary Settling Tank Model; Stichting Toegepast Onderzoek Waterbeheer (STOWa): Utrecht, The Netherlands (in Dutch).

Takács, I.; Patry, G. G.; Nolasco, D. (1991) A Dynamic Model of the Clarification-Thickening Process. *Water Res.*, **25**, 1263–1271.

Tik, S.; Vanrolleghem, P. A. (2012) Modeling and Control of a Full-Scale Chemically Enhanced Primary Treatment. *Proceedings of the International Water Association International Conference on Advances in Particle Separation*; Berlin, Germany, June 18–20.

Torres, V.; Takács, I.; Riffat, R.; Shaw, A.; Murthy, S. (2011) Determination of the Range of Anoxic Half Saturation Coefficients for Methanol. *Proceedings of the*

84th Annual Water Environment Federation Technical Exhibition and Conference [CD-ROM]; Los Angeles, California, Oct 15–19; Water Environment Federation: Alexandria, Virginia; pp 3234–3244.

Ueberl, J.; Hager, W. H. (1997) Improved Design of Final Settling Tanks. *J. Environ. Eng.*, **123**, 259–268.

Van Hulle, S. W. H.; Volcke, E. I. P.; López Teruel, J.; Donckels, B.; van Loosdrecht, M. C. M.; Vanrolleghem, P. A. (2007) Influence of Temperature and pH on the Kinetics of the Sharon Nitritation Process. *J. Chem. Technol. Biotechnol.*, **82**, 471–480.

Vanrolleghem, P. A.; De Clercq, B.; De Clercq, J.; Devisscher, M.; Kinnear, D. J.; Nopens, I. (2006) New Measurement Techniques for Secondary Settlers: A Review. *Water Sci. Technol.*, **53** (4–5), 419–429.

Vanrolleghem, P. A.; Lee, D. S. (2003) On-Line Monitoring Equipment for Wastewater Treatment Processes: State of the Art. *Water Sci. Technol.*, **47** (2), 1–34.

Vanrolleghem, P. A.; Spanjers, H.; Petersen, B.; Ginestet, P.; Takacs, I. (1999) Estimating (Combinations of) Activated Sludge Model No. 1 Parameters and Components by Respirometry. *Water Sci. Technol.*, **39** (1), 195–214.

Vanrolleghem, P. A.; Van Daele, M.; Dochain, D. (1995) Practical Identifiability of a Biokinetic Model of Activated Sludge Respiration. *Water Res.*, **29**, 2561–2570.

Wahlberg, E. J. (2002) *WERF/CRTC Protocols for Evaluating Secondary Clarifier Performance*; Water Environment Research Foundation: Alexandria, Virginia.

Walker, W. E.; Harremoes, P.; Rotmans, J.; van der Sluijs, J. P.; van Asselt, M. B. A.; Janssen, P.; Krayer von Krauss, M. P. (2003) Defining Uncertainty: A Conceptual Basis for Uncertainty Management in Model-Based Decision Support. *Integrated Assess.*, **4** (1), 5–17.

Wouters-Wasiak, K.; Heduit, A.; Audic, J. M. (1996) Consequences of an Occasional Secondary Phosphorus Release on Enhanced Biological Phosphorus Removal. *Water SA*, **22**, 91–96.

Yang, J. (2011) Convergence and Uncertainty Analyses in Monte-Carlo Based Sensitivity Analysis. *Environ. Modeling Software*, **26** (4), 444–457.

Chapter 6

Overview of Available Modeling and Simulation Protocols

1.0   INTRODUCTION

The quality of simulation studies can vary depending on project objectives, resources spent, and available expertise. Consideration should be given to model accuracy and the amount of time required to carry out a simulation project, with the goal of producing the accuracy required. A variety of approaches and insufficient documentation make quality assessment and comparability of simulation results difficult or almost impossible. A general framework for the application of activated sludge models is needed to overcome these obstacles.

During the last few years, several guidelines have been developed around the world focusing on different aspects of simulation projects. To synthesize the available experience into an internationally recognized industry standard, the International Water Association (IWA) formed a task group on good modeling practice (GMP), known as the GMP Task Group. The objective of the GMP Task Group was to prepare a scientific and technical report that describes GMP with activated sludge systems (http://iwa-gmp-tg.irstea.fr/). The report was published in 2012 (Rieger et al., 2012).

2.0   AVAILABLE PROTOCOLS

2.1   Protocols Dedicated to Wastewater Treatment Modeling

Several groups working on wastewater treatment have proposed activated sludge modeling and simulation protocols. The following four protocols have reached a level of application such that they are now considered standards in the field:

- STOWA (Hulsbeek et al., 2002; Roeleveld and van Loosdrecht, 2002)
- Water Environment Research Foundation (WERF) (Melcer et al., 2003)
- BIOMATH (Vanrolleghem et al., 2003) and extensions (e.g., Corominas et al. [2011])
- HSG (Langergraber et al., 2004)

A comparison of these four protocols can be found in works by Sin et al. (2005) and Corominas (2006). Protocols with less international scope or coverage have been proposed by Frank (2006), Japan Sewage Works Agency (2006), and Itokawa et al. (2008). Guidance for water resource recovery facility modeling can also be found in many publications presenting case studies (e.g., Meijer et al. [2002]), different books (e.g., Henze et al. [2008] and Makinia [2010]), simulator manuals, or proprietary company guidelines.

2.2 Protocols from Related Fields

In fields other than wastewater treatment, promoting the correct use of models and ensuring quality and modeling efficiency have also been studied (e.g., Scholten et al. [2000], Refsgaard et al. [2005], and U.S. EPA, 2009). A working group in The Netherlands published *Good Modeling Practice Handbook* (Van Waveren et al., 2000) for the water management field. Enhancing model credibility was also one of the objectives of the European project, HarmoniQuA (http://harmoniqua.wau.nl/), and led to the development of quality assurance guidelines and a modeling support tool, MoST (Version 3.1.5; http://harmoniqua.wau.nl/public/Products/most.htm).

2.3 Comparison of Existing Protocols

This section highlights the main features of the different protocols to identify specific items that should be included in a unified protocol.

The HarmoniQuA project suggested the following classification scheme for modeling quality assurance guidelines (Refsgaard et al., 2005):

- Type 1—internal technical guidelines developed and used internally

- Type 2—public technical guidelines

- Type 3—public interactive guidelines

The last two types are developed in a public consensus-building process. Type 3 guidelines also include organization of the interaction between the modeler and the client. Moreover, the HarmoniQuA project also discussed establishment of performance criteria and reviews of the different phases of the modeling project.

The four main activated sludge modeling guidelines (STOWA, WERF, BIOMATH, and HSG) are Type 2 and the other protocols are Type 1. All protocols have been developed in a mature scientific discipline and a mature market, as can be seen in the now widespread use of activated sludge models in engineering practice.

The main emphasis of STOWA is to help end users model their nitrogen removal facilities using Activated Sludge Model No. 1 in a systematic and standardized way. An essential part of this protocol was the development of an easy-to-use wastewater characterization procedure. As part of the development, user groups were set up and the outcome was the result of an extensive consensus-building process. The STOWA guideline is regarded as an international standard because of its ease of use and widespread application. Unfortunately, only a summary is available in English in the form of two journal publications (i.e., Hulsbeek et al. [2002] and Roeleveld and van Loosdrecht [2002]).

The WERF guidelines (Melcer et al., 2003) are based on experiences from a large market (mainly North America), with authors from consulting companies, software developers, and universities. Targeted users are municipalities and consulting engineering companies, including junior and intermediate modelers. The development consisted of research on wastewater characterization methods and a consensus-building process involving a large international reviewer group. The 575-page final report includes an extensive overview of knowledge, experience, and data and became a standard reference for wastewater characterization and simulation procedures.

BIOMATH at Ghent University (Ghent, Belgium) proposed a generic calibration procedure (Vanrolleghem et al., 2003) using state-of-the-art parameter estimation methods for step-wise calibration/validation of models, with a focus on the biokinetic model and sections on settling, hydraulics, and aeration. The protocol requires a high level of experimental results and takes advantage of systems analysis tools (see Chapter 5). The protocol summarizes the work of the BIOMATH research group and is mainly dedicated to experienced modelers. It has been applied in academic research projects to increase process understanding or during development of new models.

The HSG protocol (Langergraber et al., 2004) is a generic procedure to guide modelers through all steps of a modeling project. The HSG protocol gathers the experience of specialized researchers from German-speaking countries. The focus is on a standardized structure for modeling projects. An objective-oriented approach is encouraged, but deviations from the full procedures need to be explained and documented. The importance of data quality is highlighted. The HSG protocol targets modelers from consulting firms, water boards, and municipalities. An eight-page journal publication is publicly available.

Regional protocols (e.g., the Japan Sewage Works Agency [2006] protocol; http://www.sbmc.or.jp/) often focus on specific issues and constraints and may not allow for generalization. Company protocols (e.g., Frank [2006]) are often proprietary and not easily accessible. The focus of both types of protocols is typically on practical use. Software manuals are focused on explaining the use of respective software, but often provide a relevant source of information on how to apply models. Some software companies provide additional support to clients in their modeling work. Published case studies (e.g., Meijer et al. [2002] and Third et al. [2007]) can be used as another source of guidance, but are often too specific to be used as general guidance on activated sludge modeling.

2.4 Toward a Unified Protocol

The GMP Task Group analyzed existing protocols looking for agreements and differences to identify the strengths of each protocol, with the goal of combining them in one unified protocol. When comparing protocols, agreements outnumbered differences and, where differences were evident, they were mostly in the level of detail and foci. This can be related to the background of the authors (e.g., researchers, consulting engineers, and roundtables and the field of their expertise, e.g., process engineering or water management) and the targeted users. Differences may also be linked to the fact that the objectives of model use are different (i.e., mainly for design/redesign purposes in North America and for optimization or controller studies that require more dynamic simulations in Europe) (Hauduc et al., 2009).

All discussed protocols show some similarities in the following tasks that should be considered for inclusion in a unified protocol:

- Objectives definition
- Data collection and reconciliation strategies
- Model selection and setup
- Plant model calibration/validation, including parameter selection
- Documentation
- Interaction between modelers and end users (establishment of performance criteria and reviews of the main steps of the protocol)

The main differences of the discussed protocols are the design of measuring campaigns; experimental methods used to characterize influent, hydraulics, settling, and aeration and to estimate stoichiometric/kinetic parameters; and the procedure to calibrate and validate the plant model.

3.0 GOOD MODELING PRACTICE UNIFIED PROTOCOL

The GMP Task Group proposed the GMP Unified Protocol (Rieger et al., 2012), which combines the key aspects of the protocols discussed in the preceding sections. The GMP Unified Protocol includes interactions of stakeholders and identifies substeps in which the client should be involved in the decision-making process.

An important aspect of the GMP Unified Protocol is that every step is linked to an application matrix to assist modelers in considering the level of effort required to carry out a simulation project depending on the particular objective. The GMP application matrix provides 12 typical modeling objectives for domestic water resource recovery facilities (WRRFs) and two additional ones for industrial facilities.

3.1   Outline of the GMP Unified Protocol

The goal for the GMP Task Group was to develop a Type 3 protocol according to Refsgaard et al. (2005), that is, a protocol based on a broad international consensus and including the interaction between modelers and clients. The proposed protocol is illustrated in Figure 6.1. It comprises the following main steps, which have to be reviewed and agreed on with stakeholders before the next step is carried out (decision boxes are in black):

- Step 1—project definition
- Step 2—data collection and reconciliation
- Step 3—plant model setup
- Step 4—calibration and validation
- Step 5—simulation and result interpretation

3.2   Protocol Steps

3.2.1   Step 1—Project Definition

On the basis of requirements (why use modeling, to answer which questions, what model quality required) and of available data, the objectives of the project are defined in close cooperation with stakeholders. Boundaries and layout of the system to be modeled have to be agreed on. Performance criteria (e.g., data quality requirements or "stop criteria" for calibration) should be set at this stage, responsibilities should be defined, and required data should be identified to decide on the budget and schedule.

3.2.2   Step 2—Data Collection and Reconciliation

Data collection refers to existing (process-related and historical) and missing (measuring campaign setup) data. Because data collection and reconciliation represent the most time-consuming step (30 to 60% of the total effort) (Hauduc et al., 2009),

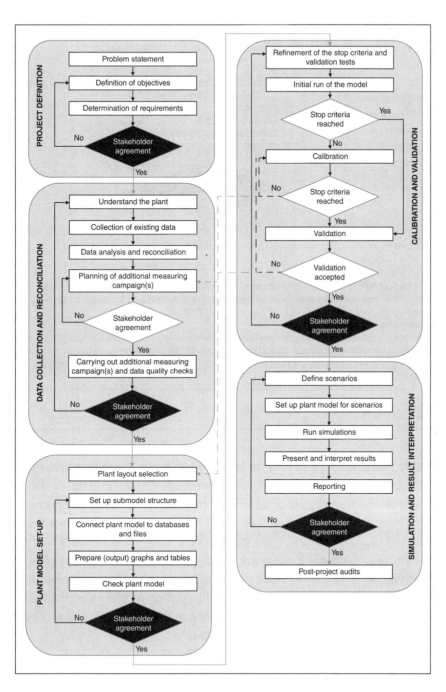

FIGURE 6.1   Proposed GMP unified protocol.

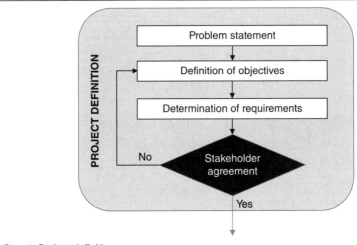

FIGURE 6.1   (*Continued*)

data quality validation should play an essential role in the planning phase (design the required data quality) (Rieger et al., 2010). Data required to perform the data validation step should be an integral part of additional measurements. Understanding process conditions based on collected data is essential to confirm the adequacy of the defined objectives and of collected data. Data validation is based on data series analysis, outlier detection, mass balances (on flows and phosphorus), and other checks (e.g., typical component ratios). From these data, plant model structure (i.e., processes included, flows, and boundaries) is defined. The data collection step finishes with an agreement between modeler and stakeholder on used data and the processes to be modeled.

3.2.3   *Step 3—Plant Model Setup*

This step includes selection of submodels to simulate selected process units. A number of physical and operational values are set at this stage, such as tank

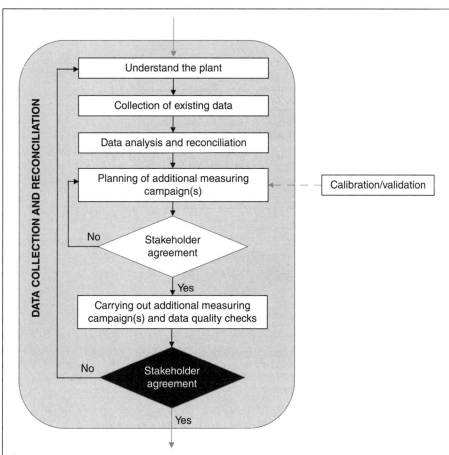

Step 2: Data collection and reconciliation
This step aims at collecting, assessing, and—if necessary—reconciling data sets for simulation projects. A stepwise procedure to analyze collected data is provided, including dedicated methods based on statistical analysis, engineering expertise, and mass balancing.
Deliverables
Reconciled data sets, which are the single data sources for all subsequent steps of the simulation project. Identify deviations from original project definition and possible modifications to project definition document before the next step: plant model setup.

FIGURE 6.1   (*Continued*)

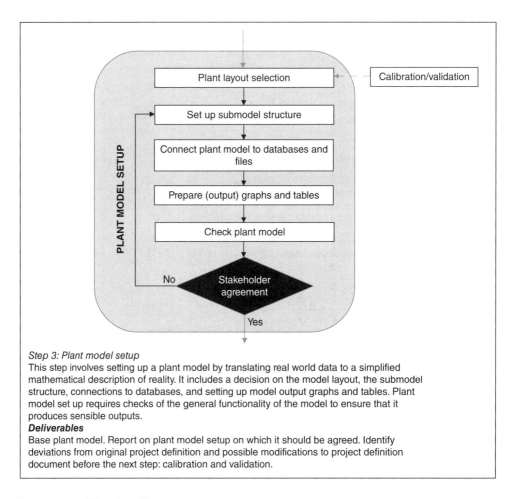

Step 3: Plant model setup
This step involves setting up a plant model by translating real world data to a simplified mathematical description of reality. It includes a decision on the model layout, the submodel structure, connections to databases, and setting up model output graphs and tables. Plant model set up requires checks of the general functionality of the model to ensure that it produces sensible outputs.
Deliverables
Base plant model. Report on plant model setup on which it should be agreed. Identify deviations from original project definition and possible modifications to project definition document before the next step: calibration and validation.

FIGURE 6.1   *(Continued)*

volumes, flows, controller setpoints, and so on. A preliminary selection of average influent data with a set of wastewater characteristics (often defaults) is used during first implementation in the simulator. A number of initial test runs allow for functionality tests and for checking mass balances, thereby better defining boundaries and critical conditions. The functional model is tested for sensible outputs, and stakeholders should agree on model adequacy.

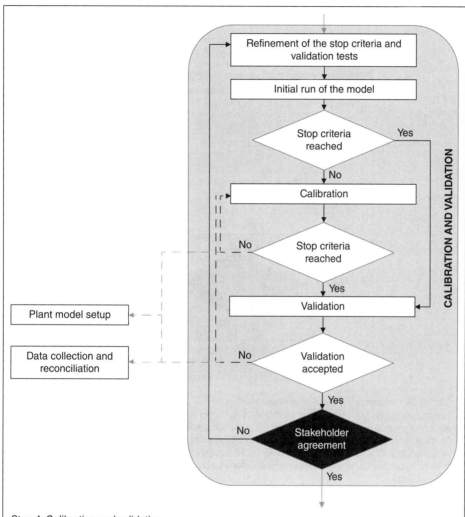

Step 4: Calibration and validation
Model calibration is the process of modifying input parameters until simulation results match an observed set of data. The process is completed when simulated results are close to experimental ones, within a previously defined acceptable margin of error. Validation tests are performed to ensure the use of the plant model with the level of confidence required to meet modeling objectives.

Deliverables
Calibrated and validated parameter sets. Report on calibration and validation on which there is agreement. Identify deviations from original project definition and possible modifications to project definition document before the next step: Simulation and result interpretation.

FIGURE 6.1   (*Continued*)

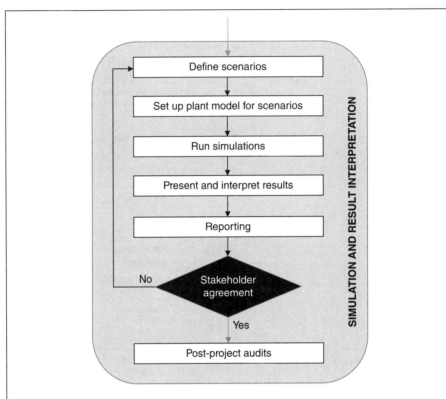

Step 5: Simulation and result interpretation
The calibrated and validated plant model is used to run simulations to meet the objective laid out in the project definition. This step includes defining scenarios, setting up the plant model for these scenarios, running simulations, presenting and interpreting results, and documenting all essential information. This step concludes with reaching an agreement with the stakeholders that the expectations of the project definition have been met.
Deliverables
Final version of the plant model, plus variants from different scenarios. A final report, including simulation and result interpretation plus information from all previous steps (typically as appendices). Agreement between stakeholders and modeler that the objectives of the project definition have been fulfilled.

FIGURE 6.1   (*Continued*)

3.2.4   Step 4—Calibration and Validation

Having defined stop criteria in Step 1, a calibration procedure is performed to get to an agreement between measured and simulated values. First, the plant model should be set up in required detail and operational settings should be adapted before changing any biokinetic parameters. The need to change biokinetic parameters often points to erroneous data or to wrong transport models (mixing characteristics).

For atypical conditions (e.g., significant industrial influent), it may be necessary to recalibrate the biokinetic model, and values may have to be assigned to specific parameters. Sensitivity analyses should be conducted to identify influential parameters (i.e., parameters that have a significant effect on results). Engineering judgment should be exercised to select few parameters to change (i.e., ones that are uncertain and have wide typical ranges) (Hauduc et al., 2011). Parameter estimation is typically carried out manually because of the over-parameterized nature of activated sludge models. However, automatic algorithms to estimate model parameters may assist in speeding up the process (see Chapter 5). It is important to define upfront how the calibration should be carried out. Good practice is to set up a table to define and track all parameter changes.

The resulting parameter set should be validated using a data set, which is independent of calibration data, but still reflects the targeted model application. Uncertainty analysis (based on data, model structure, model parameters, and other sources of uncertainty) (Belia et al., 2009) may be performed to assess the domain of validity of the model and its accuracy. Ultimately, stakeholders have to agree on the model accuracy that is reached.

3.2.5   Step 5—Simulation and Result Interpretation

The validated plant model is used to meet the objectives of the project through simulations. This typically requires definition of scenarios, setting up specific plant models, and, finally, running a set of simulations. If the original objectives cannot be met, the reasons have to be justified and discussed with the client.

All important aspects of the modeling project, including significant decisions, results, interpretation, and conclusions, have to be documented in a final report. The report should allow the client to assess the study's quality, to compare the results with similar studies from other facilities or other studies for the same facility (e.g., non-modeling studies), and to provide required information for future simulations (i.e., for reuse of the model). Finally, simulation results are presented to the client for discussion. Agreement should be sought as to whether the objectives of the project have been met.

4.0   BENEFITS AND POTENTIAL RISKS OF MODELING AND SIMULATION PROTOCOLS

4.1   Introduction

Modeling and simulation protocols can improve the quality of the results and may reduce the required effort. In addition to a direct positive effect on the simulation study, there are several additional benefits such as improved data quality for operation and design. The following are some of the main benefits of using modeling and simulation protocols:

- As a standardization process, protocols lead to better comparable, reproducible, and transferable results; comparison of simulation projects is, indeed, facilitated by use of standard procedures.

- Protocols give guidance to clearly define requirements and limitations of the obtained model and what can be reached at the beginning of the project and, therefore, help prevent misconceptions.

- Checking the quality of a project against standard procedures should lead to improved quality assurance/quality control (Shaw et al., 2011).

- Inexperienced modelers and clients are guided throughout the project.

4.2   Benefits of Using Standardized Modeling and Simulation Protocols

The benefits of using modeling and simulation protocols can be structured into the following categories: benefits for modelers, benefits for WRRFs, benefits for regulatory bodies, and general benefits:

- Benefits for modelers—a GMP survey (Hauduc et al., 2009) found through a questionnaire that the majority of model users have never received organized training in process modeling. A standardized protocol can lead modelers through all steps of a simulation project and, therefore, reduce engineer training time through one consistent source of information. A standardized procedure might highlight typical pitfalls and, therefore, save time and improve model quality. Intensive data analysis and quality checks reduce the time it takes to produce and use a high quality model, which, in turn, reduces reworking requirements. A comprehensive and consistent report format standardizes and, therefore, speeds up model documentation and communication with the client.

- Benefits to WRRFs—following a standardized protocol allows a client to check the quality of the modeler's work against standard procedures. This should lead to improved quality assurance/quality control by the client and increased confidence in the model. Detailed documentation will increase transferability of the modeling study and of the model itself so that it may be used for future applications (even by another modeler).

- Benefits to regulatory bodies—having a state-of-the-art procedure for carrying out simulation studies provides a quality measure to evaluate whether a simulation study followed good modeling practice, even for nonmodelers.

- General benefits—guidance for the interaction between modeler and clients helps define responsibilities and set clear objectives. Quality simulation projects will provide additional benefits in terms of highlighting existing (often undetected) facility problems and data inconsistencies.

4.3 Potential Risks of Standardization

One of the disadvantages of using standardized protocols is that it may block innovative and more cost-effective solutions. Therefore, a structure common to all modeling projects should be suggested, although the modeler should feel free to decide on the best methods and models available for the project's specific objectives.

Another potential problem relates to the danger of strictly following a protocol without taking into account the defined objectives and case characteristics; that is, unnecessary complex (and expensive) steps should be avoided if possible.

Modeling and simulation protocols should lead to a continuous increase in efficiency of simulation projects. Therefore, the goal should be to regularly reassess the suggested methods and to add improvements if they are commonly accepted and standardized. More simulation projects will lead to more experience with parameter values (or typical ranges), and the models will further improve in terms of completeness and applicability. Newly developed technologies might require new models, and a state-of-the-art modeling and simulation protocol should not exclude better-suited models.

TABLE 6.1   Information on protocol steps by chapter.

GMP unified protocol step	Linked chapter in MOP 31	
Step 1—project definition	Chapter 7	Project Definition
Step 2—data collection and reconciliation	Chapter 8	Building a Facility Model
	Section 2	Data Collection
	Section 3	Data Reconciliation
Step 3—plant model setup	Chapter 8	Building a Facility Model
	Section 4	Facility Model Setup
Step 4—calibration and validation	Chapter 8	Building a Facility Model
	Section 5	Calibration and Validation
Step 5—Simulation and result interpretation (including documentation)	Chapter 8	Building a Facility Model
	Section 6	Quality Assurance/Quality Control
	Chapter 9	Using Models for Design

5.0   LINK BETWEEN GOOD MODELING PRACTICE UNIFIED PROTOCOL AND MANUAL OF PRACTICE 31

This manual suggests following the five steps of the GMP Unified Protocol. Table 6.1 shows chapters in which the reader can find more information on a specific protocol step.

6.0   REFERENCES

Belia, E.; Amerlinck, Y.; Benedetti, L.; Johnson, B.; Sin, G.; Vanrolleghem, P. A.; Gernaey, K. V.; Gillot, S.; Neumann, M. B.; Rieger, L.; Shaw, A.; Villez, K. (2009) Wastewater Treatment Modeling: Dealing with Uncertainties. *Water Sci. Technol.*, **60** (8), 1929–1941.

Corominas, L. (2006) Control and Optimization of an SBR for Nitrogen Removal: From Model Calibration to Plant Operation. Ph.D. Thesis, University of Girona, Girona, Spain.

Corominas, L.; Sin, G.; Puig, S.; Balaguer, M. D.; Vanrolleghem, P. A.; Colprim, J. (2011) Modified Calibration Protocol Evaluated in a Model-Based Testing of SBR Flexibility. *Bioprocess Biosyst. Eng.*, **34** (2), 205–214.

Frank, K. (2006) The Application and Evaluation of a Practical Stepwise Approach to Activated Sludge Modeling. *Proceedings of PENNTEC 2006, Annual Technical Conference and Exhibition*; State College, Pennsylvania, July.

Hauduc, H.; Gillot, S.; Rieger, L.; Ohtsuki, T.; Shaw, A.; Takacs, I.; Winkler, S. (2009) Activated Sludge Modeling in Practice—An International Survey. *Water Sci. Technol.*, **60** (8), 1943–1951.

Hauduc, H.; Rieger, L.; Ohtsuki, T.; Shaw, A.; Takács, I.; Winkler, S.; Héduit, A.; Vanrolleghem, P. A.; Gillot, S. (2011) Activated Sludge Modeling: Development and Potential Use of a Practical Applications Database. *Water Sci. Technol.*, **63** (10), 2164–2182.

Henze, M.; van Loosdrecht, M. C. M.; Ekama, G. A.; Brdjanovic, D. (2008) *Biological Wastewater Treatment—Principles, Modeling and Design*; IWA Publishing: London.

Hulsbeek, J. J. W.; Kruit, J.; Roeleveld, P. J.; van Loosdrecht, M. C. M. (2002) A Practical Protocol for Dynamic Modeling of Activated Sludge Systems. *Water Sci. Technol.*, **45** (6), 127–136.

Itokawa, H.; Inoki, H.; Murakami, T. (2008) JS Protocol: A Practical Guidance for the Use of Activated Sludge Modeling in Japan. *Proceedings of the International Water Association World Water Congress and Exhibition*; Vienna, Austria, Sept 7–12.

Japan Sewage Works Agency (2006) *Report on the Technical Evaluation of Activated Sludge Models for Practical Application*; Report number 05-004 [CD-ROM]; Japan Sewage Works Agency: Tokyo, Japan.

Langergraber, G.; Rieger, L.; Winkler, S.; Alex, J.; Wiese, J.; Owerdieck, C.; Ahnert, M.; Simon, J.; Maurer, M. (2004) A Guideline for Simulation Studies of Wastewater Treatment Plants. *Water Sci. Technol.*, **50** (7), 131–138.

Makinia, J. (2010) *Mathematical Modeling and Computer Simulation of Activated Sludge Systems*; IWA Publishing: London.

Meijer, S. C. F.; van der Spoel, H.; Susanti, S.; Heijnen, J. J.; van Loosdrecht, M. C. M. (2002) Error Diagnostics and Data Reconciliation for Activated Sludge Modeling Using Mass Balances. *Water Sci. Technol.*, **45** (6), 145–156.

Melcer, H.; Dold, P. L.; Jones, R. M.; Bye, C. M.; Takacs, I.; Stensel, H. D.; Wilson, A. W.; Sun, P.; Bury, S. (2003) *Methods for Wastewater Characterization in Activated Sludge Modeling*; Water Environment Research Foundation: Alexandria, Virginia.

Refsgaard, J. C.; Henriksen, H. J.; Harrar, B.; Scholten, H.; Kassahun, A. (2005) Quality Assurance in Model Based Water Management—Review of Existing Practice and Outline of New Approaches. *Environ. Modeling Software*, **20**, 1201–1215.

Rieger, L.; Gillot, S.; Langergraber, G.; Ohtsuki, T.; Shaw, A.; Takács, I.; Winkler, S. (2012) *Guidelines for Using Activated Sludge Models*; IWA Scientific and Technical Report No. 22; IWA Publishing: London.

Rieger, L.; Takács, I.; Villez, K.; Siegrist, H.; Lessard, P.; Vanrolleghem, P. A.; Comeau, Y. (2010) Data Reconciliation for WWTP Simulation Studies—Planning for High Quality Data and Typical Sources of Errors. *Water Environ. Res.*, **82** (5), 426–433.

Roeleveld, P. J.; van Loosdrecht, M. C. M. (2002) Experience with Guidelines for Wastewater Characterization in The Netherlands. *Water Sci. Technol.*, **45** (6), 77–87.

Scholten, H.; Van Waveren, R. H.; Groot, S.; Van Geer, F.; Wösten, H.; Koeze, R. D.; Noort, J. J. (2000) Good Modeling Practice in Water Management. *Proceedings of HydroInformatics2000*; International Association for Hydraulic Research: Cedar Rapids, Iowa.

Shaw, A.; Rieger, L.; Takács, I.; Winkler, S.; Ohtsuki, T.; Langergraber, G.; Gillot, S. (2011) Realizing the Benefits of Good Process Modeling. *Proceedings of the 84th Annual Water Environment Federation Technical Exhibition and Conference* [CD-ROM]; Los Angeles, California, Oct 15–19; Water Environment Federation: Alexandria, Virginia.

Sin, G.; Van Hulle, S. W. H.; De Pauw, D. J. W.; van Griensven, A.; Vanrolleghem, P. A. (2005) A Critical Comparison of Systematic Calibration Protocols for Activated Sludge Models: A SWOT Analysis. *Water Res.*, **39**, 2459–2474.

Third, K. A.; Shaw, A. R.; Ng, L. (2007) Application of the Good Modeling Practice Unified Protocol to a Plant Wide Process Model for Beenyup WWTP Design Upgrade. *Proceedings of the 10th International Water Association Specialized Conference on Design, Operation and Economics of Large Wastewater Treatment Plants*; Vienna, Austria, Sept 9–13; pp 251–258.

U.S. Environmental Protection Agency (2009) *Guidance on the Development, Evaluation, and Application of Environmental Models*; EPA-100/K-09-003; U.S. Environmental Protection Agency, Council for Regulatory Environmental Modeling: Washington, D.C.

Vanrolleghem, P. A.; Insel, G.; Petersen, B.; Sin, G.; De Pauw, D.; Nopens, I.; Weijers, S.; Gernaey, K. (2003) A Comprehensive Model Calibration Procedure for Activated Sludge Models. *Proceedings of the 76th Annual Water Environment Federation Technical Exhibition and Conference* [CD-ROM]; Los Angeles, California, Oct 11–15; Water Environment Federation: Alexandria, Virginia.

Van Waveren, R. H.; Groot, S.; Scholten, H.; Van Geer, F. C.; Wösten, J. H. M.; Koeze, R. D.; Noort, J. J. (2000) *Good Modeling Practice Handbook*; STOWA Report 99-05, Utrecht, RWS-RIZA; Lelystad, The Netherlands.

Chapter 7

Project Definition

1.0 INTRODUCTION

The first step in any application of wastewater process simulations is to define the objectives of the model. Each model must have a purpose, and the purpose must be defined clearly. It is important that all stakeholders are involved in this process. Objectives are typically identified during the proposal phase of a project and further defined during the first phase of project execution. Expectations, with respect to budget, data collection needs, and level of simulation effort required, should be established when the modeling objective is defined. The objectives will be used as the basis for model selection, data collection, calibration steps, evaluation of alternatives, and conclusions.

This chapter reviews uses for models, then describes how a modeling project should be scoped so that all stakeholders agree on objectives and deliverables and concludes with a brief discussion on modelers' training needs and expertise. This information could be used by model developers, consultants, and municipality/utility staff when scoping a project.

Project definition encompasses the first steps in the Good Management Practice Unified Protocol, which are

- Problem statement
- Objectives
- Requirements
- Client agreement

Together, these constitute the project definition, which is an important process to achieve agreement on for the overall project prior to expending significant effort.

2.0 DEFINING PROJECT PURPOSE

It is important to clearly define model purpose, which is ultimately defined by model use. This section describes different purposes based on project phases and model end use. For example, most projects begin with the planning phase and end with optimization.

It is also important to set project boundaries. Project boundaries determine where the model begins and ends. For a water resource recovery facility (WRRF), the boundaries are typically the fence wall (i.e., not to include the sewer collection system); however, sometimes the boundaries can be individual processes such as secondary treatment.

2.1 Planning

Planning models occur in the early stages of a project. The model goal for planning purposes typically is screening options during planning of the capital improvement program, conceptual studies, and proposals. Additionally, models can be used to evaluate current WRRF capacities and forecast future system capacities and effluent quality to understand the effect of discharge on receiving waterbodies. Modeling can also help forecast the effects of wet weather.

Models for planning purposes typically have limited data available, and assumptions are made when needed. Model results are used to determine the most suitable option.

2.2 Design

Models are often used in the design of facility upgrades (Section 2.2.1) or completely new facilities (also known as *Greenfield designs*; see Section 2.2.2). The design process can be conducted on a design-bid-build project or a design-build project. In design-bid-build projects, all design is completed before a construction bid, where a contractor bids to build the project. In a design-build project, designers are partnered with contractors either in the same firm or multiple firms on the same team. In both project execution environments, designers can use models to design facilities. Typically, design-build projects demand much of the project work be completed during the pursuit phase to propose on the project. This is especially true in a design-build type of performance contracting project. In all instances, modelers must take care to obtain accurate information needed to model the facility. Project definition may need to occur with the design-bid team internally so that appropriate expectations are set for the pursuit. The reader is referred to Chapter 9 for additional direction on using models for facility design.

2.2.1 *Water Resource Recovery Facility Upgrades*

Typically, current data such as influent and recycle flows, chemical oxygen demand (COD), biochemical oxygen demand (BOD), total suspended solids (TSS), loads, and so on from the existing facility are used to assess flows and loads that can be treated to existing or future effluent limits with existing process tankage.

Many facilities require simulation during the design of upgrades of existing facilities. Typically, there are data available for the existing facility, such as influent loads and concentrations, return stream flows and loads, and biological kinetics. However, special attention is needed if unit process changes are to be implemented as part of the upgrade. Any safety factors being applied to the design must be recognized, and decisions must be made about how the simulation results and sensitivity analysis should interface with these safety factors. Decisions on safety factors should be made during the conceptual or predesign level. Decisions made during design have significant implications for clients, including, but not limited to, cost, operational flexibility, and effluent quality.

2.2.2 Greenfield Design

Water resource recovery facility process models are widely used at the conceptual and predesign level. At the end of predesign, most important decisions should have been made and agreed on by the parties involved. Detailed design of new facilities is completed after predesign once a process has been determined for a new facility. Data may not exist because the facility is not yet in operation; thus, typical data or data from a neighboring facility, such as influent TSS, BOD, COD, total Kjeldahl nitrogen (TKN) loads, and concentrations, may be used with caution. *Design of Municipal Wastewater Treatment Plants* (WEF et al., 2009) provides typical influent constituent concentrations at various influent flowrates that can also be used; in addition, Chapter 8 includes tables of typical values.

2.3 Operations

Using a model to operate an actual facility system possibly requires the highest level of modeling accuracy and involves using simulations on a regular basis to assess and modify facility operations. For such an application, periodic calibration of kinetic and stoichiometric parameters may be valuable. This could help improve model accuracy in different seasons, for example, or as influent loads change.

Another valuable use of models is to test different control strategies in principle, before applying them to full-scale conditions. For this situation, a calibrated model can be used to assess different control strategies and to give the control engineer a better understanding of system dynamics. An example of this is Fairey et al. (2006), who tested a novel control system for a sequencing batch reactor in a process model before implementing a full-scale trial.

Wastewater process models can also be used for existing facilities to make the most of operations for chemical dosages, power requirements, aeration requirements, reactor configurations, mixed liquor suspended solids (MLSS) concentrations, clarifier loads, and solids retention time (SRT). Models can be used to quantify these costs for operations and to make the most of the facility process to minimize costs.

Use of models to develop probability distributions of treatment levels is becoming increasingly important with the low level nutrient values being proposed by regulatory agencies. In this situation, models can be used to determine the ability of an existing facility to meet a lower nutrient standard by, for example, adjusting aeration or using swing zones differently. This may require rerating facility capacity, but could minimize capital investments.

2.4 Training

Training engineers and WRRF operators can use wastewater models to understand the dynamic behavior and process interactions of their facility. Models are great training tools because they help communicate knowledge about the behavior of the system, especially highly interrelated and complex wastewater systems. A whole-facility model can be developed for a specific facility to show interactions between processes, single models can be developed for individual processes on a site, or generic "fictional" models can be used in a classroom setting to provide training on general principles. For example, models can be used to illustrate effects of increasing or decreasing MLSS inventory on sludge blanket height, waste activated sludge rate, and aeration.

2.5 Research

Models help researchers understand new and developing technologies. For example, models can be used to help interpret and understand pilot test data; this information can then be used to design a full-scale system. An example of this is given by Phillips et al. (2007), who used process models to investigate pilot- and full-scale performance of an integrated fixed-film activated sludge system and then used that knowledge to design other facilities. Models are a key part of discussions on how to develop and design emerging and new processes.

3.0 SCOPING A MODELING PROJECT

Scoping the modeling project is a critical step to ensuring project success because it gives project managers and modelers the opportunity to look at and assess the modeling project from the beginning. The main purpose of scoping the project is to specifically establish how it needs to be organized and managed. Questions to ask when scoping a project include

- What are the goals and objectives?
- What are the boundaries and limits?

- What are the needs, requirements, and possible difficulties?

- What features are and are not included in the model?

- How should the project be organized and tackled?

Typically, it is best to develop the scope using discussions with all key stakeholders and potential team members and contributors. The bulk of the work involves discussions with relevant parties to determine the most important goals or objectives and then some structured thinking and evaluation to establish priorities.

Depending on the level of experience with project stakeholders, modelers should describe typical model terminology and concepts early in the project. This includes simulator and model linkages, state variables versus facility data, and dynamic versus steady-state modeling (discussed in Chapter 9). Appendix E contains an example of scope language for reference purposes.

3.1 Define Project Objectives

Defining project objectives is the first step toward developing modeling goals. After defining project objectives, the modeler and client must clearly state the list of tasks to be tackled.

When scoping a project with all stakeholders, it is important to understand and communicate how the model will help to achieve the objectives of the project. Some project objectives can be addressed by facility inspection, mass balance, or modification of facility operation, rather than by modeling the facility. Modelers should assist clients to determine if a model is the right tool to achieve their project objectives.

Model objectives should define statements such as "this is what we should do" and "this is how the results should be presented." For many projects, the available budget constrains the modeling objectives. However, the objectives for the model determine the time, effort, and data required to complete the project. In a broad context, defining the objectives is an interactive process where the client states the project needs, the modeler describes the amount of effort and data required, including data cleanup/conditioning, and both agree on the specific objectives (outputs) the model is to deliver.

Examples of project objectives are to determine aeration demand on seasons, minimize sludge production, and evaluate the effect of sidestreams on aeration basins.

Preferably, defining project objectives should be obtained through an interactive workshop session with stakeholders, where model ideas are brainstormed and prioritized. The definition process should include a thorough description of project objectives, with some indication of how the model will be able to provide solutions.

3.2 Define Project Boundaries

Model objectives should also determine the model scope or boundaries, such as the following:

- Include only liquid train.

- Include liquid and solids trains.

- Describe only the secondary treatment.

- Steady-state vs dynamic modeling.

For most applications, it is preferable to carry out a whole-facility model with liquid and solids, which includes sufficient detail of all process interactions. However, there are applications where modeling individual processes may be sufficient for the project objectives.

3.3 Define Project Scope

The project scope describes project objectives and specifies budget and schedule. It is important to have a detailed written scope that stakeholders agree to and can be further expanded to a work plan that identifies which individual or organization will complete and fund each activity.

Like the project objectives, the project scope can be defined during a workshop with all key stakeholders. This can be combined with a site visit to further review facility operation, including recycle stream returns, sampling procedures, and solids processing.

3.4 Understanding Uncertainties

It is important to understand uncertainties and the level of accuracy for a modeling effort. The level of accuracy for any modeling effort depends on the objectives, availability of data, type of process to be modeled, budget, and schedule. For any modeling effort, there will be a level of uncertainty. During any stage, it is important to understand the level of effort to remove that uncertainty versus the reduction in risk if uncertainty remains.

When dealing with stakeholders and uncertainty, transparency and credibility are required. Modelers need to document where safety factors are used in the model, identify risk, and quantify the benefits and limitations of modeling. Further discussion on uncertainty is provided throughout this manual and, specifically, in Chapter 9. The required level of accuracy is discussed in Section 5.0 in Chapter 8.

3.5 Specify Project Requirements

Once objectives have been determined, the modeler should identify project requirements. Project requirements should be defined for each step of the modeling project. Examples of specific requirements to define include, but are not limited to, the definition of calibration level and the number and type of alternatives to analyze.

If requirements are not completely defined and described and if there is no effective change control in a project scope or requirement, the designer may be asked to do more work than was originally envisioned or budgeted for; this is commonly termed *scope creep*. Scope creep management is important for effective project execution. Proper definition of specific requirements at the modeling step can help prevent scope creep. The following subsections address specific project requirements for discussion with all stakeholders.

3.5.1 Meeting Regulatory Stipulations

Models may be required to meet regulatory requirements. This requirement is dependent on the location of the facility. It is important to work with regulators early in the project definition phase so that all regulatory (i.e., local, state, and federal) requirements are met.

3.5.2 Selecting Unit Processes to Model

Process model selection should be isolated within project boundaries that were identified when the objectives were set. As a first step to selecting unit processes to model, the modeler must determine if both liquids and solids processes must be modeled (i.e., whole-facility model) or if specific processes only are needed (i.e., secondary treatment only, liquids only, digester only, etc.). This determination is based on project objectives, but is more specific in terms of what will be in the modeling software and different models that might need to be built. Next, the modeler must determine if each individual component of the process must be modeled separately or if the components can be combined. For example, if there are four secondary clarifiers in the facility, these are commonly modeled as one clarifier unit with the surface area to account for all four clarifiers. Combining components typically results in faster model run speed, but can be inaccurate if each unit does not receive the same flow or if there are other differences in the processes.

3.5.3 Selecting Steady-State versus Dynamic Model

It is important to determine if steady-state versus dynamic modeling is appropriate for the model application. A comparison of steady-state and dynamic modeling is described in Section 4.3 of Chapter 2. As stated in Chapter 2, steady-state models are useful for evaluations and may be useful for initial planning purposes. However, steady-state models do not consider actual facility flows, loads, and operating conditions throughout the day, week, month, or year. A primary purpose of dynamic modeling is to capture and mitigate transient conditions. These can include intermittent solids processing operations, wet weather events, diurnal variation in operating setpoints for energy and chemical cost curtailment, and so on. Depending on facility variability, a dynamic model may be more appropriate.

3.5.4 Determining Sampling Needs

To accurately represent the facility, modelers typically accumulate data from the following sources:

- Historical data collected at the facility (operator logs, laboratory logs, supervisory control and data acquisition, and operator interviews)

- Special samples obtained from a sampling program

- Direct parameter measurements from dedicated tests

When scoping a project, it is important to conduct a preliminary review of historical facility records to determine any gaps in available information. If modelers identify critical missing information, they should inform the client about the special sampling needs and the time and effort that is required to collect additional information. Specific data needs are further defined in Chapter 8.

When dealing with project stakeholders, modelers and WRRF operators should discuss the overall sampling plan, including

- Availability and quality of historic data

- Type, quantity, and cost of special sampling

- Entity/personnel that will take and analyze the additional samples

Modelers should inform WRRF staff that there is the possibility that additional parameter measurements may be required following the initial sampling program. During facility modeling, it is common for additional questions to arise that can only be answered by obtaining additional samples. Modeling is an iterative process and these additional samples can be useful to learn more about overall facility performance.

During this stage, it is also important for modelers to differentiate project requirements from "wish-list" items or unnecessary data. This will promote focus on the main objectives of the model rather than the academic exercise of exactly matching facility performance. For example, it may be interesting to do testing for the actual nitrification rate at the facility being modeled. However, if the facility is fully nitrifying, not expected to expand, and only proposing upgrades to final clarifiers, an estimated nitrification rate based on influent and effluent may be sufficient

3.5.5 *Specifying Project Quality Assurance/Quality Control*

There are three main components to quality assurance/quality control (QA/QC) for wastewater modeling projects:

- Hand calculations check

- Model review

- Reality check

During a simulation review, the reviewer should check the following:

- Quality of data collected during any special sampling and analyses and fractionation of COD, nitrogen, and phosphorus

- Influent characteristics for both dynamic and steady-state simulations, which should be compared with loadings in the design criteria

- Mass balances of inert suspended solids, TSS, COD, and total phosphorus around unit processes to check data consistency

- Comparison of model sludge production with typical yields

- Kinetic and stoichiometric parameters, especially those that have been changed from defaults during calibration

- Selected models for each unit process

- Process layout and sizes, including tank sizes and depth of aeration tanks

- Process operation, including percent solids removal for clarifiers, dissolved oxygen setpoints in aeration tanks, recycle flowrates, yield, MLSS, and SRT

In addition to checking model files, the QA/QC reviewer should review the documentation and conclusions of wastewater simulations against professional experience and knowledge of the facility. The reviewer should also check any hand calculations performed by the modeler.

3.5.6   Determining Calibration Methodologies

Chapter 8 describes calibration and validation for building a facility model. This describes the overall goal of calibration as minimizing the error between measured data and model predictions. Typically, the effort for a modeling project (cost, time, etc.) is directly related to the level of calibration defined in Chapter 8. It is important to define the level of calibration early so that the time and budget is available for discussion and consideration during the project scoping.

3.5.7   Specifying Sensitivity Analysis ("What if" Scenarios)

A typical facility modeling project will include a sensitivity analysis for unknown or assumed parameters that were used in the modeling effort. For example, for a new Greenfield facility design, influent TKN may be assumed based on typical wastewater for a new development. The modeler may ask, "What if the influent TKN was higher or lower than the assumption?" Additional scenarios can be run in the model to determine the effect on influent constituents. These types of "what if" scenarios should be discussed in project definition to accurately scope which scenarios will be modeled. Chapter 9 describes additional example sensitivity analysis process for design of facilities.

3.5.8   Defining Deliverables

At the earliest stage of the project, it is useful to define the deliverables. Deliverables are the products, services, or results obtained by executing the project. Examples of deliverables are

- Workshops
- Technical memoranda
- Draft and final modeling report
- Training
- Model files
- Detailed documentation (described in Chapter 8)
- Analysis of facility data

All stakeholders should agree on project deliverables as part of project scoping.

3.5.9   Specifying Model Outcome Communication Plan

The process for communicating the model output should be defined during the project definition stage. Model outcome is typically described in a technical memorandum or report that includes data tables, graphs, screen shots of the model, and summary tables and figures to convey final results and recommendations. Presentations and workshops with model stakeholders are also effective means to communicate model results. Frequently, model files can be provided for those needing more detailed information on the model. While most stakeholders may not have access to the model, those with modeling capabilities can review model files, if requested. It is important for the modeler to fully document the model setup, calibration, and scenario analysis in reports and presentations.

3.6   Develop a Schedule and Budget

The work plan for a modeling project can be defined using the protocol defined in Chapter 6. Modelers should identify the level of effort (time and budget) required for each step in the protocol. An additional tool to develop a work plan is the work breakdown structure described by Haugan (2002).

When developing the work plan, it is important to always consider modeling objectives so that these are met but effort is not expended beyond the agreed scope, which would result in excessive additional costs. The work plan should also identify other work areas or activities in the modeling project to make sure that 100% of the work is covered and to identify areas that cut across the deliverables, represent intermediate outputs, or complement the deliverables. Appendix E includes an example of a scope of work that can be part of the work plan.

Development of a schedule should consider the overall project schedule, especially the special sampling program defined in Chapter 8. A special sampling program can add considerable time to the project based on the length of the sampling period and turnaround time from the laboratory completing the analysis. Laboratories often follow their own QA/QC procedures before releasing analytical results, which can expand the turnaround time for results.

3.7   Agree on Model Scope

Obtaining context agreement with all stakeholders requires a general understanding of how the model fits into the overall project. For design, models are typically used at the beginning of the project for process selection and design criteria. Therefore, it is important to agree on the scope of modeling early to begin data collection and screening and be able to provide model results in the early project phases. Appendix E includes an example of a scope of work for reference purposes.

It is critical to review the facility model and operating procedures built into the model with operations staff. Operating strategies and facilities often are modified from the original design and, often, operations staff is the only entity aware of the changes. Information collected during the workshop and site visits will be used during modeling to more accurately represent the facility's processes. Expected model outputs should be clearly identified, understood, and communicated.

4.0 EXPERTISE AND TRAINING FOR MODELERS

Successful application of wastewater process modeling techniques requires highly trained and skilled personnel experienced in the art of the process and in the science and engineering of wastewater processes and treatment. Different types of model training include:

- University-level coursework
- On-the-job training, such as design activities and operational experience gained during facility startups
- Model developer training

Many commercial simulators are user-friendly, but this can be dangerous if engineers try to run them "out-of-the-box" without proper training and supervision. Changing parameters within the simulator packages requires a strong understanding of fundamentals in the wastewater model. Modification of these parameters can be useful as a training tool to understand their effects on the overall process. However, using simulators for design requires careful consideration of model parameters.

5.0 REFERENCES

Fairey, A. W.; Shaw, A.; McConnell, O.; Cook, J. (2006) Development and Field Testing of a New Intelligent Sequencing Batch Reactor (iSBR) Control System. *Proceedings of the 79th Annual Water Environment Federation Technical Exhibition and Conference* [CD-ROM]; Dallas, Texas, Oct 21–25; Water Environment Federation: Alexandria, Virginia.

Haugan, G. T. (2002) *Effective Work Breakdown Structures*; Management Concepts: Vienna, Virginia.

Phillips, H. M.; Shaw, A.; Sabherwal, B.; Harward, M.; Lauro, T.; Rutt, K. (2007) Modeling Fixed Film Processes: Practical Considerations and Current Limitations from a Consulting Viewpoint. *Proceedings of the 80th Annual Water Environment Federation Technical Exhibition and Conference* [CD-ROM], San Diego, California, Oct 13–17: Water Environment Federation: Alexandria, Virginia.

Water Environment Federation; American Society of Civil Engineers/ Environmental & Water Resources Institute (2009) *Design of Municipal Wastewater Treatment Plants,* 5th ed.; WEF Manual of Practice No. 8; ASCE Manuals and Reports on Engineering Practice No. 76; McGraw-Hill: New York.

Chapter 8

Building a Facility Model

1.0 INTRODUCTION

Once the problem statement has been clearly defined and modeling objectives have been agreed on by all stakeholders (Chapter 7), the facility model is ready to be built. This exercise is best initiated by holding a modeling "kickoff" workshop, where the modeler meets with a small group of facility personnel including the facility superintendent, facility operators, and laboratory manager. Before holding the meeting, the modeler should collect all pertinent information to discuss with the group. Table 8.1 provides a checklist of typical information that is needed to build a facility model. The modeler should use this information during the meeting as discussion points for the proposed model layout or representation of the water resource recovery facility (WRRF). The intent of the meeting is to confirm all model inputs and expectations of model outputs before the modeler begins the model calibration and validation steps.

After the kickoff workshop, the modeler should have stakeholder agreement on facility model setup and can begin the model calibration and validation steps. This chapter provides guidance on the second, third, and fourth steps of the International Water Association (IWA) Good Modeling Practice (GMP) Unified Protocol that is, respectively, data collection and reconciliation, facility model setup, and calibration and validation. The biggest challenges to successfully executing these steps are managing stakeholder expectations and model complexity and modeling the unknown. However, clear communication throughout each step of the project, beginning with a well-planned kickoff meeting, can help ensure that a high quality product can be delivered on time and within budget.

2.0 DATA COLLECTION

It is important to involve facility operators and laboratory managers in any discussions on data collection and analysis because they can provide insight to what is typical or unusual for the facility being modeled. The following sections describe how data are typically used in the model, including historical data (influent, effluent, and process control data), online data (if available), and special sampling data. The cost of sampling vs benefit is also discussed, and an example of a sampling program is included. Additional information on analytical techniques is provided in Chapter 5.

TABLE 8.1 Checklist of typical information needed to build a facility model.

Information	Purpose
Most recent facility plan document	Confirm information is still current.
Recent facility influent data plots, and analyses of peaking factors	Discuss recent data trends vs observed long-term historical trends; discuss proposed modeling approach.
Additional sampling results	Determine if sampling occurred during a "typical" period or if additional sampling is required.
Recent facility operational data	Discuss trends and typical operating strategies.
Current PFD	Determine locations of samplers, recycle flows, etc.
Current P&ID	
Current as-built plan drawings	Determine surface area, volume, and configuration of all basins being modeled.
Current as-built hydraulic profiles	Determine sidewater depth of all basins being modeled.
Current inventory of equipment (blowers, diffusers, mixers, pumps, etc.)	Determine firm capacity and discuss typical operation of equipment (constant speed or variable frequency drives for pumps, etc.).
List of calibration parameters	Discuss expectations and agree on level of accuracy.
Sample of model output and report	Discuss content and format; adjust based on feedback.

When modeling a facility, it is impossible to have a complete set of data that describes all flows and concentrations throughout the entire facility. In reality, the modeler typically must deal with significant gaps in data. One powerful feature of process models is that they can bridge these gaps as they complete the mass balance. This is particularly useful in estimating flows and loads of sidestream returns, although the modeler must be careful in applying this information. It is always better to take real measurements, if possible, using good sampling and measurement techniques, than to blindly trust the output of a model. The IWA Task Group on Design and Operations Uncertainty (DOUT) was formed to develop a transparent method of addressing unknown factors in wastewater treatment modeling and to link efforts in academia with practice (Belia, Amerlinck et al. 2009; Belia, Benedetti et al., 2009).

2.1 Historical Facility Data

Historical data collected at WRRFs are discussed in resources such as *Design of Municipal Wastewater Treatment Plants* (WEF and ASCE, 2009). Facilities typically monitor and record influent and effluent biochemical oxygen demand (BOD), total suspended solids (TSS) concentrations, and flow. They also may analyze for nutrients, pH, and dissolved oxygen. However, most models are based on, among other parameters, chemical oxygen demand (COD) and inert suspended solids (ISS). Often, no historical data exist for these parameters. Therefore, the modeler must clearly state his or her assumptions for the model calibration/validation efforts and his or her assumptions for the process evaluations—because they may differ—to account for future flow and load projections. This includes stating assumptions for solids processing sidestream contributions, including flow, quality, and return location.

2.1.1 Influent Data

Selecting the basis of the design is often a critical milestone in the project schedule, which means that the project manager may expedite the task to keep design efforts moving forward. There is significant risk in completing this task before special sampling results are in, however, because an incorrect assumption of wastewater characteristics can result in design failure. Although historical data are used to give a basic characterization of influent wastewater, there is often no record of COD, ISS, and even total Kjeldahl nitrogen (TKN) or total phosphorus data. When designing nutrient removal facilities, in particular, the engineer must insist that all necessary analyses be completed before any significant design efforts take place or communicate the risk and associated cost of rework if data are not yet available. The engineer must also consider future changes in the service area when developing the design flow and loads, which serve as the basis for process evaluation.

When applicable, it is often possible to use historical BOD and TSS concentrations to estimate historical COD and ISS concentrations, which are required model inputs. These parameters provide more detailed and useful characterization of influent wastewater needed in modeling and, as such, it is preferable to measure them directly, if possible. Care must be taken when using BOD data on nitrifying facilities with return liquors in the influent because the presence of nitrifying bacteria can cause significant errors in BOD measurement. The carbonaceous BOD (CBOD) test can be used on influent to overcome this problem; however, the inhibitor that is added as part of the CBOD test can also affect carbonaceous bacteria and cause it to measure erroneously low.

There are several methods for analyzing flow and load data, such as statistical methods, seasonal analyses, or applying published per capita factors in the case of a Greenfield facility. It is common to size secondary treatment and solids processing facilities based on maximum month conditions or the highest 30-day average loads for a given set of data (Metcalf and Eddy, 2003). Model calibration efforts might focus on a month of special sampling data, but a validation effort must be conducted using a different and wider set of data. Others' efforts may calibrate a model using an entire year of data to account for dry and wet weather seasons. Regardless of the method used, it is important to analyze wastewater loads independent of concentration because wet weather dilution can skew apparent observations in data. The modeler must also acknowledge that "worst-case" flows and loads in terms of capacity likely do not represent the worst case for nutrient removal evaluations because BOD is required for denitrification. Tables 8.2, 8.3, and 8.4 summarize typical wastewater flowrates, per capita factors, and concentrations, respectively. For service areas that have significant commercial, institutional, recreational, or urban contributions, additional references should be consulted (e.g., Metcalf and Eddy, 2003).

2.1.1.1 Developing Seasonal Peaking Factors

Monthly average or seasonal influent peaking factors are often developed for modeling efforts because facility capacity is dictated by different factors at different times of the year, especially facilities with combined sewer systems. For example,

TABLE 8.2   Typical wastewater flowrates from urban residential sources in the United States (from Metcalf and Eddy [2003], with permission).

Household size, no. of persons	Flowrate, gal/capita·d		Flowrate, L/capita·d	
	Range	Typical	Range	Typical
1	75–130	97	285–490	365
2	63–81	76	225–385	288
3	54–70	66	194–335	250
4	41–71	53	155–268	200
5	40–68	51	150–260	193
6	39–67	50	147–253	189
7	37–64	48	140–244	182
8	36–62	46	135–233	174

a facility may receive high constituent loadings during the winter, but the highest flows are more likely to occur during the spring when inflow and/or infiltration are high. Either of these seasonal conditions could limit facility capacity. On the other hand, peak aeration demands could occur during the summer if high loads coincide with warmer temperatures (and lower oxygen transfer efficiencies). The effects of seasonal influent flows and loads must be considered in design modeling efforts, especially if the service area includes a college town, significant tourist attractions, or other causes of population fluctuations. Attention should also be given to seasonal industrial loadings, which may have significant effects on the overall facility loading.

It is customary to size facilities based on influent loads rather than concentrations. However, Figure 8.1 shows an example of a facility that routinely experiences increasing flows and decreasing BOD concentrations during spring months; decreasing flows and increasing BOD concentrations during summer/fall months; and relatively stable, but low flows and high BOD concentrations during winter months. For this particular facility, the BOD loads increased steadily for the time period evaluated, but the concentrations varied seasonally, thereby affecting nutrient removal. For facilities that experience regular influent trends, the engineer may select representative months or seasons to develop influent peaking factors

TABLE 8.3 Typical per capita factors of untreated domestic wastewater (adapted from Metcalf and Eddy [2003], with permission).

| Constituent | Value, lb/capita·d | | | Value, g/capita·d | | |
	Range	Typical without ground-up kitchen waste	Typical with ground-up kitchen waste	Range	Typical without ground-up kitchen waste	Typical with ground-up kitchen waste
BOD$_5$	0.11–0.26	0.180	0.220	50–120	80	100
COD	0.30–0.65	0.420	0.480	110–295	190	220
TSS	0.13–0.33	0.200	0.250	60–150	90	110
NH$_3$ as N	0.011–0.026	0.017	0.019	5–12	7.6	8.4
Organic N as N	0.009–0.022	0.012	0.013	4–10	5.4	5.9
TKN as N	0.020–0.048	0.029	0.032	9–21.7	13	14.3
Organic P as P	0.002–0.004	0.0026	0.0028	0.9–1.8	1.2	1.3
Inorganic P as P	0.004–0.006	0.0044	0.0048	1.8–2.7	2.0	2.2
Total P as P	0.006–0.010	0.0070	0.0076	2.7–4.5	3.2	3.5
Oil and grease	0.022–0.088	0.0661	0.075	10–40	30	34

to apply to future modeling scenarios. Other facilities may periodically experience high flows at the same time as high loads, and this must also be considered. For nutrient removal facilities, it is important to evaluate the timing of high nutrient loads because the facility will likely produce the highest effluent nutrient concentrations when influent nutrient loads are high and the BOD load is low (especially if the temperature is also low). Recycle streams generated by biosolids processing are typically not included in influent sampling, but can have a larger effect on biological nutrient removal (BNR) performance than diurnal fluctuations and, therefore, must be included in modeling efforts. Daily and hourly peaking factors should also be evaluated for hydraulic and aeration calculations and to evaluate effluent quality if the facility has daily or weekly permit limits.

TABLE 8.4   Typical composition of untreated domestic wastewater (adapted from Metcalf and Eddy [2003], with permission).

| Contaminants | Unit | Concentration[a] | | |
		Low strength	Medium strength	High strength
Total solids	mg/L[c]	390	720	1230
Total dissolved solids	mg/L	270	500	860
Fixed	mg/L	160	300	520
Volatile	mg/L	110	200	340
TSS	mg/L	120	210	400
Fixed	mg/L	25	50	85
Volatile	mg/L	95	160	315
Settleable solids	mg/L	5	10	20
BOD_5, 20 °C	mg/L	110	190	350
Total organic carbon (TOC)	mg/L	80	140	260
COD	mg/L	250	430	800
Nitrogen (total as N)	mg/L	20	40	70
Organic	mg/L	8	15	25
Free ammonia	mg/L	12	25	45
Nitrites	mg/L	0	0	0
Nitrates	mg/L	0	0	0
Phosphorus (total as P)	mg/L	4	7	12
Organic	mg/L	1	2	4
Inorganic	mg/L	3	5	10
Chlorides[b]	mg/L	30	50	90
Sulfate[b]	mg/L	20	30	50
Oil and grease	mg/L	50	90	100
Volatile organic compounds	mg/L	< 100	100 to 400	> 400
Total coliform	no./100 mL	10^6 to 10^8	10^7 to 10^9	10^7 to 10^{10}
Fecal coliform	no./100 mL	10^3 to 10^5	10^4 to 10^6	10^5 to 10^8
Cryptosporidium oocysts	no./100 mL	10^{-1} to 10^0	10^{-1} to 10^1	10^{-1} to 10^2
Giardia lamblia cycts	no./100 mL	10^{-1} to 10^1	10^{-1} to 10^2	10^{-1} to 10^3

[a]Low strength is based on an approximate wastewater flowrate of 750 L/cap·d (200 gal/cap·d). Medium strength is based on an approximate wastewater flowrate of 460 L/cap·d (120 gal/cap·d). High strength is based on an approximate wastewater flowrate of 240 L/cap·d (60 gal/cap·d).

[b]Values should be increased by the amount of constituent present in the domestic water supply.

[c]Note that mg/L = g/m^3.

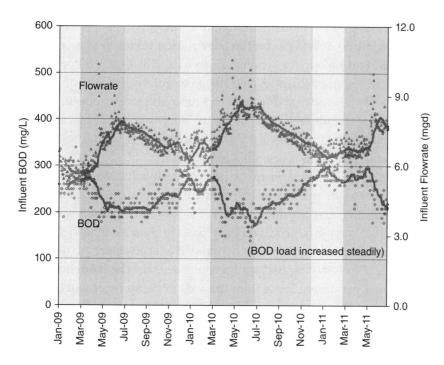

FIGURE 8.1   Example of seasonal influent flow and BOD concentration variability.

2.1.1.2   Developing Diurnal Peaking Factors

For dynamic modeling, it is important to develop diurnal peaking factors for influent flow and concentrations because variable loads throughout the day can affect aeration, hydraulic design, and effluent quality. For most facilities, diurnal patterns do not vary significantly throughout the year. However, many facilities experience different patterns on weekends compared to weekdays, so it may be applicable to model these differences. The modeler should also confirm that there is nothing upstream of the influent sampler that might affect diurnal patterns; for example, if biosolids processing sidestreams are returned upstream of the sampler, influent data and model results will be significantly affected.

2.1.2   Operational Data

It is important to closely evaluate historical operating data before building a facility model because calibration efforts should focus on a period in which the facility was operating in a stable and consistent mode. While this is not always possible, modeling efforts that attempt to calibrate around a time period for basins being taken in and out of service for maintenance, intermittent use of chemicals, or wide variations in recycle or wasting rates (solids retention time [SRT]) will likely not yield success-ful results. In addition, it is important to have a good understanding of how opera-tors control aeration unless the process is run with effective aeration controls and dynamic data can be used directly. Dissolved oxygen readings may only be taken once or twice a day, and only at one location within the aeration basin. The modeler must have an understanding of how and when process control data are collected to accurately apply them to modeling efforts. A summary of typical operational data needed for model calibration is provided in Table 8.5.

It is important to consider biosolids processing in addition to liquid treatment performance because the two are directly related. It is common for facilities to dewa-ter solids in shifts (8 hours per day, 5 days per week) and, if the filtrate (or centrate) is not equalized, it can introduce significant load variations on the facility that must be included in the modeling effort. Furthermore, if a facility stores digested biosolids in the winter and uses drying beds in the summer only, the modeling effort must address this because nutrient loads from the beds in the summer will be substantial. Even if the facility hauls sludge from the facility and sidestreams are not a concern, it is important to collect and evaluate waste activated sludge (WAS) data to ensure that the activated sludge model is predicting accurate sludge yields. Reconciling sludge data is discussed further in Section 3.2.

2.1.3   Effluent Data

For facilities that have very stable operating conditions and effluent quality, model calibration efforts can be relatively easy. Facilities that must perform to low effluent nutrient limits are often difficult to calibrate because the effluent concentration may be at or below the half-saturation constant and such instances may require additional batch testing. For facilities that have unexplained spikes in effluent quality, modeling can be a tool to ascertain the cause and help make the most of facility performance. Arriving at the solution may add model complexity, however, because effluent spikes can result from variable influent flows and loads, variable operations, or both. In nutrient removal facilities, spikes in effluent nitrogen or phosphorus are often the result of sidestream return loads. It is common practice for smaller plants to dewater digested solids on weekday shifts only; as a result, effluent quality may be low and stable on the weekend, but may spike on Monday afternoon after sidestream loads have passed through the facility. A second significant cause of effluent variability is blower operation because many facilities choose to over- or under-aerate at different times of the day to avoid surcharges associated with turning blowers on and off.

TABLE 8.5 Example checklist of operational data needed for model calibration.

Process	Minimum data needed	Importance of modeling
Primary clarifiers		
Performance	% TSS and % BOD (or COD)	Defines loads to secondary processes
Operational	Removal	Defines loads to biosolids processing
	Primary sludge flowrate, % TSS	
Trickling filters		
Performance	Effluent BOD (or COD), TSS, NH_3-N	Calibrating biofilm model
Operational	Recirculation rate (if applicable)	Affects performance
Aeration basins		
Performance	Effluent NH_3-N, NO_x-N, PO_4-P	Calibrating activated sludge model
Operational	MLSS, MLVSS, dissolved oxygen, MLR flowrate	Nutrient and soluble BOD profiles also useful
Secondary clarifiers		
Performance	SVI, effluent TSS, BOD (or COD)	Calibrating settling model
Operational	RAS, WAS flowrates, % TSS	WAS data defines sludge yield
Filters		
Performance	Effluent TSS, BOD, nutrients	Calibrating filter model
Operational	Backwashing routine, chemicals	Defines sidestream return loads
Sludge thickening		
Performance	Thickened solids flow, % total solids	Defines loads to biosolids processing
Operational	% total solids removal, chemical doses	Defines sidestream loads
Anaerobic digestion		
Performance	SRT, % VSr, gas production	Defines loads to dewatering, biogas to cogeneration, sidestream nutrient loads
Operational	temperature, mixing type	
Aerobic digestion		
Performance	SRT, % VSr	Defines loads to dewatering, sidestream nutrient loads
Operational	temperature, dissolved oxygen, airflow	
Dewatering		
Performance	Dewatered solids flow, % total solids	Defines cake
Operational	% total solids removal, chemicals, nutrients	Defines sidestream return loads

All effluent data should be evaluated closely, especially when selecting a data set for model calibration and validation efforts.

2.2   Online Data

With increased acceptance of online instrumentation by wastewater operators, it is common to see several devices at WRRFs. Online instrumentation provides a wealth of data to modelers, filling in gaps of missing information or providing insight to why the facility (and the model) may have performed unexpectedly at any given time.

Many larger facilities have dozens of online instruments. Before entering all of these dynamic inputs to the model, modelers should first determine which parameters are important to their modeling objectives and which will actually make a difference in model output. For example, a facility may feed an anaerobic digester at slightly different rates throughout the day, but, because the digester has a long retention time, digester effluent quality will not vary significantly and online data can be neglected in the model. On the other hand, if the facility dewaters digested sludge and sends centrate to the headworks only 8 hours a day, 5 days a week, the facility effluent will be affected and online centrate data should be included in the model. Keeping the model as simple as possible to meet project objectives will minimize unnecessary work (and guesswork) if the model and multiple data sets do not reconcile.

2.2.1   Flow Meters

Flow meters are a common online instrumentation seen at WRRFs. There are several types of flow meters and each is designed to measure flow of a medium (water or air) in open or closed pipe. Sources of error include improper maintenance or calibration and potential issues with incorrect installation (e.g., a meter too close to a bend or valve).

Use of online data from an influent flow meter can be used as flow input to a dynamic model. Online data can be used to model a single dynamic event such as a wet weather scenario or the data can be used to generate hydraulic diurnal peaking factors. The influent flow diurnal peaking factors can be used in other dry weather scenarios such as a process upgrade at future flows.

2.2.2   Analyzer and Sensor Data

Recent improvements have allowed online analyzers and sensors to be used as a rich source of dynamic data through continuous measurement of state variables. However, care must be taken to prevent error due to the following sources:

- Lack of maintenance and calibration
- Poor location of sensor

- Turndown range
- Response time

Data gathering using analyzers requires proper planning to ensure relevant data collection. The type and locations of analyzers must be carefully selected to ensure the data collected can be used to verify the simulation. Common types of data from nutrient analyzers are ammonium (NH_4-N), nitrate (NO_3-N), and soluble phosphorus for representative BNR processes. Location of analyzers is important; analyzers need to be placed where the data received can be used in the model. Properly located analyzers can provide information such as diurnal ammonium loading entering and leaving the aerobic zone or diurnal nitrate loading entering and leaving the anoxic zones. Nutrient analyzers are not limited to a short duration of intensive sampling; multiple days of dry and wet weather or even the change of season can be sampled.

Online data can be useful for calibrating dynamic models if instruments are maintained and calibrated regularly. The modeler should always compare analyzer data with laboratory data to ensure agreement before adjusting model kinetics. Data reality checks using nutrient mass balances can also be used to verify the data. The accuracy of the analyzer needs to be considered during low range measurements where inaccuracy can be a high percentage of the measurement. Problems can occur with online analyzers, such as power outages or sampling tube plugging, which can produce inaccurate data if instruments are not regularly maintained (and sometimes even if they are). If samples are conveyed any distance from the sampling point to the analyzer, biological growth can accumulate on the sample line walls and significantly affect analytical results. Therefore, routine cleaning of piping is critical. Alternatively, direct insertion of the probe in the tank can eliminate such issues. In the example shown in Figure 8.2, online analyzer data validated the effluent nitrate pattern predicted by the model for the first few days, but there were problems with the analyzer on days 26 through 28 where the model aligned better with daily composite data.

2.3 Special Sampling Data

For modeling studies, it is not uncommon to perform detailed facility profiling over a short period of time to obtain the requisite information for modeling. The amount of time for this will vary depending on project scope and budget; however, a period of 2 weeks is typically sufficient. This is an extremely important step in developing the database to be used as input to the model. This step allows fractionation of COD, TKN, and phosphorus. It also allows determination of ISS and confirmation with historical data if historical TSS and volatile suspended solids (VSS) data were collected.

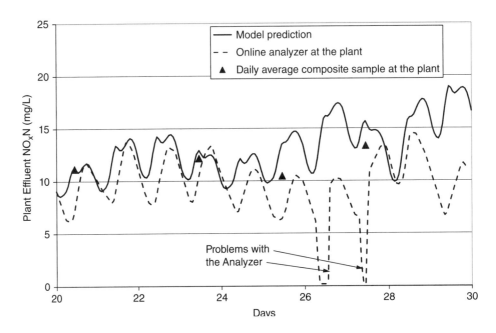

FIGURE 8.2 Dynamic modeling using laboratory data and online analyzer data.

Additionally, this step allows a correlation between COD and BOD to be developed to convert historic BOD data to COD approximations. Finally, it allows a correlation between TKN and ammonia to be developed and also allows a link between BOD/COD and TKN and phosphorus, if historic nitrogen and phosphorus data have never been collected at the subject facility. Aspects of special sampling are described in the following sections; Chapter 5 also provides details about relevant analytical methods.

2.3.1   Influent Fractionation

A critical step to calibrate any model is to fractionate the influent wastewater COD into its various components. Daily 24-hour flow-weighted composite samples of the influent are most appropriate for influent COD fractionation, which typically involves estimating the following COD fractions:

- Particulate—un-biodegradable particulate fraction (X_u) and slowly biodegradable particulate fraction (X_b)

- Colloidal—colloidal biodegradable fraction (after coagulation does not pass through 0.45-µm filter) or C_b
- Soluble—un-biodegradable soluble fraction (S_u) and soluble readily biodegradable COD (RBCOD) fraction or S_b

In the laboratory, influent fractionation typically involves performing the following four types of COD analyses:

- Raw COD (unfiltered) (COD_{XX})
- Glass-fiber-filtered COD (1.2-m filter) (COD_{XG})
- Membrane-filtered COD (0.45-m filter) (COD_{XM})
- Flocculated and membrane-filtered COD (0.45-m filter) (COD_{XF})

Fractionating the six components of COD also requires two additional unknowns, which are often derived through trial-and-error approximations. The additional unknowns include the following:

- Fraction of un-biodegradable particulate material (F_{up}) in the influent
- The COD_{XF} concentration in the facility effluent, which is assumed to be equal to the un-biodegradable soluble concentration of COD (S_u)

The six components of COD can be fractionated as follows:

$$X_u = COD_{XX} \times F_{up} \tag{8.1}$$

$$X_b = COD_{XX} - COD_{XG} - X_u \tag{8.2}$$

$$C_b = COD_{XG} - COD_{XF} \tag{8.3}$$

$$S_u = COD_{XF\text{-facility effluent}} \tag{8.4}$$

$$S_b = RBCOD = COD_{XF} - S_u \tag{8.5}$$

The aforementioned discussion is one example of influent characterization, and nomenclature and fractionation methods typically vary between models. In some models, C_b is not defined, and an assumption must be made if the substrate is considered to be slowly biodegradable (X_b) or readily biodegradable (S_b). Melcer et al. (2003) contains further details and information on influent fractionation. Table 8.6 provides influent characterization data for several facilities; site-specific sampling should always be conducted because data can vary significantly.

TABLE 8.6 Typical wastewater ratios for facility data (from Hauduc et al. [2010]).

	Ratio	Unit	N[a]	Mean	Std. %[b]	Median	Min.	Max.
Raw influent	Total nitrogen/COD	g N/g COD	12	0.095	17%	0.091	0.050	0.150
	NH_3-N/TKN	g N/g N	13	0.684	8%	0.670	0.500	0.900
	Total phosphorus/COD	g P/g COD	12	0.016	22%	0.016	0.007	0.025
	PO_4-P/total phosphorus	g P/g P	12	0.603	16%	0.600	0.390	0.800
	COD/BOD_5	g COD/g BOD	12	2.060	11%	2.050	1.410	3.000
	fCOD[c]/COD	g COD/g COD	13	0.343	29%	0.350	0.120	0.750
	TSS/COD	g TSS/g COD	12	0.503	18%	0.500	0.350	0.700
	XCOD[d]/VSS	g COD/g VSS	11	1.690	12%	1.600	1.300	3.000
	VSS/TSS	g VSS/g TSS	12	0.740	20%	0.800	0.300	0.900
	BOD_5/BOD_∞	g BOD/g BOD	7	0.655	7%	0.650	0.580	0.740
	Alkalinity	Mol_{eq}/L	11	5.173	35%	5.000	1.500	9.000
Primary effluent	Total nitrogen/COD	g N/g COD	9	0.134	35%	0.120	0.050	0.360
	NH_3-N/TKN	g N/g N	11	0.755	4%	0.750	0.430	0.900
	Total phosphorus/COD	g P/g COD	9	0.023	25%	0.023	0.010	0.060
	PO_4-P/total phosphorus	g P/g P	10	0.741	12%	0.750	0.500	0.900
	COD/BOD_5	g COD/g BOD	9	1.874	31%	1.900	0.500	3.000
	fCOD/COD	g COD/g COD	10	0.449	31%	0.495	0.150	0.750
	TSS/COD	g TSS/g COD	9	0.380	21%	0.400	0.180	0.560
	XCOD/VSS	g COD/g VSS	9	1.718	14%	1.700	1.400	3.500
	VSS/TSS	g VSS/g TSS	9	0.794	7%	0.800	0.700	0.909
	BOD_5/BOD_∞	g BOD/g BOD	6	0.644	10%	0.656	0.533	0.760
	Alkalinity	Mol_{eq}/L	9	5.711	40%	6.000	1.500	9.000
Activated sludge	COD/VSS	g COD/g VSS	9	1.434	7%	1.420	1.266	1.600
	Total nitrogen/COD	g N/g COD	7	0.073	35%	0.060	0.045	1.116
	Total phosphorus/COD	g P/g COD	7	0.020	64%	0.015	0.010	0.044
	VSS/TSS	g VSS/g TSS	10	0.739	8%	0.750	0.650	0.900

[a]Number of answers.

[b]Standard deviation in %.

[c]fCOD = filtered COD.

[d]XCOD = particulate COD.

In practice, influent characterization is often limited by schedules, budgets, or both. For nutrient removal projects, however, quantifying the amount of RBCOD in the influent wastewater can mean the difference between success and failure. When designing a sampling protocol, the engineer must consider the following factors:

- Raw influent and primary effluent—primary clarifier operation can greatly affect RBCOD entering secondary treatment. If sludge is stored within the primary clarifiers for a long period of time, fermentation can potentially occur, which increases the RBCOD concentration. Such conditions are difficult to model, but additional sampling will result in a better calibration.

- Seasonal effects—temperature within the collection system and changes in retention time caused by seasonal flow fluctuations can also affect influent characteristics. It is common for WRRFs to receive increased inert solids loads during and after a wet weather event because higher flows scour settled solids from the collection system. Odor control practices in the collection system can also affect RBCOD concentrations; nitrate salt is commonly added to create anoxic conditions in the sewer to prevent sulfide productions, although anoxic conditions also prevent volatile fatty acid (VFA) formation (Kobylinski et al., 2008).

- Changing wastewater characteristics—sampling provides a snapshot in time of wastewater characteristics, but many designs have been affected by changes in the service area between design and startup. For example, an industry that typically discharges high RBCOD quantities may decide to relocate to another town. These circumstances cannot always be foreseen, but it is in the engineer's best interest to have a full understanding of the collection system and its connectors. In addition, wastewater fractions such as the NH_3-N to TKN ratio or the RBCOD to total COD ratio may change daily or diurnally and the engineer must have an understanding of this, especially if discharge limits are stringent.

2.3.2 *Diurnal Profiles*

Diurnal profiles are used to model variations in influent loads throughout the day. Diurnal profiles are developed through a series of samples at regular time intervals over a 24-hour period. Typically, it is best to sample from facility influent and influent of the activated sludge process at a frequency of 1 to 2 hours. This information will be used to develop a diurnal loading pattern for design. Figure 8.3 presents an example of diurnal COD variations obtained from a 114-ML/d (30-mgd) facility in Virginia. The loads are high during the day and drop off overnight.

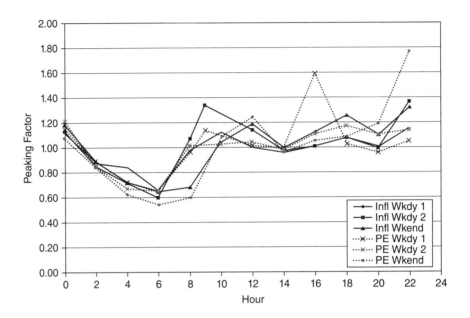

FIGURE 8.3   Example diurnal COD variation for a 114-ML/d (30-mgd) facility.

2.3.3   Settling Tests

If a model clarifier is being used, settling parameters can be determined using a variety of published methods (see Chapter 5). The challenge of conducting settling tests is that the data represent a snapshot in time and, therefore, may not represent typical settling characteristics. At a minimum, most facilities measure sludge volume index (SVI) routinely so the results of settling tests should be compared to historical SVI to determine the applicability of the test data. Depending on the modeling objective, more advanced testing and modeling may be required (computational fluid dynamic [CFD] modeling, tracer studies, etc.) if clarifier capacity is being significantly stressed. In many instances, settling tests are not conducted because it is known that there is excess clarifier capacity for the flows and loads being evaluated (facility planning efforts, etc.).

2.3.4   Aeration Tests

Calibrating the aeration model is often the most difficult design task because many of the factors that influence oxygen transfer are not easily quantified nor are they specifically addressed within the aeration model. Clean water shop tests provide very reproducible data for a specific diffuser grid design, but translating this data into process water performance requires careful selection of the alpha factor (α), which is a function of the following factors:

- Specific oxygen uptake rate

- Mixed liquor suspended solids (MLSS) concentration

- Solids retention time (endogenation)

- Air intensity and turbulence

- Water chemistry, the presence of surfactants, and oil and grease

- Presence of moving bed biofilm reactor (MBBR) media or other equipment

- Fouling (fine bubble diffusers)

In many instances, such as designs with MBBR media carriers, field tests are conducted to confirm design efforts as part of the equipment supplier guarantee. While these tests are much more costly than shop tests, they account for full-scale conditions not addressed within a small test tank (basin geometry, the presence of MBBR media, etc.). Offgas testing can also provide insight to calibrating a model for existing aeration systems.

2.3.5 Kinetic Studies

Default or current published values for kinetic parameters can usually predict the performance of most domestic WRRFs. However, in some instances, wastewater characteristics and the resulting bacterial community in the biological treatment process are so unusual compared to typical domestic wastewater that the system performance cannot be explained using typical kinetic parameters. An example of such an instance may be a WRRF receiving large contributions of industrial wastewater. In such instances, it is beneficial to conduct kinetic studies to better understand the reason for deviations from model predictions. Kinetic studies can reveal possible causes for unusually slow (or fast) nitrification or denitrification rates. Phosphorus uptake and release tests are also very useful in calibrating BNR models, especially if inline fermentation or other uncommon design approaches are being taken. Chapter 5 summarizes the variety of options available for kinetic studies. In all instances, kinetic rate test results should be calibrated and validated in the model.

2.3.6 Facility-Wide Sampling

Special sampling is used to enhance understanding of the facility's process performance and assist in model calibration. During the special sampling period, it is recommended that 24-hour flow-weighted composite samples be collected at the following locations:

- Raw influent, fully upstream from any recycle flows
- Primary clarifier influent (if recycle streams are introduced upstream of the primary clarifier)
- Primary clarifier effluent
- Activated sludge treatment influent (if recycle streams are introduced downstream of primary clarifier)
- Secondary effluent, upstream of disinfection
- Facility effluent (if the facility has filters; otherwise, it should be the same as secondary effluent)
- Sidestream contributions

The purpose of this detailed facility profile is to enhance historical data with a record of how each process performs. These data can be used in the model calibration for process troubleshooting and/or to verify the calibrated model. Example sampling programs are presented in Section 2.5.

In addition to composite sampling, the modeler should also evaluate the need for the following complementary types of sampling:

- Grab sampling at various locations for temperature, dissolved oxygen, and pH
- Mixed liquor/process sampling
- Dissolved oxygen profiling
- Recycle stream sampling

2.3.6.1 *Preliminary Treatment Considerations*

Facility-wide models often do not include the headworks facility because screening and degritting processes typically have minimal effect on nutrient removal, which is the primary focus of most modeling activities. However, depending on screen type and size, a relatively significant portion of grit (ISS) material can be removed during preliminary treatment. In addition, hauled waste, oil and grease, or in-facility sidestreams may be introduced in the headworks facility; as such, it is important for the modeler to have a good understanding of the process flow diagram and any potential effect that preliminary treatment may have on model results. By conducting fractionation studies on both raw influent and primary influent samples, any effect of preliminary treatment on wastewater quality can be quantified.

2.3.6.2 *Primary Treatment Considerations*

The primary clarification process has the potential to significantly complicate modeling efforts because current settling models assume equal settling velocities for all particulate COD fractions. In addition, if clarifiers are used to thicken primary sludge or co-thicken primary sludge with WAS, biological reactions (hydrolysis and fermentation) are likely to take place, which greatly increases model complexity. The use of a biological model in conjunction with a settling model complicates calibration efforts, but is necessary in many circumstances. By characterizing both the primary influent and primary effluent streams, uncertainty can be minimized, at least for the operating conditions occurring during the sampling program. In the absence of detailed influent characterization data, calibration of the model to primary clarifier performance can serve as a means for approximating influent COD fractions. Sludge depth measurements can provide insight to the potential for biological reactions, and they can help close mass balances for clarifiers in which intermittent sludge pumping is practiced.

Lastly, when using metal salts in primary clarification or upstream of it, colloidal material can be removed across the primary unit, thus complicating mass balance efforts. Not all simulators have coagulation capabilities built into their primary settling model.

2.3.6.3 *Secondary Treatment Considerations*

It is often sensible to measure secondary influent constituents in addition to primary effluent because sidestream loads may affect concentrations. It may also be necessary to measure the influent to individual treatment trains to confirm that each train receives approximately the same flow and loads. Some facilities have long secondary influent channels and it is not uncommon for solids to settle in the channel, causing the first train to receive higher loads than the downstream trains. The same is true for return activated sludge (RAS) and mixed liquor recycle (MLR) distribution, so it is important for the modeler to have a good understanding of how and where flows are introduced to the trains and to recommend additional sampling when in doubt.

Once it is confirmed that individual treatment trains are receiving the same flow and loads, their operational conditions must be confirmed. This includes collecting airflow and dissolved oxygen data for each diffuser grid, confirming similar MLSS and mixed liquor volatile suspended solids (MLVSS) concentrations in each train, and conducting nutrient profiles, if possible. Knowing the effluent nutrient concentrations may not always be enough for model calibration efforts. It is helpful, and often necessary, to measure ammonia-nitrogen (NH_3-N), nitrite (NO_2-N), phosphate (PO_4-P), and SCOD concentrations within each reactor (anaerobic, anoxic, and oxic) to properly calibrate a model. These data are also useful to facility operators; if it is determined that nitrite concentrations are nondetectable halfway through the anoxic zone, the operators could increase the MLR flowrate to further improve effluent quality. However, care must be taken in collecting mixed liquor samples because, if they are not immediately filtered, reactions will continue to take place and results will be invalid.

It is also important to understand clarifier design. Depending on the type of sludge collection mechanism (orificed header vs center withdrawal), the RAS concentration can vary significantly over a diurnal timeframe. This coupled with return strategy can significantly affect the instantaneous food-to-microorganism ratio (F/M).

2.3.6.4   Biosolids Processing Sidestreams

It is always preferable to conduct facility-wide modeling that includes both the liquid and biosolids processes as opposed to including a separate model input for sidestream loads. Sidestreams generated during biosolids processing, particularly those generated through dewatering of digested sludge, can introduce significant nutrient loads to the liquid treatment facility. Anaerobic digestion can product significant NH_3-N and PO_4-P sidestream loads and aerobic digestion can product significant NO_3-N and PO_4-P sidestream loads. A properly calibrated model should be able to predict these loads, but sampling is necessary to confirm the flows and concentrations, especially if these loads are intermittent (shift-dewatering). In addition, struvite formation in the digester affects the amount of NH_3-N and PO_4-P (and magnesium) recycled in dewatering sidestreams. Good modeling practice applies a facility-wide model and sampling results to confirm sidestream predictions from the model.

2.4   Cost of Sampling versus Benefit

While sampling programs require additional effort and expense on the part of the owner, they can significantly reduce uncertainty in the model, which can reduce capital and operating costs. Utilities can reduce expenditures by doing some of the analyses "in house"; however, it is often more convenient to send samples to a commercial laboratory. Analysis costs vary from individual laboratories and depend on the parameter and method used; a sample of approximate analytical costs is provided in Table 8.7.

TABLE 8.7 Approximate commercial laboratory costs of common wastewater analyses.

Parameter	Approximate cost per sample, U.S. dollars
COD (or soluble COD)	$20 to $40
Filtered flocculated COD	$35 to $50
VFAs	$75 to $125 for individual acids
BOD (or soluble BOD)	$30 to $45
TSS	$15 to $20
VSS	$20 to $30
TKN	$30 to $45
Ammonia	$15 to $30
Total phosphorus	$25 to $40
Orthophosphorus	$15 to $30
Alkalinity	$10 to $20
pH	$10 to $20
Magnesium	$10 to $30
Calcium	$10 to $30

The financial benefits of additional sampling can be difficult to quantify, but several case studies report significant savings. The Suomenoja Wastewater Treatment Plant in Espoo, Finland, reported a savings in methanol costs of U.S. $40,000 to $66,000 (30,000 to 50,000 Euros) per year based for their sampling program and modeling efforts (Phillips et al., 2009). The Metro Wastewater Reclamation District in Denver, Colorado, has been using models for nearly a decade; facility operators have quantified at least $2 million in capital savings (Phillips et al., 2009) and as much as $250,000 in operating costs in 2011 because of modeling efforts and dissolved oxygen optimization.

2.5 Example Sampling Programs

A typical sampling program for composite samples to gather sufficient data for model calibration is shown in Table 8.8. If the facility typically analyzes for CBOD in its raw wastewater, analysis of uninhibited BOD and CBOD must be performed

TABLE 8.8 Sample composite sampling program.

Sample location and guidelines*	Parameter									
	TSS	VSS	BOD$_5$	COD	VFA	TKN	NH$_3$-N	NO$_x$-N	TP	PO$_4$-P
Raw influent										
XX	X	X	X	X	X	X			X	
XG			X	X		X	X	X	X	X
XM				X						
XF				X						
Primary influent										
XX	X	X	X	X		X			X	
XG			X	X		X	X	X	X	X
Primary effluent										
XX	X	X	X	X	X	X			X	
XG			X	X		X	X	X	X	X
Secondary effluent										
XX	X	X	X	X		X			X	
XG				X		X	X	X	X	X
XF				X		X				

*XX = not filtered, XG = filtered with glass fiber filter (1.2 µm), XM = filtered with 0.45-µm membrane filter, and XF = flocculated and filtered with 0.45-µm membrane filter.

in parallel. At the end of all uninhibited BOD tests, analysis of nitrate should also be performed to ensure that there is no effect from nitrifying bacteria. Parallel analysis of CBOD and uninhibited BOD should be performed to identify the effect of nitrification inhibitors on heterotrophs in the BOD test. A typical sampling program for other parts of the facility would follow a similar logic and approach.

3.0 DATA RECONCILIATION

Data reconciliation is the final step of the data collection procedure, and it must be conducted before the model is set up (Step 3 of the GPM Unified Protocol). The principle of data reconciliation is to apply one or several independent checks to verify the consistency of the data. Depending on original data quality and quantity, the reconciliation process can be time consuming. On the other hand, data that have not been reconciled have the potential to provide erroneous simulation results. Some examples of data reconciliation independent checks that can be performed on typical wastewater process data are listed in the following sections.

3.1 Fundamental Checks

Fundamental checks, such as flow balances or mass balances on inert components or phosphorus, should be carried out. Solids can be considered inert around clarifiers, assuming that biological transformations are not occurring in the clarifier. When performing an inert mass balance (i.e., nonvolatile inorganic solids or total phosphorus) around the entire process or facility, all inputs and outputs should be considered, including chemical dosages. If a mass balance error is detected, it should be rectified before using data in the model. It is impossible for the mass balance to be perfect; as such, judgment must be used to specify a suitable tolerance for the data. In many applications, a 10% margin of error is deemed to be reasonable for most measurements, although the actual value used depends on many factors, including data quality, measurement methods, modeling purpose, and effluent permit. It is important for all stakeholders to agree on the acceptable error early in the project. The modeler must then be prepared to give a justifiable reason for any errors exceeded. The error can be the result of several incorrect measurements, which should all be adjusted in proportion to their reliability and noisiness (standard deviation). In practice, it is common that one flow variable (i.e., recycle or wastage flow) needs to be modified to maintain the mass balance. It is important to discuss these errors with facility staff because they often have information on which measurements have the best and least confidence; this information can be used to decide which values can be adjusted in the model to give a better mass balance.

3.2 Empirical Checks Based on Engineering Knowledge

Experience-based "empirical" knowledge of the activated sludge process should be used to check the reliability of the data set. For example, observed sludge production figures can be calculated and checked against typical values before trying to force a fit to data, which might be outside the typical range. Use of histograms and descriptive statistics provide a way to visually and mathematically see the normal distribution of an influent parameter loading such as BOD and TSS. Figure 8.4 presents a histogram

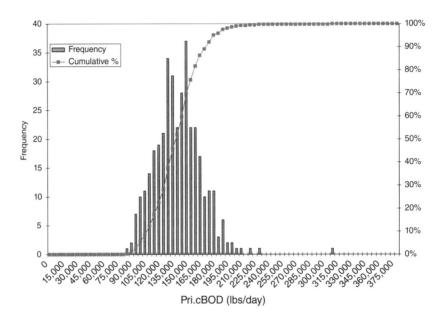

FIGURE 8.4   Example histogram of primary effluent BOD loading.

of primary effluent BOD loading. The loading data fit a normal distribution curve, which is expected; however, to the far right of the figure, there are a few data points outside normal distribution. The data points could be considered outliers and need to be evaluated more closely to determine if they are erroneous. Outliers may also be detected by examining the median and standard deviation of the data. Data more than approximately 3 standard deviations may be considered suspect or atypical. Examination of chronological data before and after the suspect point is also beneficial because, if the data smoothly build up to the apparent outlier and/or smoothly drop off from the outlier, the data point may be valid.

3.3 Historical Checks

In assessing the validity of historical or other data, it often is useful to examine the ratios of certain parameters rather than the absolute values of the parameters themselves. Some common ratios to check are influent COD/BOD, TSS/BOD, TKN/COD, VSS/TSS, particulate COD/VSS, and mixed liquor VSS/TSS. For example, there are typically considerable day-to-day variations in concentration values, but the relation of COD/5-day BOD (BOD_5) should not show large fluctuations. Therefore, it is often helpful to identify outliers or invalid data by looking at ratios. For example, the COD-to-BOD_5 ratio for typical domestic wastewaters should be in the range 2.0 to 2.2 for raw wastewater and 1.9 to 2.1 for primary effluent. Some of these ratios can be calculated from raw wastewater characteristics defined in sources such as *Design of Municipal Wastewater Treatment Plants* (WEF and ASCE, 2009). Significant deviations from these values would indicate that the data are suspect or that some influent wastewater characteristic is different from typical values. Typical ratios are referenced in Table 8.6.

3.4 Identification and Repair

The aforementioned methods help identify and isolate potential erroneous data from the data set. The next step is to determine if the data are truly erroneous. Obviously, some data may be wrong, such as NH_3-N being higher then TKN, but other data may need further investigation such as a possible uncalibrated flow meter. Additional samples may be required to fill in the holes left by erroneous data.

4.0 FACILITY MODEL SETUP

Managing model complexity and client expectations are some of the biggest challenges that a modeler faces when setting up a facility model (Step 3 of the GMP Unified Protocol [Rieger et al., 2012]). Common methods of managing complexity include merging parallel treatment trains that operate similarly or disabling add-on features for cursory simulations. For example, some models provide the ability to model or not model pH, dissolved oxygen, or precipitation reactions. Enabling these features nearly always increases run time. Modeling pH and precipitation of metal salts may not be necessary for initial calibration runs, such as the initial solids balance. Selectively enabling and disabling these features can affect how quickly one is able to complete projects without negatively affecting the desired outcome. It is important to select a model that is appropriate for the objective; if phosphorus is not measured or regulated for a given facility, it may be appropriate to select a model that simulates carbon and nitrogen transformations only. The following sections provide step-by-step checklists for model development.

4.1 Influent Model

Model inputs should be included for all influent streams, including raw wastewater, chemicals, sidestreams (if modeled separately), and wash water for solids processing. These items are addressed in the following sections.

4.1.1   Wastewater

Influent wastewater is the most important aspect of any modeling effort because it ultimately defines sludge production (and facility capacity) and effluent quality. If good data are unavailable, the modeling task itself may be useless because "garbage in equals garbage out." With good data, the modeler can achieve calibration and validation of data without adjusting yield or kinetic parameters from their peer-reviewed published values.

The modeler must first decide how many model inputs are necessary to characterize the raw influent wastewater. Often, there are different parts of the collection system that enter the facility separately and that are measured independently. It is common for industrial contributions to be measured separately because the utility charges them rates for treatment accordingly. Significantly different influent streams should always be modeled separately, especially if flow or characteristics are going to change in the future. For example, industrial contributions will often remain constant for several years then change suddenly when the industry modifies a process. At the same time, domestic flow contributions may grow steadily with population, but the waste strength may increase as a result of water conservation efforts. Modeling efforts often evaluate several long-term planning milestones. Therefore, depending on the complexity of the collection system, it may be easiest to model significant influent contributions separately.

Most of the time setting up a model will likely be spent on influent characterization and aligning primary effluent concentrations with measured data. If biological reactions occur in the primary clarifiers, efforts will be significantly complicated. The modeler must specify how much of the influent COD entering the facility (and the model) is soluble or particulate and biodegradable or non-biodegradable in order for the model to determine its fate. Figure 8.5 shows how influent characterization affects primary clarifier removal efficiencies. It is critical to examine primary clarifier performance during special sampling and to compare it to historical performance (and future performance projections). A common pitfall with modeling is to assume unrealistic primary clarifier removal efficiencies. For example, a modeler might apply relatively conservative primary clarifier removal efficiencies for long-term planning purposes (e.g., 60% TSS and 30% BOD). However, if the clarifiers perform much better than this during startup (e.g., 70% TSS and 40% BOD), there may not be enough carbon to support BNR.

Influent data used for model calibration and validation often differs from data used for simulation and interpretation of results. Sometimes, no matter how much planning went into the special sampling program, unforeseeable events such as wet

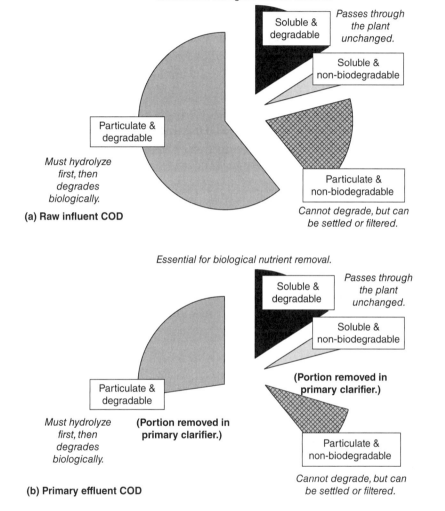

FIGURE 8.5 Example of how wastewater COD fractions change from raw influent (a) to primary effluent (b) after clarification.

weather may affect results, causing the modeler to use completely different sets of data for calibration, validation, and other modeling efforts. In these instances, the modeler must justify influent flows and loads for all model simulations and clearly state the assumptions.

4.1.2   Chemical Addition

If chemicals are added at the facility for phosphorus precipitation, denitrification, or even struvite or odor control, they will affect sludge production. During the site visit and before model setup, the modeler should take inventory of all chemicals used at the facility, where they are used, and how they are used, including doses. It is common for facilities to use ferric chloride for odor control in the headworks or primary clarifiers. While iron will initially react with sulfide, excess dosing will precipitate phosphorus and, potentially, colloidal COD. Although phosphorus removal may not be a modeling objective, it may be necessary to model ferric chloride addition to calibrate the model and reconcile the sludge balance. A list of common chemicals used in wastewater treatment and associated modeling considerations are provided in Table 8.9.

4.1.3   Sidestream

In general, it is always recommended to build a facility-wide model that includes both liquid and biosolids processes as opposed to including a separate model input for sidestreams generated through biosolids processing. However, there are circumstances that may warrant separate model inputs for sidestreams, such as the following examples:

- The facility influent sampler is downstream of the location where sidestreams are returned, and all historical data includes sidestream loads (in this instance, it is recommended that the sampler be moved for regulatory reasons).

- Biosolids processing will change in the future (e.g., the facility will install digestion and dewatering and the engineer wants to account for worst-case sidestream loads).

- A new sidestream load will be generated (e.g., the facility will abandon liquid hauling and install dewatering equipment, and the engineer wants to account for worst-case sidestream loads).

- A new waste will change sidestream quality (e.g., co-digestion will import nitrogen and phosphorus, and volatile solids to boost biogas production).

While each of the scenarios can be modeled, in instances where data are not available for future facility configurations, the modeler may find it easier to use a separate sidestream input with data from a similar facility.

TABLE 8.9 Modeling considerations for chemicals used in wastewater treatment.

Reason for chemical addition	Typical chemical(s) used	Modeling considerations
Odor control (in the liquid)	• Nitrate salts or other oxidants (collection system) • Iron salts	• Nitrate salts and oxidants can change influent characteristics, preventing fermentation in the collection system. • Overdosing iron salts can precipitate phosphorus and colloidal COD and increase sludge production.
Chemically enhanced primary treatment	• Iron and aluminum salts • Polymer	• Can significantly increase primary clarifier removal efficiencies, precipitate phosphorus and colloidal COD, and increase sludge production. • pH and alkalinity may be lowered.
Denitrification	• Acetic acid • Glycerol • High fructose corn syrup • Methanol • Sucrose • Waste products • Other proprietary products	• Each chemical has a specific substrate utilization efficiency and sludge yield. • Methylotrophs should be modeled separately because their kinetics are much different than ordinary heterotrophic organisms and an acclimation period is required. • When modeling waste products, make sure to account for any N or P associated with the waste (or metals, total dissolved solids, or other inhibitory compound). • Model inputs are generally expressed as COD, so convert from mg/L of chemical accordingly.
Biological phosphorus removal	• Acetic acid • Propionic acid	• Model inputs are generally expressed as COD, so convert from mg/L of chemical accordingly. • Volatile fatty acids can be fermented onsite, often unknowingly at the bottom of the clarifier, at the corner of anaerobic zones, or during sludge storage and processing. • Sludge production will increase.
Chemical phosphorus removal	• Iron salts • Aluminum salts	• pH and alkalinity will be lowered. • Sludge production will increase.
Filament control	• Chlorine • Polyaluminum chloride	• Chlorine overdose could inhibit nitrification. • Polyaluminum chloride overdose could precipitate phosphorus.
Struvite control	• Iron salts	• Will precipitate phosphorus (and magnesium), affecting sidestream quality.
Settling, filtration, thickening, or dewatering aid	• Polymer	• Can affect clarifier model assumptions. • May increase sludge production.

4.2   Physical Configuration

When setting up facility model layout, it is important to have accurate and current piping and instrumentation diagrams (P&ID) and process flow diagrams (PFD) and to interview facility staff that work with each process on a daily basis. For each sub-model, the modeler will need to specify the following physical information:

- Number of treatment units in parallel

- Number of treatment units in series

- Reactor volume (or surface area)

- Reactor depth

- Additional process-specific information (elevation, number of diffusers, etc., for aeration models; depth of feed point for some clarifier models; media surface area for biofilm processes; etc.)

Model complexity can be reduced easily by merging parallel trains into one unit. For example, if a facility has four parallel 1000-m^3 aeration tanks, these could be represented by one 4000-m^3 aeration basin instead of four separate 1000-m^3 aeration tanks. However, there may be valid reasons for representing the trains separately. For example, if all the aeration basins have different dissolved oxygen concentrations as a function of fouling or aeration system design, it may be preferable to model each train separately. Furthermore, if there is an uneven flow distribution among each of the trains, it may be required to model them separately. Whenever parallel treatment units are operated significantly differently or they are physically configured differently (clarifiers with different depths), they must be modeled separately.

Selecting the appropriate number of reactors in series to simulate plug flow conditions is also a topic to carefully consider because an adequate number is necessary to simulate performance, while too many adds unnecessary model complexity. Section 4.0 of Chapter 5 describes hydraulic characterization methods that can be used to determine the appropriate number of tanks in series.

4.3   Selection of Submodels

The theoretical details of activated sludge, biofilm, clarification, and other submodels are presented in Chapter 3. The following sections provide brief checklists for the modeler to use when specifying submodel input. The reader is referred to Chapter 3 for more detailed information.

4.3.1 Biokinetic Models

There are many biokinetic models for a modeler to choose from for a given modeling assignment, and selection of biokinetic models should be based on model objectives. If a facility has effluent limitations for BOD, TSS, and nitrogen (but not phosphorus), then it may be acceptable to select a carbon/nitrogen biokinetic model for simplification purposes. The same can be said for modeling pH and alkalinity. When modeling large facilities or complex processes (i.e., biofilms), in particular, the modeler must select a biokinetic model that will meet objectives with the available project budget and schedule, factoring in solving time. However, a risk with selecting simpler biokinetic models is that phosphorus is a necessary nutrient for all biological activity and there are circumstances in which it can be a limitation. If a simpler biokinetic model is selected, the modeler must collect data to ensure that there is sufficient phosphorus (or alkalinity) to support the biological reactions being modeled. Some simulators have the ability to switch off certain reactions; however, the modeler could switch them back on for the final solution, which can expedite solving time tremendously. When deciding which biokinetic model to use, the modeler should ask the following questions:

- What biokinetic reactions do I need to model to meet my objectives?
- Do I have data to calibrate the model for these biokinetic reactions?
- Where in the facility are biokinetic reactions taking place (aeration basin, trickling filter, clarifier, sludge holding tank, or digester) and do I need to model all of them?

4.3.2 Clarifier Models

Clarifier modeling is another example of when the modeler must make a choice between obtaining increasing levels of accuracy and reducing run time. For example, the modeler may choose between using a fixed-percent TSS removal to model secondary clarifier performance ("ideal clarifier") or using a one-dimensional clarifier model that accounts for storage in the clarifiers ("model clarifier"). The latter requires an increased run time, but more accurately models clarifier performance. If a facility historically has not experienced problems with its secondary clarifiers, or if secondary clarifier performance is not the focus of the study, then perhaps it would be an appropriate situation to use the less detailed representation of the clarifiers. It is typically preferred to model the clarifier, especially if a state point analysis indicates that the capacity will be stressed. In some circumstances, more advanced CFD modeling is required to support clarifier design. The modeler must also decide whether to include biological reactions within the clarifier, which is often necessary when modeling clarifiers or gravity thickeners that have relatively deep sludge blankets.

4.3.3   Aeration Models

When modeling aeration in activated sludge facilities, the modeler often has the choice of specifying a dissolved oxygen concentration and letting the model predict the airflow needed to achieve that setpoint or specifying the airflow and letting the model predict the resulting dissolved oxygen concentration. Either approach requires the modeler to have knowledge of the diffuser equipment, its transfer efficiency at given operating conditions, and, most importantly, a correct assumption for the alpha factor (α), the ratio of process-to-clean-water mass transfer. Section 2.3.4 summarizes the factors that affect α and provides guidance on incorporating aeration test data to modeling efforts.

4.3.4   Other Models

Some model layouts may require additional submodels to simulate filters, disinfection, dissolved air floatation, sidestream treatment, or other processes. It is common for manufacturers to conduct detailed modeling of these processes independently of the facility-wide model because each product can behave differently depending on design, chemical dose, and so on. Manufacturers often use their own proprietary models to design such systems. However, the modeler must include any process that affects the overall solids balance (filtration, dissolved air floatation, or other thickening or dewatering processes) in the facility-wide model to account for sidestreams. It is common to use an empirical solids separation device for such purposes, where, for example, the modeler specifies a percent TSS capture for a filter.

4.4   Operational Parameters

When setting up the facility model layout, it is important to have accurate accounts of all operating conditions; this is best accomplished by interviewing facility staff after reviewing available process data. For each submodel, the modeler will need to specify the following operational parameters:

- Number of treatment units in service (this affects physical model parameters)
- Operational setpoints

- Method of control (manual, online, etc.) and frequency of process adjustments
- Additional process-specific information

4.4.1 Operational Setpoints

Operational setpoints will likely vary from facility to facility, even if the processes are similar. It is important to interview operators because there is more than one way to operate a facility to achieve the same results. Table 8.10 provides a checklist of common questions to ask an operator to obtain the necessary information for model development.

TABLE 8.10 Questions to ask an operator when developing a facility model.

Process	Question to ask	Why ask the question?
Headworks	Do you equalize flow anywhere?	The model should mimic the approach.
Primary clarifiers	Do you pump the sludge continuously?	The model should mimic the approach.
	Do you notice a lot of odors or bubbling?	Could indicate biological activity.
	Do you measure sludge blanket depth?	Could indicate biological activity.
	Do you add any chemicals?	Could affect wastewater characteristics.
	When do you notice a difference in performance?	Operators' observations can explain the data.
Aeration basins	When do you take basins out of service?	
	How do you control dissolved oxygen?	
	How do you control SRT?	
	Are the MLR pumps flow-paced or other?	The model should mimic the approach.
	Do you add any chemicals?	
	When do you notice a difference in performance?	Operators' observations can explain the data.
Biofilm processes	Do you monitor biofilm growth, or take photographs throughout the year?	Biofilms behave very seasonally.
Secondary clarifiers	Are RAS and WAS a common line?	The model should mimic the approach.
	Is the RAS flow-paced or other?	The model should mimic the approach.
	Do you waste continuously?	The model should mimic the approach.
	Do you measure sludge blanket depth?	Could indicate biological activity.
	Do you notice a lot of odors or bubbling?	Could indicate biological activity.
	Do you notice a lot of algae growth?	Will affect effluent TSS.
	When do you notice a difference in performance?	Operators' observations can explain the data.

TABLE 8.10   Questions to ask an operator when developing a facility model. (*continued*)

Process	Question to ask	Why ask the question?
Sludge storage	How long is the sludge stored?	Biological activity could occur.
	Is the storage tank aerated or mixed?	Defines the type of biological activity.
Sludge thickening	Do you thicken sludge continuously?	The model should mimic the approach.
	What type of chemicals do you use?	
	When do you notice a difference in performance?	Operators' observations can explain the data.
Anaerobic digestion	How are the digesters mixed?	Short-circuiting could affect performance.
	Are the digesters all heated?	
	Are the units run in series or parallel?	
	Are chemicals added?	The model should mimic the approach.
	When do you notice a difference in performance?	Operators' observations can explain the data.
Aerobic digestion	Is the air cycled on and off, and how?	The model should mimic the approach.
Dewatering	Do you dewater continuously?	
	What type of chemicals do you use?	The model should mimic the approach.
	When do you notice a difference in performance?	Operators' observations can explain the data.

4.4.2   Controllers

Controllers are commonly used for process management at WRRFs, and the model should reflect the same strategy and algorithms implemented by the full-scale controllers. Common applications for controllers include

- Recycle flows
- Solids retention time
- Aeration
- Influent flow equalization and bypassing

5.0   CALIBRATION AND VALIDATION

Step 4 of the GMP Unified Protocol (Rieger et al., 2012) is model calibration and validation. Process models for existing WRRFs must be calibrated to accurately describe the performance of the existing treatment process. Well-calibrated models can be used with confidence to investigate the effect of changes in process configuration, implementation of external carbon addition, and changes in wastewater composition and/or strength, among other possibilities. Applying models without calibration and validation can be considered engineering negligence, and can result in designs that fail to meet effluent criteria.

Calibration is typically based on matching facility performance during the time period in which characterization data were collected. Validation of the model occurs by using the calibrated model to match facility performance over a different time period or during specific operating conditions, such as high-flow or low-temperature periods. Calibration may include modifying model parameters such as kinetics, but validation serves to confirm these modifications. Modification of kinetic parameters should always be avoided unless biokinetic rate studies were conducted or no other factors (influent characterization, hydrodynamics, etc.) can reconcile model output with facility data. The calibration and validation procedure is shown in Figure 8.6.

Before initiating model calibration, the modeler and the client must agree on target variables and stop criteria; that is, which parameters must be calibrated to support the model objectives (sludge production or effluent quality) and what is the acceptable variance between model predictions and facility data (i.e., when can the modeler stop calibrating). Table 8.11 provides a list of common target variables and stop criteria for model calibration. The discussion on model calibration in the following sections is based on detailed discussions presented in Melcer et al. (2003) and in the GMP Unified Protocol (Rieger et al., 2012). The reader should consult these reports for further details on recommended model calibration practices.

5.1 Target Variables

Model calibration efforts typically focus on three target variables: (1) solids production (basin sizing and facility capacity), (2) effluent quality (permit compliance), and (3) aeration (diffuser and blower design and operating costs). The modeler should select his or her calibration parameters or target variables based on project-specific objectives defined in Chapter 7.

5.1.1 Solids Production

A steady-state period of operation is typically selected for calibrating the overall solids production. In some instances, however, the facility may never operate under steady-state conditions because of construction activity or other factors so the modeler must do their best to account for dynamic conditions when calibrating the model.

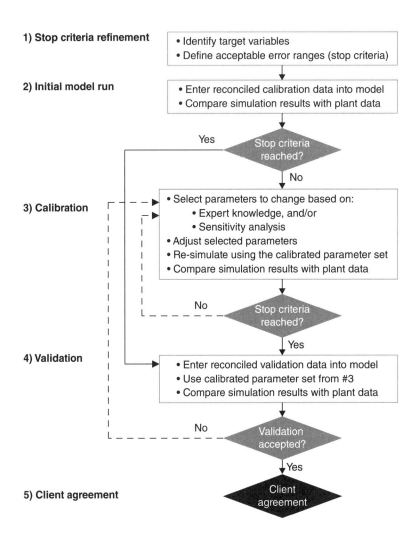

FIGURE 8.6 Model calibration and validation protocol.

The modeler may need to simulate a long period of time for calibration efforts and calculate the long-term average solids production for comparison purposes.

The influent wastewater stoichiometry has the biggest effect on solids yield; as such, the modeler should never adjust yield parameters within the model unless all other attempts to reconcile data with model predictions fail. The modeler should have first identified any unusual data as part of the data reconciliation step and

TABLE 8.11 Common target variables and stop criteria for model calibration (adapted from Rieger et al. [2012]).

Modeling task	Averaging period	Target variable	Acceptable error range (±)
Calculate sludge production	Monthly average	MLSS	10%
		MLVSS/MLSS	5%
		WAS mass load	5%
		Effluent TSS	5 mg/L
		SRT	1 day or 20% for SRT < 5 days
Design aeration system	Monthly average	Airflow rate	10%
	Daily average	Dissolved oxygen (profile)	0.3 mg/L
	Hourly peaks	Oxygen uptake rate	10 mg/L/h
Develop a process configuration for nitrogen removal	Monthly or annual average	NH_3-N	
		NO_x-N	1.0 mg/L
		Total nitrogen	
Design a treatment system to meet peak effluent nitrogen limits	Instantaneous values	NH_3-N	
		NO_x-N	0.5 mg/L
		Total nitrogen	
Develop a process configuration for phosphorus removal	Monthly or annual average	PO_4-P	0.5 mg/L
		Total phosphorus	
Assess plant capacity for nitrogen removal	Monthly or annual average	NH_3-N	
		NO_x-N	1.0 mg/L
		TN	
Make the most of aeration control	Hourly average	Airflow rate	10%
		Dissolved oxygen	0.5 mg/L
		Effluent NH_3-N	0.5 mg/L
Test effect of taking tanks out of service	Monthly average	NH_3-N	1.0 mg/L
		NO_x-N	1.0 mg/L
		Total nitrogen	1.0 mg/L
		PO_4-P	0.5 mg/L
		Total phosphorus	0.5 mg/L

TABLE 8.11 Common target variables and stop criteria for model calibration (adapted from Rieger et al. [2012]). (*continued*)

Modeling task	Averaging period	Target variable	Acceptable error range (±)
Use model to develop sludge wasting strategy	Weekly average Daily average	WAS mass load SRT NH_3-N PO_4-P	10% 1 d mg/L 0.5 mg/L
Develop a strategy to manage storm flows	Hourly average	MLSS Effluent TSS NH_3-N NO_x-N PO_4-P	10% 10 mg/L 1.0 mg/L 1.0 mg/L 0.5 mg/L
Develop a site-specific model for operator training	Monthly	MLSS WAS mass load Effluent TSS NH_3-N NO_x-N PO_4-P Airflow rate Dissolved oxygen	10% 5% 5.0 mg/L 1.0 mg/L 1.0 mg/L 1.0 mg/L 10% 0.5 mg/L

avoided using that data for model calibration purposes. For example, if the modeler calculated the sludge yield (the sum of WAS and effluent TSS mass divided by the secondary influent BOD mass) for 1 month to be 1.7, when all other months were 0.8 (a reasonable value), the month with the data anomaly should be avoided for calibration efforts. There are many factors that affect solids yield and the modeler should have first carefully evaluated flow measurement, sampling, and analytical techniques as part of the data reconciliation step (Rieger et al., 2010).

5.1.2   *Effluent Quality*

Model predictions should accurately predict effluent nutrient levels including ammonia, nitrate, nitrite, and phosphorus. Calibrating effluent TSS and BOD is more difficult because this depends on the level of effort put into the clarifier model. Calibration of effluent nutrients should first be attempted without modifying biokinetic parameters because there are many other factors that affect nutrient removal.

Influent stoichiometry and primary clarifier performance affect denitrification and biological phosphorus removal more than any other factors. Nitrate and nitrite concentrations are influenced by the carbon available for denitrification (and also the mixed liquor return rate, the anoxic volume, and other factors). Phosphorus removal is influenced by the concentration of RBCOD and VFAs available. Ammonia concentrations are governed by aerobic SRT of the system, dissolved oxygen concentrations, tanks-in-series selection, and available alkalinity. Because SRT is a function of the solids yield, influent stoichiometry plays a direct role in all nutrient transformations.

Nutrient removal is also significantly affected by facility operations, especially dissolved oxygen control. The modeler may have difficulty calibrating ammonia concentrations because they do not have dissolved oxygen data for a period of time in the day when effluent ammonia was high. Denitrification is also affected by dissolved oxygen concentrations through carryover in MLR or RAS, overmixing, or hydraulic conditions. Carryover of dissolved oxygen into anaerobic zones creates anoxic conditions that are detrimental to biological phosphorus removal. Model calibration efforts must, therefore, mimic facility operating conditions, especially if they vary throughout the day, which is often the case.

The physical configuration of basins can also affect nutrient removal. It is not uncommon for the anoxic zone influent streams to enter the anoxic zone over weirs, cascading down several inches or feet (although this is not preferred design practice). As a rule of thumb, approximately 1 mg/L of dissolved oxygen is entrained for every 0.3 m (1 ft) of drop in water level. Screw pumps entrain a tremendous amount of dissolved oxygen in the liquid.

5.1.3 Aeration

Predicted airflow should be calibrated with observed airflow and dissolved oxygen concentrations in the aeration basins. Oxygen-transfer-efficiency parameters can be used to accurately calibrate the aeration process, as discussed in Chapter 3. The biggest challenge of calibrating aeration models is lack of data. Many facilities do not measure airflow to individual drop legs and may only monitor dissolved oxygen concentrations once or twice per shift. The modeler, therefore, must use available data and interview operations staff to help fill in missing information. The operator might not be able to quantify the airflow entering each diffuser grid, but if he or she can describe the valve positions and how often they are changed, the modeler can approximate airflow distribution. Calibrating the aeration model is often difficult because of a lack of site-specific data. However, the modeler cannot ignore this task because undersizing and oversizing aeration equipment can result in performance failure.

5.2 Stop Criteria

Depending on available information and the ultimate goal of the modeling effort, different levels of calibration can be conducted, each having different levels of expected accuracy (or stop criteria). Stop criteria are typically quantitative, such as variance in the MLSS concentration as a percent or variance in effluent concentration in milligrams per liter. The expectation for accuracy for each target variable must be agreed on early in the project because they define the amount of additional sampling required. At times, stop criteria are based on other factors such as the maximum number of model runs when project budgets are limited. It may even be acceptable to not meet some of the criteria if most criteria are being met and further improvement appears to be impractical. Table 8.12 summarizes typical stop criteria for common target variables.

TABLE 8.12   Example checklist for minimum QA/QC tasks.

Point in schedule*	Modeler's tasks	Quality assurance engineer's task	Quality control engineer's task
Scope development	All parties agree on QA/QC tasks and schedule before work begins.		
Data collection	• Check process flow diagram. • Verify sample points and probe locations. • Confirm physical facility data, operational conditions.	• Compare data collection plan with standard industry practice.	• Cross reference other sets of drawings and reports. • Review data collection plan. • Obtain stakeholder feedback.
Data reconciliation	• Fundamental checks. • Empirical checks based on engineering knowledge. • Historical checks. • Identification and repair.	• Comment on all assumptions, especially any data repair. • Comment on the design conditions with respect to service area growth or changes.	• Review modeler's data analysis report and confirm all assumptions. • Check spreadsheets. • Obtain stakeholder feedback.

TABLE 8.12 Example checklist for minimum QA/QC tasks. (*continued*)

Point in schedule*	Modeler's tasks	Quality assurance engineer's task	Quality control engineer's task
Facility model setup	• Check influent stoichiometry balance. • Check each submodel, splitter, combiner, etc., for accurate connections. • Check input/output spreadsheets for functionality.	• Comment on choice of submodels.	• Review model layout, input/output spreadsheets. • Obtain stakeholder feedback.
Calibration and validation	• Validate with second data set if budget allows. • Document all model inputs, outputs, results.	• Comment on kinetic parameter adjustments. • Comment on applicability of calibration/validation conditions to future design conditions.	• Review model development report. • Obtain stakeholder feedback.
Using the model	• Conduct sensitivity analyses and/or Monte Carlo simulations to validate design. • Check model results with parallel modeling, hand calculations, or vendor recommendations. • Document all model inputs, outputs, results, and archive files.	• Compare design parameters to typical published values. • Compare design to other designs similar in size and location. • Suggest other "what if" modeling scenarios such as changes in service area or more stringent regulatory limits.	• Compare design parameters to state/regulatory policies for minimum values. • Confirm that startup and minimum turndown conditions are satisfied. • Obtain stakeholder feedback.

*QA/QC tasks should be conducted upon completion of significant project milestones before moving on to the next tasks in the schedule.

5.3 Model Maintenance

After the model has been calibrated and validated for a specific modeling task, the project may not be over. Indeed, the model is often used to design a facility expansion and then later to assist with construction sequencing. Eventually, after the facility is constructed, the model may be used again to help decide how many treatment trains are needed for startup or to troubleshoot problems that might arise. Ideally, the model should be used to train operators on how to run the new facility. It is becoming increasingly common for utilities to purchase simulator software to and assume ownership of the model for training and daily operations support, particularly when a facility is changing treatment processes or implementing nutrient removal for the first time; there is a lot of value in using a desktop tool to answer "what if" questions when running a new process.

The utility must decide who is going to retain ownership of the model. Because much time and money likely were spent on the special sampling program and model development, it is in the utility's best interest to retain a copy of the model and all associated data and reports. Then, when it is time for the next facility expansion, the model can be shared with the next engineering firm and hopefully lessen the time and effort spent on model development. Ideally, after startup, either the design engineer or the utility would conduct a model validation to confirm design. If the utility uses the model for day-to-day operations, it is recommended that routine validations be conducted, especially if there are significant changes in operations. Any time that physical changes are made to a facility, the facility model must be validated or recalibrated, if necessary. Figure 8.7 shows a model maintenance schedule for a utility that conducts routine operations modeling.

The most important component of model maintenance is documentation. As part of a design, it is the engineer's responsibility to document all model inputs, outputs, and assumptions. Often, this information must be submitted to the local regulatory agency for a design to be approved. The utility should retain copies of modeling reports, and routinely write their own modeling reports, if the model is used for facility operations or other in-house analyses. Good documentation and routine validations will help prevent modeling rework and extend the life of the model.

6.0 QUALITY ASSURANCE/QUALITY CONTROL

Quality assurance should be included for all modeling projects by a qualified modeler who was not directly involved in model development. This person serves as an objective reviewer of all procedures used, the model itself, and model results

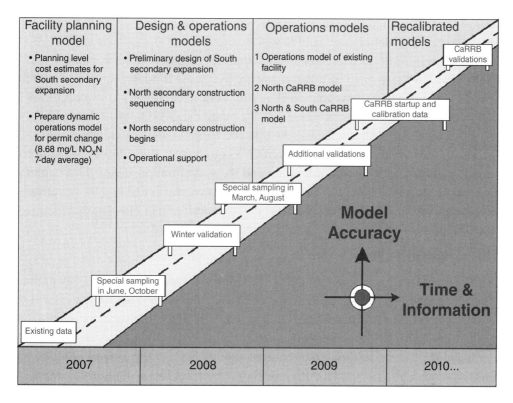

FIGURE 8.7   Example of a model development and maintenance schedule.

and reports, and should be able to quickly spot mistakes because this person is not directly involved in day-to-day details. The quality assurance modeler often works within the same company, although her or she could be from an outside firm (e.g., through a value engineering effort). Quality assurance should include a planned system of review that is agreed on early in the project and documentation of when and how the quality assurance work was conducted.

Quality control should be conducted throughout the modeling project. Quality control includes tasks such as data reconciliation, verification of model layout, and model validation. Quality control does not need to be conducted by a third party. Often, the engineering manager or other process engineers working on the project will provide quality control, interjecting relevant questions as needed throughout the project. The modeler should conduct his or her own quality control as well and can use parallel models or spreadsheet calculations to confirm results. If good quality control is implemented throughout the project, it is less likely that the quality assurance modeler will identify any fatal flaws.

Many significant decisions on an engineering project are made based on model predictions, and care must be taken to ensure that these model results are accurate. Most engineering firms have standard quality assurance/quality control (QA/QC) protocols for design projects; however, specific modeling QA/QC tasks also need to be identified for the project to stay on schedule and to avoid fatal flaws. While the QA/QC procedure may differ somewhat from project to project, there are several uniform steps that should be taken at a minimum and these are listed in Table 8.13. Chapter 7 also includes a discussion of QA/QC. Specific data reconciliation and model calibration checks are discussed in Sections 3.0 and 5.0, respectively, of this chapter.

7.0 EXAMPLES

Development of a process model will be different for every project, and the amount of effort spent on sampling, calibration, and validation should be directly related to modeling objectives and associated project risk. The following sections summarize model development steps that were taken for a basic planning-level design model, a detailed design model, and an operations model. Each of the models described is a facility-wide dynamic model with modeling objectives focused on facility capacity and/or nutrient removal.

7.1 Basic Planning-Level Design Model Development

The Big Dry Creek Wastewater Treatment Facility (BDCWWTF) is owned and operated by the City of Westminster, Colorado, and has a maximum monthly treatment capacity of 45 ML/d (11.9 mgd) and 8437 kg/d (18 600 lb/d) of BOD_5. During the past few years, flows have increased steadily at anticipated rates. However, organic loading to the facility has increased at rates higher than anticipated and disproportionate to the flow increase. Recent organic loadings to the facility exceeded 80% of the permitted capacity, which serves as a regulatory trigger to evaluate the need for rerating and/or expanding the facility. City officials selected an engineer to develop both a hydraulic and process model to evaluate facility capacity through a comprehensive analysis of each unit process and to eventually apply for facility rerating with the Colorado Department of Public Health and Environment. The BDCWWTF has a number of unique features, including primary clarifiers converted into anaerobic zones and the use of coarse bubble diffusers in the anoxic zones.

TABLE 8.13   Basic planning-level design model calibration and validation results.

Parameter	June 2011 data	Model prediction	Variance	Jan 2011 data	Model prediction	Variance
Facility influent						
Flow, mgd	7.73	7.73	0%	6.47	6.47	0%
BOD$_5$, mg/L	214	214	0%	278	278	0%
TSS, mg/L	224	224	0%	264	264	0%
TKN, mg/L	32.2	32.2	0%	-	38	no data
Total phosphorus, mg/L	5.4	5.4	0%	-	5.9	no data
Alkalinity, mg CaCO$_3$/L	262	262	0%	226	226	0%
Activated sludge						
MLSS, mg/L	2750	2967	8%	2732	2999	10%
Total SRT, days	11.5	11.5	0%	12.6	12.2	−3%
Airflow, scfm	7605	7800	3%	7555	7870	4%
MLR, % influent	300%	300%	0%	300%	300%	0%
RAS, % influent	75%	75%	0%	75%	75%	0%
WAS, ppd	11 600	12 000	3%	10 640	12 000	13%
Effluent quality						
BOD$_5$, mg/L	3.1	1.7	−1.4 mg/L	4.1	1.6	−2.5 mg/L
TSS, mg/L	3.7	4.6	0.9 mg/L	6.0	3.8	−2.3 mg/L
NH$_3$N, mg/L	0.1	0.1	0.0 mg/L	0.5	0.3	−0.2 mg/L
NO$_x$N, mg/L	7.6	7.7	0.1 mg/L	7.6	8.2	0.6 mg/L
Total phosphorus, mg/L	2.0	1.8	−0.2 mg/L	0.6	1.0	0.5 mg/L
Orthophosphorus, mg/L	1.5	1.6	0.0 mg/L	0.1	0.9	0.8 mg/L
Biosolids processing						
Digester feed, % total solids	3.2%	3.2%	0.0%	3.7%	3.3%	−0.4%

TABLE 8.13   Basic planning-level design model calibration and validation results. (*continued*)

Parameter	June 2011 data	Model prediction	Variance	Jan 2011 data	Model prediction	Variance
Digester feed, % volatile solids	79%	75%	−4.0%	83%	73%	−10%
Digester volatile solids destruction, %	37%	27%	−10%	41%	30%	−11%
Biosolids thickener feed, % total solids	2.2%	2.5%	0.3%	2.1%	2.5%	0.4%
Biosolids thickener feed, % volatile solids	72%	72%	0.5%	74%	69%	−5%
Thickened sludge, % total solids	6.5%	6.6%	0.1%	5.7%	6.5%	1%

The facility-wide dynamic process model was developed using BioWin software (Envirosim, Version 3.1). To support model development, intensive sampling efforts were conducted between June 12 and June 25, 2011. The sampling included parameters that are not typically measured at the facility, including diurnal patterns, influent and effluent COD, TKN, anaerobic and anoxic zone profiles, and sidestream quality. Two outside laboratories were used to assist with sample analyses.

In general, influent COD fractions measured at the BDCWWTF are typical for domestic wastewater. One exception is RBCOD, which is higher than values typically observed. The high values are not surprising, however, because the facility has demonstrated excellent BNR. Diurnal influent flow and concentration patterns are typical for domestic wastewater. Sidestream sampling confirmed that high concentrations of nutrients (123 mg/L total phosphorus and 402 mg/L TKN) are recycled during biosolids thickening (approximately 40 hours per week), and the effect of these sidestreams was detected in anaerobic and anoxic zone profile sampling. Sampling also confirmed that "secondary" phosphorus release (in the anoxic zone) did not occur on either day of sampling, which is important for successful biological phosphorus removal. Dissolved oxygen spot sampling confirmed that the dissolved oxygen setpoints may not always be precisely maintained, which is difficult because levels are adjusted manually.

Default values were used for all model kinetic parameters with the exception of the simultaneous nitrification/denitrification switching function or the heterotrophic dissolved oxygen half-saturation constant. This parameter was increased from its default value of 0.05 to 0.25 mg/L to align airflow and effluent nitrate plus nitrite predictions with facility data. The model was calibrated using June 2011 data and validated using winter (January 2011) data to confirm that the selected kinetic constants were accurate for multiple conditions. Model predictions aligned well with facility data (Table 8.13) and, therefore, it was concluded that the model could be used to assess facility capacity. The biggest challenge with model calibration was simulating shift dewatering because it had a tremendous effect on effluent total phosphorus results, as shown in Figure 8.8. As part of the project, the engineer conducted operator training for both the process and hydraulic models.

7.2 Detailed Design Model Development

The City of St. Cloud, Minnesota, owns and operates the St. Cloud water resource recovery facility and has accomplished biological phosphorus removal through unconventional means for several years by operating the front of the aeration basins at low dissolved oxygen concentrations. City officials hired an engineer to evaluate ways to implement permanent BNR facilities while expanding facility capacity. As part of a 2006 facility planning effort, the city conducted extensive sampling knowing

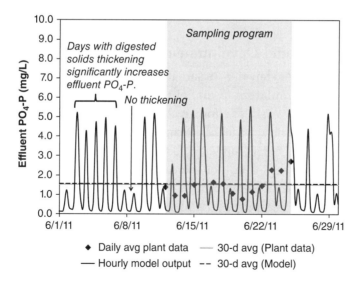

FIGURE 8.8 Effects of digested solids thickening on effluent phosphate (PO_4-P) concentrations.

that the results would be used to model the selected facility design alternative. By conducting sampling early in the planning stages, the city was able to have several years of data to support facility design. Planning/design model development was based on the following analyses:

- Reconciliation of historical influent data. Data collected prior to October 2004 was believed to be inaccurate because the composite sampler collection line was long and likely contained bacteria that affected influent concentrations. After a thorough comparison of available data from other studies and contributions from individual communities, correction factors were developed for the incorrect data.

- Influent characterization, including diurnal profiles and COD fractions.

- Low F/M batch nitrification tests to confirm the absence of inhibitory substances in the influent. Initial tests were conducted by the engineer and the city conducted subsequent tests to confirm results on a seasonal basis.

- Settling tests to quantify Vesilind settling parameters (Wahlberg, 2003). Initial tests were conducted by the engineer and the city conducted subsequent tests to confirm results on a seasonal basis.

- Aeration shop tests were conducted for detailed design.

7.3 Operations Model Development

Since 2003, the Metro Wastewater Reclamation District (Metro District) in Denver, Colorado, has used simulators to model the Robert W. Hite Treatment Facility (RWHTF) to support a variety of efforts including long-term planning, design, and daily operations. The facility model has exchanged hands and evolved over the years and, in 2007, the Metro District implemented an intensive sampling program to calibrate the model for dynamic conditions since a 7-day average discharge limitation of 8.68 mg/L nitrite began being enforced.

To ensure frequent and proper use of the model, the Metro District formed an internal task group, the Model User Group (MUG), which meets monthly to discuss modeling assignments either internally or with hired consultants for specific projects. The Metro District regularly conducts projections modeling to predict future operating conditions and effluent quality. Each year, MUG conducts at least one detailed model validation and issues an annual model report that documents all modeling tasks, sampling results, or other related information. In 2011, the Metro District made

the decision to change simulators and the annual model reports made it easy to recalibrate the facility-wide model. The model layout and selected calibration results are
shown in Figures 8.9 and 8.10, respectively.

In general, regardless of the simulator used, the Metro District noticed the following similar trends in modeling efforts when comparing model output to facility data:

- The default kinetics in the current versions of the simulators (GPS-X 6.0 and
  BioWin 3.1) accurately predicted facility performance.

- Rather than adjusting kinetic parameters or yield factors, influent stoichiometry is adjusted on a monthly basis to account for seasonal primary clarifier performance and differences in the two influent streams. Special sampling data
  cannot be used for every scenario, and the modeler must carefully balance the
  influent stoichiometry for every monthly simulation.

- The models tend to overpredict effluent nitrite concentrations if conservative
  influent peaking factors are selected. Planning-level flow and load projections
  cannot be used to predict near-term effluent quality because they include
  safety factors that are intended for design purposes (i.e., the carbon-to-nitro-
  gen ratio must evaluated on a monthly basis for operations modeling).

- Models tend to underpredict effluent NH_3-N concentrations slightly, but the
  models also assume the same diurnal pattern for every day of the simulation and a constant dissolved oxygen concentration. If increased accuracy is
  desired, then hourly dissolved oxygen data should be added as model input.

- The models are not capable of predicting effluent TSS and CBOD concentrations unless settling parameters are entered seasonally to simulate upset conditions (e.g., *Microthrix parvicella* is common during the winter). Section 5.2
  discusses reasonable expectations for model calibration efforts, including a
  stop criterion of ± 5 mg/L for TSS.

- The models tend to underpredict the MLSS concentration and WAS production
  for the BNR facility even though the SRT often aligns well with facility data.

- The models tend to overpredict high-purity oxygen WAS production.

- There are often discrepancies between thickening and dewatering data vs
  model predictions; however, there is often not closure in the mass balance of
  the data itself. Because there is difficulty in collecting representative sludge
  samples, many of these discrepancies are ignored provided that the overall
  facility-wide mass balance reconciles well.

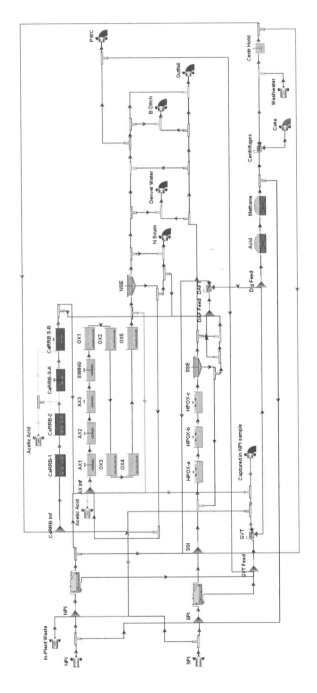

FIGURE 8.9   Second example layout of an operations model.

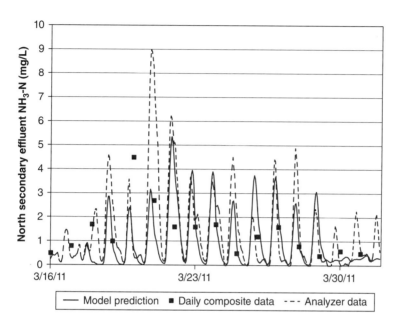

FIGURE 8.10   Operations model calibration results for effluent ammonia-nitrogen (second example).

- The models consistently predict the digester feed total solids mass within 10% or less and, therefore, these data are used as a barometer to gauge the accuracy of the overall solids mass balance.

- The models tend to underpredict digester VSS.

- The models are not consistent at predicting centrate quality; however, water chemistry and struvite formation play an important role in centrate concentrations (in addition to digester VSS).

8.0   REFERENCES

Belia, E.; Amerlinck, Y.; Benedetti, L.; Johnson, B.; Sin, G.; Vanrolleghem, P. A.; Gernaey, K. V.; Gillot, S.; Neumann, M. B.; Rieger, L.; Shaw, A.; Villez, K. (2009) Wastewater Treatment Modeling: Dealing with Uncertainties. *Water Sci. Technol.*, **60** (8), 1929–1941.

Belia, E.; Benedetti, L.; Johnson, B.; Murthy, S.; Vanrolleghem, P. A.; Gernaey, K. V.; Gillot, S.; Neumann, M. B.; Weijers, S.; Copp, J. B. (2009) Wastewater Treatment Modeling: Discussing Uncertainty. *INFLUENTS*, **4**, 81–83.

Hauduc, H.; Rieger, L.; Takács, I.; Héduit, A.; Vanrolleghem, P. A.; Gillot, S. (2010) A Systematic Approach for Model Verification: Application on Seven Published Activated Sludge Models. *Water Sci. Technol.*, **61** (4), 825–839.

Melcer, H.; Dold, P.; Jones, R. M.; Bye, C. M.; Takacs, I.; Stensel, H. D.; Wilson, A. W.; Sun, P.; Bury, S. (2003) *Methods for Wastewater Characterization in Activated Sludge Modeling, Report No. 99-WWF-3*; Water Environment Research Foundation: Alexandria, Virginia.

Metcalf and Eddy, Inc. (2003) *Wastewater Engineering: Treatment and Reuse*, 4th ed.; McGraw-Hill: New York.

Phillips, H. M.; Sahlstedt, K. E.; Frank, K.; Bratby, J.; Brennan, W.; Rogowski, S.; Pier, D.; Anderson, W.; Mulas, M.; Copp, J. B.; Shirodkar, N. (2009) Wastewater Treatment Modeling in Practice: A Collaborative Discussion of the State of the Art. *Water Sci. Technol.*, **59** (4), 695–704.

Rieger, L.; Gillot, S.; Langergraber, G.; Ohtsuki, T.; Shaw, A.; Takács, I.; Winkler, S. (2012) *Guidelines for Using Activated Sludge Models*; IWA Scientific and Technical Report; IWA Publishing: London.

Rieger, L.; Takács, I.; Villez, K.; Siegrist, H.; Lessard, P.; Vanrolleghem, P.A.; Comeau, Y. (2010) Data Reconciliation for Wastewater Treatment Plant Simulation Studies—Planning for High-Quality Data and Typical Sources of Errors. *Water Environ. Res.*, **82**, 426–433.

Water Environment Federation; American Society of Civil Engineers/ Environmental & Water Resources Institute (2009) *Design of Municipal Wastewater Treatment Plants*, 5th ed.; WEF Manual of Practice No. 8; ASCE Manual and Reports on Engineering Practice No. 76; McGraw-Hill: New York.

Chapter 9

Using Models for Design, Optimization, and Control

(continued)

(continued)

1.0 INTRODUCTION

During the definition phase of the project (Chapter 7), project scope and objectives are defined. As discussed in Chapter 8, the model is then developed and calibrated using operating data and any additional sampling data collected. The level of calibration effort will depend on the objectives of the modeling project and available data.

Once a calibrated model is developed, it can be used to assist with a wide range of activities throughout the entire life-cycle of a water resource recovery facility (WRRF), such as

- Treatment process selection
- Facility design and retrofitting
- Sensitivity and uncertainty analysis
- Control system design and tuning
- Operator training
- Facility startup planning
- Optimization
- Troubleshooting and "what if" analyses
- Facility capacity analysis and rerating
- Facility auditing and benchmarking
- Planning and scheduling
- Biosolids management
- Design of experimental programs
- Data and sensor validation

Modeling scenarios can be created within the calibrated model that can be used to study different loading and operational scenarios that relate to the goals defined for the project. For example, when selecting a treatment process for a Greenfield facility, a number of different treatment processes can be evaluated without any concrete being poured. Of course, benefits achieved through modeling are dependent on model setup, the scenarios selected, and the correct interpretation of results.

In this section, use of models for design, optimization, and control are discussed. The format of the steps follow the International Water Association (IWA) Good Management Practice Unified Protocol discussed in Chapter 5.

2.0 MODEL SCENARIOS

The goal of the modeling task is identified early in the project plan (Chapter 7) and the basic model is developed and calibrated so that it can be used to answer key project questions (Chapter 8). A WRRF is a complex system that undergoes changes in loading, operation, controls, and available facilities. The model can be used to determine performance in various process configurations, loading considerations, upset conditions, and operational changes.

Defining modeling scenarios requires consideration of key project objectives, operational conditions, and process configurations that are possible within the context of the project. This section will provide background on whole-facility modeling, steady-state and dynamic modeling, scenario development, sensitivity analysis, handling uncertainty, and benchmarking of modeling results.

Whole-facility modeling refers to modeling of the entire WRRF including biosolids treatment, thickening, dewatering, and the recycling of return liquors to the WRRF headworks. Before detailed anaerobic digestion and sidestream treatment process models were available in process simulators, the return streams from biosolids treatment were typically modeled as separate influent streams with assumed flows and concentrations or were ignored. With the realization that these streams can introduce a significant nutrient loading to the WRRF and with the advancement of WRRF process models, the trend in recent years has been toward modeling the entire facility.

There are two general categories of wastewater treatment simulation: steady state and dynamic. A steady-state model does not include time variant inputs, while a dynamic model does. Most commercial simulators have the capability to do both.

Both steady-state and dynamic simulations have their use in modeling projects. Selecting which method to use depends on specific requirements of the project.

For example, common design practice typically uses the maximum month flows and loads as the design basis for a facility, which would be a steady-state simulation. In contrast, a dynamic simulation would be warranted if the facility under design was required to achieve a daily limit on a nutrient component or if further information was needed on transient behavior in the facility, such as peak aeration demands or slug load effects of a dewatering liquor return. A modeler's decision about which type of simulation is best suited to meet project goals should be given careful consideration. In general, dynamic simulation requires significantly more time and effort to achieve a required level of accuracy than steady-state simulation.

Both steady-state and dynamic simulations require appropriate consideration of model uncertainty. Typically, this involves application of safety factors to model inputs, outputs, or both when using the models for design. Safety factors for steady-state-only simulations are larger than those required for dynamic simulations for a given level of calibration. Therefore, careful consideration must be given to what the appropriate safety factors are in a particular design. A key part of any simulation effort is knowledge and documentation by the user of how these safety factors are applied in the design procedure and their effect on simulation results. The following sections discuss use of simulators and models in both steady-state and dynamic modes.

2.1 Whole-Facility Modeling

Simulators can be used for tasks ranging from unit-process evaluations to whole-facility studies. Models for biological unit processes and clarification/thickening unit operations are typically mechanistic models, while those for dewatering and wastewater solids filtration unit operations are often empirical models based on solids removal efficiency. Simulators typically are not required for most non-biological treatment unit processes, but are useful for evaluating the dynamic behavior of settling tanks. Flow equalization units, although not biological treatment units, can have a significant effect on process performance by storing wet weather flow, leveling diurnal flow and load variations, and storing dewatering sidestream recycle flows for return during low-load periods. Most wastewater treatment modeling tasks involving nitrogen or phosphorus should consider the need to model the entire facility, especially when the facility uses anaerobic digestion and has dewatering recycle sidestreams. This approach captures the effects of recycling solids processing liquors within the system. The need to use a whole-facility model must be carefully balanced against project needs because whole-facility models are more complex and require additional time to solve.

2.2 Steady-State Modeling

Steady-state simulation is typically used to analyze averaged process performance over an extended time period (e.g., three solids retention times [SRTs]) over which the facility can be considered to be operating under steady conditions. For design, looking at averaged long-term performance is appropriate when developing the basic process configuration and tank sizes. Dynamic simulations can be used to further refine design, make the most of operating conditions, and ensure that the facility can handle anticipated peak loading conditions. One advantage of steady-state simulation analysis is its speed in both model solutions and in the time required by the user to produce needed results. The time required to achieve a model solution is directly proportional to the complexity of the model implemented in the simulator and the method the simulator uses to achieve the steady-state solution. There are two basic approaches to determining a steady-state solution. The first executes the dynamic solution routine until a steady state is achieved (based on a defined tolerance on the derivative terms in the dynamic mass balances). Unless an efficient integration algorithm (designed for stiff systems of equations) is used, this method is typically slower than the method of directly solving for the steady-state solution.

2.2.1 Steady-State Data Requirements

Data required to set up and run a steady-state simulation are identical to a dynamic model, except that the time-variant nature of the data is not required. This simplifies data needs in most instances, but adds an additional need to develop the data set to be used for design that is consistent across the facility. Daily data from the facility must be averaged over the desired period. Desired periods typically are

- Average annual—the 365-day average value for various facility parameters.
- Average monthly—the 30-day average value for the various facility parameters.
- Shorter periods—careful consideration must be given by the modeler with the use of shorter periods of data. As an example, maximum weekly data are sometimes used. However, it is likely that the facility is not at steady state for such a short period. In general, a facility is at steady state over periods of approximately three SRTs. In most facilities, the SRT of the bioreactor system is between 5 and 15 days, so a reliable steady-state period might range between 15 and 45 days. Shorter averaging periods can be used if desired, but the modeler needs to be aware that this is not a true steady-state application. Results from such runs can be either higher or lower than the

simulator indicates. Dynamic simulation may be required if more definition is required.

2.2.2   Steady-State Simulation Limitations

As previously discussed, steady-state simulation is not valid when data from short-term periods are used. Water resource recovery facilities are inherently dynamic; however, depending on project needs, much of the dynamic behavior can be ignored in design. For those areas that are not captured in steady-state modeling, there are two basic approaches to design: use of peaking factors or multipliers built into the design procedure or use of dynamic modeling.

Dynamic modeling can provide additional information that can be beneficially used for design. The following are three significant areas where steady-state models are not appropriate and dynamic modeling should be considered instead:

- Design of aeration and blower systems—the appropriate size of aeration piping and blowers requires that a maximum airflow be defined. Dynamic modeling of maximum loads to the bioreactor system can be used to determine expected maximum air demands. It is important to note that the dynamic approach does not eliminate the use of safety factors (either on the air rate or on expected loads) because there typically is significant uncertainty in model parameters.

- Secondary clarifier design—secondary clarifiers are inherently dynamic processes and should be designed with consideration given to maximum flows and expected return activated (RAS) sludge rates. Design tools for this process may be as simple as one-dimensional flux models or as complicated as three-dimensional computational fluid dynamic models. Most simulators offer a range of secondary clarifier modeling options, which are selected by the modeler according to the needs of the project. In steady-state designs, the dynamic aspect is often captured in the use of large safety factors defined by separate clarifier modeling tools. Dynamic simulators capture these effects as part of the simulation.

- Design for short-term effluent limits—short-term effluent limits, such as maximum daily or weekly limits, cannot be accurately designed with steady-state simulations. Therefore, dynamic simulation is called for in these situations.

- Design of noncontinuous processes—processes that do not operate continuously require dynamic modeling. This includes operation of batch processes such as a sequencing batch reactor (SBR) or scheduled thickening and dewatering processes where the return streams are not equalized.

2.3 Dynamic Modeling

Wastewater treatment processes are inherently dynamic and experience significant variations in flows and concentrations on an hourly, daily, weekly, and seasonal basis. To truly model the process performance and observe interactions between processes, it is necessary to run a dynamic simulation. Results from a dynamic simulation can be different from those obtained from a steady-state simulation, particularly for nutrient removal systems. This is because biological nutrient removal (BNR) processes behave differently when they are subject to substrate and nutrient concentrations that repeatedly vary compared to behavior when they are subject to stable "average" conditions. Some processes, such as SBRs and biological activated filters (BAFs), which change their mode of operation with time, cannot be modeled using steady-state simulations and must be modeled dynamically if short-term control/performance information is needed.

Before running a dynamic simulation, it is useful to first run a steady-state simulation to obtain the approximate model variable values. Operating conditions used to run the initial steady-state simulation should be representative of the sludge inventory in the facility during an extended period preceding dynamic simulation (e.g., a period of 3 times the SRT that precedes the dynamic period being simulated). Alternatively, the model must be run for an extended period of time to reach a pseudo steady state (typically 3 times the SRT for suspended growth processes) until model outputs have a repeatable pattern that matches measured or expected response in the real facility.

For time-varying processes such as SBRs, it is advisable to run the model dynamically until a pseudo steady state is reached before running any investigative dynamic simulations. Simulation outputs typically are graphed, and averages, maximums, and minimums are obtained from the graphs. An infinite number of dynamic simulations can be run; some common simulations that are useful for design are described in the following subsections.

2.3.1 Diurnal Modeling

One of the simplest dynamic simulations is to run the model over several days with a repeating diurnal flow and concentration pattern to see how the system responds to a typical diurnal variation. An example of this is shown in Figure 9.1 (input flows) and Figure 9.2 (output effluent ammonia and nitrate). Figure 9.3 shows dynamic output for a 1-week simulation showing the effect of dewatering during weekdays only.

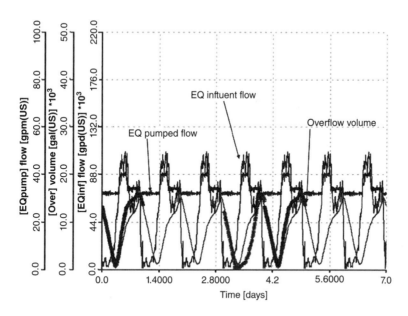

FIGURE 9.1 Dynamic input (flows) for process model showing the same repeated influent diurnal pattern over 7 days (EQ = equalization) (gal × 3.785 = L; gpd × 0.004 = m³/d; and gpm × 3.785 = L/min).

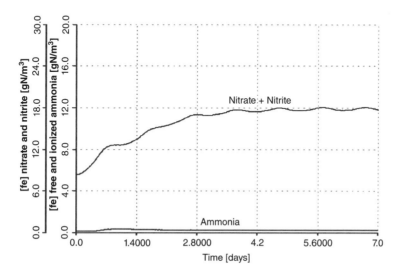

FIGURE 9.2 Dynamic output showing model effluent ammonia and nitrate gradually reaching a pseudo-steady-state condition.

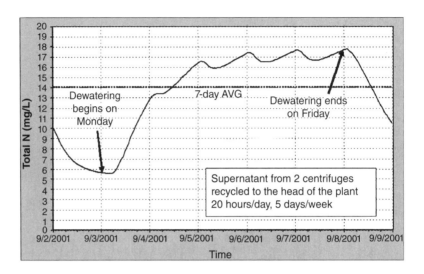

FIGURE 9.3 Dynamic output for l-week simulation showing effect of dewatering during weekdays only.

2.3.2 Weekly, Monthly, and Seasonal Variations

An extension of diurnal modeling is to model a weekly pattern or to extend this to monthly or even seasonal variations. This can provide a better picture of the effect of different flows and loads or operating practices on weekends. An example of this is shown in Figure 9.4, which illustrates the 7-day output from a model that had different sludge liquor returns on the weekend than during the week; these had a direct effect on the effluent total nitrogen concentration. This model was subsequently used to demonstrate the benefit of running sludge dewatering operations 7 days a week.

2.3.3 Maximum Month/"Birthday Cake" Analysis

A useful dynamic analysis for design is to run a simulation over several weeks and include the design loadings that are typically used to size equipment, including annual average, maximum monthly average, and peak day events. Ideally, long-term simulations would use actual facility data to describe the variation in influent characteristics. If this data are unavailable, an artificial loading pattern can be used that combines the annual average, maximum monthly average, and peak-day loading. The resulting flow input to the model resembles a birthday cake. From this simulation, average and peak load conditions can be modeled, and the process response can be examined. Figures 9.4, 9.5, and 9.6

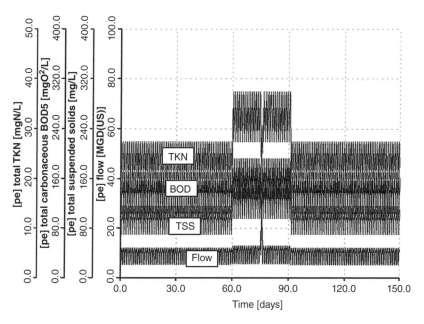

FIGURE 9.4 Example "birthday cake" input (mgd × 3.785 = ML/d).

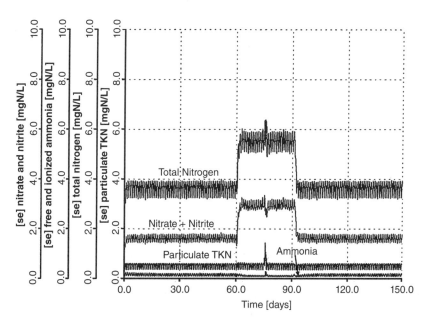

FIGURE 9.5 Modeled effluent nitrogen species output response to "birthday cake" input shown in Figure 9.4.

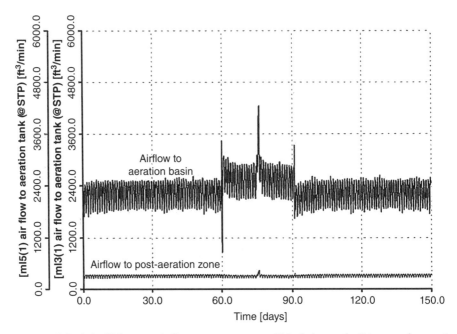

FIGURE 9.6 Modeled blower airflow responses to "birthday cake" input shown in Figure 9.4 (cu ft/min × 0.471 9 = L/s).

show typical birthday cake analysis inputs from a model. From these types of figures, the expected average and peak performance can be deduced.

2.3.4 Dynamic Responses to Storms and Other Events

In addition to examining typical flow and load patterns throughout a facility, simulations can be run to observe how the model responds to storms or other unusual events. The simulation typically should be run for several days before the event and several days after the event to observe how long it takes for the outputs to return to "normal." Figures 9.7 and 9.8 show the dynamic input and response of a model that was subjected to 3 days of abnormal "special event" flows.

As allowances for frequency and volume of combined sewer overflows and sanitary sewer overflows are reduced or eliminated by regulatory agencies, these wet weather flows are routed to WRRFs. A common solution to modeling wet weather flows is to have a separate stormwater input, as shown in Figure 9.9.

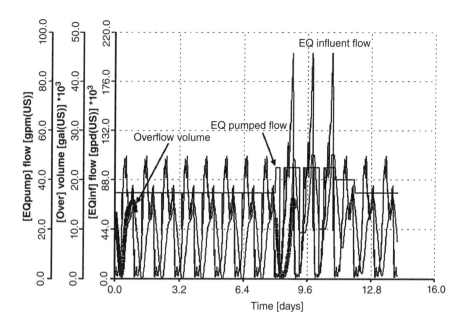

FIGURE 9.7 Simulation of a 3-day special "event" after 7 days (EQ = equalization) (gal × 3.785 = L; gpd × 0.004 = m³/d; and gpm × 3.785 = L/min).

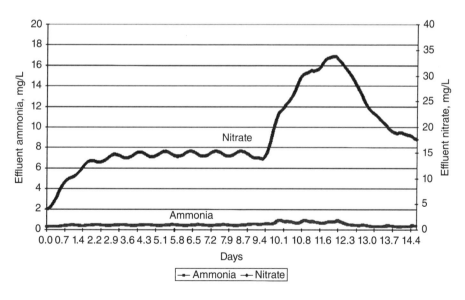

FIGURE 9.8 Graph showing dynamic response of model effluent ammonia and nitrate to the 3-day special event shown in Figure 9.7.

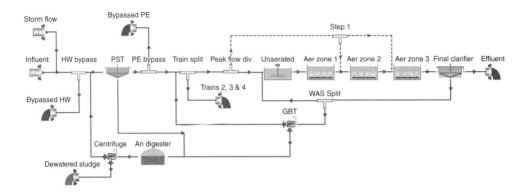

FIGURE 9.9　Example wet weather peak flow diversion configuration.

2.4　Scenario Development

Once a model has been developed and calibrated, there are many ways to use the model to answer "what if" questions. The modeler may choose to investigate several scenarios or model conditions to make the most of operations at an existing facility or to make the most of design of a new facility. Common scenarios used for design and operations are summarized in Table 9.1.

Evaluating different influent loading conditions is a relatively straightforward process; however, investigating some of the other scenarios listed in Table 9.1 can be time consuming without an organized plan. Therefore, it is best to select design flows, loads, SRT, and approximate volumes before evaluating the effects of minor changes to recycle rates and other operating parameters.

2.5　Sensitivity Analysis

Sensitivity analysis involves determining how sensitive the model outputs are to changes in the model inputs, which can include model parameters and initial conditions for a given scenario. Sensitivity analysis can be used to better understand system behavior, to determine the identifiability of model parameters, and to quantify uncertainty in modeling results. Sensitivity studies may include scenario analysis, local sensitivity analysis, and global sensitivity analysis.

Scenario analysis (i.e., scenario-based sensitivity analysis) is used to help understand the behavior of the model under different operating conditions. This may involve specific "what if" studies, manually running a series of simulations over a

TABLE 9.1 Examples of modeling scenarios.

Influent loading conditions	Effect on design
Design flow and loads, winter temperature	To size the aeration basins and clarifiers Determine effect on biological process at coldest temperature
Design flow and loads, summer temperature	To determine maximum oxygen demand at lowest oxygen transfer efficiency Determine effect on biological process at warmest temperature
Average flow and loads, average temperature	To determine average operations and maintenance costs
Minimum flow and loads, minimum	To determine minimum mixed liquor, aeration demand, pump turndown requirements, and options for taking basins out of service
Seasonal loading conditions	To determine the effect of seasonal loading from industrial sources
Operating conditions	
Basin out of service (maintenance or redundancy)	To determine capacity effect To determine effects on effluent quality To determine effect on aeration system (diffused air)
Higher (or lower) SRT	To determine effects on effluent nutrient concentration and capacity
Higher (or lower) RAS flowrates	To determine effects on secondary clarifier solids loading and effects to biological process performance
Higher (or lower) mixed liquor recycle rates	To determine effects on effluent quality
Digested sludge dewatering recycle return	To determine effects on effluent quality
Facility sizing scenarios	
Smaller (or larger) anoxic zones	To determine effects on effluent nitrates or the potential for swing zone operation
Smaller (or larger) anaerobic zones	To determine effects on effluent phosphorus or the potential for swing zone operation
More (or less) chemical addition	To make the most of performance, size pumps, and potential storage requirements

range of parameters, or having the simulator automatically run a set of predefined simulations. These types of simulations are useful for assessing the relative importance of design criteria, loading conditions, operating conditions, facility capacity studies, development of operating strategies, and selection of sampling locations for model calibration because they will highlight the most sensitive locations.

To conduct a scenario-based sensitivity analysis, the modeler should change one model input (e.g., SRT) and observe changes in model output (e.g., effluent ammonia). The process is repeated for several different values of the input until a curve can be generated that shows the sensitivity of model output to changes in model input. It is important to vary only one parameter at a time to accurately show cause and effect. For example, by using a sensitivity analysis to compare effluent ammonia with SRT, the engineer can quantify the risk of nitrifier washout and select a design SRT with an appropriate safety factor, as shown in Figure 9.10. This figure was generated with a dynamic model, and the daily average, minimum, and maximum ammonia concentrations are shown as functions of SRT.

An important application of sensitivity analysis is to study the effect of important kinetic parameters on simulation results. For example, if the nitrifier growth rate was not experimentally measured at a particular WRRF and an upgrade to the facility was being designed to handle an increased influent ammonia loading, it would be important to study the effect of this parameter on the proposed design.

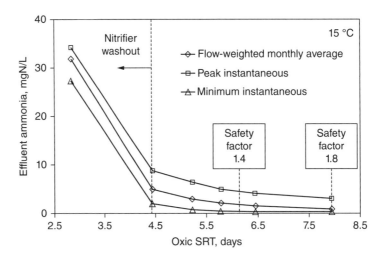

FIGURE 9.10   Sensitivity analysis of effluent ammonia versus solids retention time.

The aforementioned analysis of effluent ammonia vs SRT could be repeated over a range of nitrifier growth rates at the minimum expected system temperature to generate a series of effluent ammonia curves on the same plot. The range of the effluent ammonia curves at a given SRT shows the potential uncertainty in effluent ammonia at the selected loading conditions.

Another example analysis is to run a series of dynamic simulations with projected daily and diurnal variations over an entire year at a series of different nitrifier growth rates. This procedure could be repeated at a number of different target SRTs. Plots could then be generated at each SRT showing the dynamic effluent ammonia curves for different nitrifier growth rates. This analysis would help demonstrate how the upgraded facility would perform under dynamic loading conditions given potential uncertainty in the nitrifier growth rate, thereby helping to select appropriate SRT and tank sizes.

Sensitivity analyses can also be used for process optimization. Most activated sludge models have the ability to track residual dissolved oxygen throughout the process, which is essential for BNR facilities. Whether the engineer is selecting the size of mixed liquor recycle pumps or an operator is setting the pump speed, sensitivity analyses can be useful in predicting the points of diminishing return when recycled dissolved oxygen hinders denitrification, as shown in Figure 9.11. Local and global sensitivity analysis methods are discussed in detail in Chapter 5.

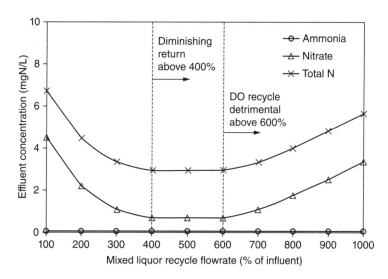

Figure 9.11   Sensitivity analysis of effluent ammonia versus mixed liquor recycle rate (DO = dissolved oxygen).

2.6   Dealing with Uncertainty

2.6.1   *Scenario/Sensitivity Analysis of Design*

Table 9.2 summarizes the types of parameters available in WRRF models and provides examples of which parameters might be used in a sensitivity analysis. Physical parameters for study include reactor configuration (e.g., number of tanks in series,

TABLE 9.2   Parameters for sensitivity analysis in facility models (adapted in part from Melcer et al. [2003]).

Parameter type	Examples
Physical	• Reactor configuration (e.g., number of tanks in series, effect of trains out of service, use of swing zones, and anaerobic, anoxic, aerobic volumes/mass fractions) • Wastewater temperature
Influent	• Flow • Concentrations • Diurnal pattern • COD, nitrogen, and phosphorus fractionation among model state variables (e.g., readily biodegradable fraction of COD, fraction of un-biodegradable particulate COD)
Operational	• Recycle flows • Wastage flow • Chemical dosage • Pumping schedules • Airflows • Aeration system constraints • Oxygen transfer efficiency • Aeration alpha factor • Controller setpoints (e.g., dissolved oxygen) • SRT, MLSS
Biological	• Maximum specific growth rates (e.g., nitrifier growth rate, anoxic hydrolysis rate factor, polyphosphate-accumulating organism maximum specific growth rate) • Half-saturation coefficients (e.g., half-saturation coefficient for poly-hydroxy-alkanoate) • Stoichiometric (e.g., nitrogen and phosphorus fractions in biomass)
Settling	• Vesilind and double-exponential parameters
Initial conditions	• Initial biomass concentrations

tanks out of service for maintenance, or use of swing zones), anaerobic/anoxic/aerobic mass fractions, and wastewater temperature. These parameters would typically be studied during design-based analyses.

Influent variables for study can include influent flow, diurnal patterns, concentrations, and composition (i.e., chemical oxygen demand [COD], nitrogen, and phosphorus fractionation among state variables). These parameters all relate to facility loading and are important when studying facility design, performance, and capacity.

Specification of operational variables and strategies is an important aspect of troubleshooting and optimization studies. A sensitivity analysis can help identify the most important optimization variables. Potential parameters for analysis include recycle flows, wastage flows, airflows, chemical dosage, and controller setpoints (e.g., for dissolved oxygen, ammonia, SRT, and mixed liquor suspended solids).

As mentioned in Section 2.5 of this chapter, sensitivity of kinetic and stoichiometric parameters in activated sludge models has been well studied. In most instances, the majority of these parameters are left at published default values during modeling studies because of the difficulty in identifying them using typical facility data. Certain parameters may be site-specific and their effect can be studied using a sensitivity analysis. For example, Melcer et al. (2003) suggested that the following parameters may require adjustment during calibration: nitrifier growth rate, anoxic hydrolysis rate factor, polyphosphate accumulating organism maximum specific growth rate, and the half-saturation coefficient for poly-β-hydroxyalkanoate. These parameters would be good candidates for further study using sensitivity analysis. Other parameters that are candidates for sensitivity analysis include the nitrogen and phosphorus fractions in biomass (which may need to be adjusted for industrial WRRFs) and settling model parameters (Vesilind and double-exponential parameters).

2.6.2 Probabilistic Modeling

Monte Carlo simulations can be used to quantify uncertainty in WRRF design. The basic Monte Carlo method was described in Section 2.5, but some additional background is provided here. Monte Carlo methods use statistical sampling techniques to obtain a probabilistic approximation to the solution of a mathematical model or problem. One of the earliest documented applications of the basic Monte Carlo method, also known as *statistical sampling*, was in determination of the value of π (Hall, 1873). The term *Monte Carlo* was coined in the 1940s by researchers working on nuclear fusion at the Los Alamos National Laboratory, Los Alamos, New Mexico (Metropolis and Ulam, 1949).

Monte Carlo methods allow for the solution of the integral of the product of two variables. Many problems can be formulated in this context such as finding the mean of a stochastic variable, which is defined as the integral of the variable multiplied by its probability density function. Monte Carlo methods are often used to evaluate difficult multidimensional integrals with complicated boundary conditions.

In the context of wastewater treatment, Monte Carlo methods can be used to determine the mean and variance of the probability distribution(s) of the output(s) of a WRRF simulation. Monte Carlo simulation involves iteratively solving a deterministic model using sets of random numbers (generated from probability distribution functions) as inputs. As described in Section 2.5, a series of stochastic simulations are run using random or pseudo-random values of the WRRF model input variables and parameters to generate probability distributions of model output variables. These probability distributions can be analyzed to determine the probability of effluent variables exceeding effluent limits.

The basic Monte Carlo method can require a large number of samples to converge. The uncertainty in Monte Carlo simulations is proportional to $1/\sqrt{n}$ (Eckhardt, 1987), where n is the number of samples. This means that every decimal point of extra accuracy requires 100 times the number of samples. As a result, Monte Carlo simulations could require hundreds or thousands of simulations to converge, depending on the required accuracy.

To reduce the number of simulations that must be run, methods have been developed to more efficiently generate the sets of random numbers required as model inputs. These include Markov Chain Monte Carlo (Metropolis et al., 1953), stratified sampling methods such as Latin Hypercube Sampling (LHS), and Quasi-Monte Carlo (see Torvi and Hertzberg [1998]). In LHS, the range of the input variable distribution is divided into subintervals with equal probability. One value is selected from each subinterval (either a random value of the median) and this process is repeated for all input variables. The generated input variable values are then paired randomly to a sequence to generate a sequence of input samples for use in Monte Carlo simulations. Quasi-Monte Carlo methods construct deterministic sequences such as the Halton, Sobol, or Hammersley sequences that share properties of random or pseudo-random sequences. These methods are found to have less error than random Monte Carlo methods and require fewer samples to converge, although the advantage may be slight in large problems (Morokoff and Caflisch, 1995).

Probability density functions used for model input variables and parameters depend on available data. In instances where data are available, distribution of data

can be determined using statistical techniques. For variables for which little information is known except for expected minimum and maximum values, a uniform distribution is often used. A triangular distribution is used if a most likely value and
minimum and maximum values are known.

Inputs to a simulation model may be correlated in some way (e.g., influent biochemical oxygen demand [BOD] will be larger when the influent total suspended
solids [TSS] is larger), and this should be accounted for when running Monte Carlo
simulations. One approach is to develop a multivariable distribution for model
inputs that accounts for correlation between the variables and then select random
points from this distribution as inputs into the Monte Carlo simulations. This can
be achieved by decomposing the correlation matrix of the distribution using the
Cholesky decomposition. The resulting lower triangular matrix is then multiplied by
a vector of uncorrelated random values of input variables to create a correlated input
vector for the simulation problem. In this manner, Monte Carlo analysis is not biased
by unreasonable combinations of variable values (e.g., high BOD and low TSS and
total Kjeldahl nitrogen [TKN]).

An example of a Monte Carlo analysis is shown in Figure 9.12. A conventional
plug flow activated sludge facility with primary clarification was studied using
Monte Carlo analysis, with a WRRF simulator with nitrifier growth rate, wastewater temperature, influent flow, and influent TKN used as stochastic input variables.
Normal distributions were used for all variables except liquid temperature, and 1000
steady-state simulations were run with random samples taken from the distributions.
The nitrifier growth rate distribution was centered at 0.7 d^{-1} (to be conservative) and
the liquid temperature was allowed to be as low as 8 °C. The reader is referred to
Benedetti et al. (2011) for more information on using Monte Carlo methods to study
uncertainty in the context of WRRF modeling.

2.7 Benchmarking Results

Models of existing facilities can be benchmarked against typical facility performance for similar facilities within the same region. The facility data (and model)
can also be checked for consistency using basic data checks such as checking the
influent flow expected based on population, influent concentration ratios (e.g.,
BOD/COD, volatile suspended solids/TSS, TKN/COD, and total phosphorus/
COD), facility sludge production (based on flow, per capita, or BOD removed),
clarifier solids removal efficiency, clarifier solids mass balances, total phosphorus
mass balances, oxygen consumption, digester volatile solids destruction, digester

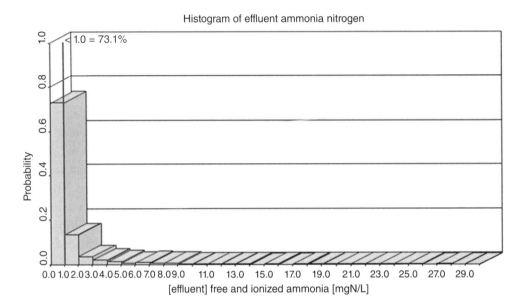

FIGURE 9.12   Histogram of effluent ammonia nitrogen concentrations produced using a Monte Carlo analysis using nitrifier growth rate, wastewater temperature, influent flow, and influent TKN as inputs.

gas production (per capita, facility flow, or theoretical sludge production), and the hauled sludge concentrations. Models used for design can be benchmarked against performance for similar process configurations published within design manuals and for similar operating facilities in addition to the aforementioned basic data/ model checks.

3.0   APPLICATION OF MODELS

Chapter 8 discussed model setup and calibration. As the model is applied to design and optimization, each unit process should be evaluated to determine the validity of the model setup under a new set of conditions. In Section 2.4, development of scenarios was discussed to evaluate various conditions. In this section, the model established in Chapter 8 is used to simulate the scenarios previously defined. During evaluation of alternative process scenarios, simulations should strive to conserve consistency in unit processes.

3.1 Preliminary Treatment

In most modeling practices, preliminary treatment is not included in the overall facility model. Solids that are removed in the grit basin and screens typically are not included in any sampling used to characterize influent waste. The washing and compacting steps return most of the organic waste to the influent stream. The decision to include preliminary treatment steps should be based on the various scenarios considered in the process. Conditions in which preliminary treatment might be included are

- Evaluation of scenarios where preliminary treatment currently does not exist at the treatment system. Initial calibration of the model may have included a greater mass of inert solids in the influent waste stream.

- Known effects from the preliminary treatment process such as dissolved oxygen effects from aerated grit removal facilities.

3.2 Primary Clarification

In full-facility models, the primary clarification process is modeled on the basis of the percentage of TSS removal. Actual performance of the primary clarifiers may change based on changes in loading, wastewater characteristics, chemical addition, or co-thickening. The scenarios considered under this arrangement cannot be field-measured unless a pilot test is conducted. Particular care must be taken for configuration that includes more sophisticated unit modeling than solids removal.

3.3 Activated Sludge Systems

The majority of models available are centered on the operation and design of an activated sludge system. The various scenarios allow the designer to determine the comparative performance of process configurations, effect from changes to zone sizing, and variations in operation options. In general, the process model will be used to

- Develop sizing of anaerobic, anoxic, and aerated zones.

- Predict solids generated for solids system sizing.

- Characterize aeration requirements.

- Evaluate alternate process configuration (e.g., step feed/Johannesburg/3-stage Phoredox) and swing zone use (i.e., to switch functions of zones between anaerobic, anoxic, and aerobic for seasonal variation).

- Determine effects from operation with a unit out of service.
- Determine effects of changes to RAS and mixed liquor recycle flowrates.

The validity of results from scenario evaluation is dependent on the accuracy of the data that are available and the fit of the data during initial setup and calibration of the model. The models can only be as accurate as the conditions defined in each unit process. Several areas that should be considered during design or can be evaluated as part of troubleshooting are

- Actual flow splitting
- Areas of possible air entrainment
- Back-mixing potential
- Requirements for air and mixing throughout the aerated zones
- Diffuser grid airflow control

3.4 Fixed-Film and Hybrid Processes

Fixed-film processes can be modeled using commercially available models as described in previous chapters. For optimization, similar parameters described under the calibration scenarios should be observed to maintain that model predictions under various scenarios are equivalent between options.

In general, equipment suppliers should be consulted concerning operating characteristics to determine if the predicted results match installation experience. Scenarios to evaluate during design or optimization of a hybrid system are media fill fraction and location of media in the treatment process. Scenarios to evaluate during design or optimization of a trickling filter system are media height, filters in series operation, and recycle ratios.

3.4.1 Predictive Ability of Current Models

While fixed-film treatment is not new to the industry, full-scale examples of integrated fixed-film activated sludge (IFAS) and moving bed bioreactor (MBBR) installations are limited. Parameters such as film thickness and detachment rates are generally empirically derived. The predictive ability of the models is directly tied to the quantity of full-scale operational data or pilot testing results where the biofilm thickness, detachment rates, and distribution of biomass population can be assessed.

3.4.2   *Approach to Process Sizing/Evaluation Using Fixed-Film Modules*

Using a fixed-film process requires the modeler to consider metrics such as mass of biomass per surface area to help assess the validity of model results. The approach to modeling has typically been to define maximum biofilm thickness and detachment rates to match the mass loading on the media to reported parameters. Modelers should note the purpose of the unit process as different mass loads are expected on media placed in different zones of the process.

3.5   Solids Handling and Recycle Stream Simulation

Recycle streams from solids processing can have a big effect on the overall treatment process. Simulating these streams provides loading characteristics to the rest of the process and allows the engineer to evaluate potential sidestream treatment options.

   The solids processing simulation is dependent on liquid process stream characteristics. While developing scenarios, it is important to balance the conservatism in sizing needed for the solids system with actual return streams. The following actions should be taken for each simulation:

- Review solids concentrations after each unit process.
- Watch solids processing flow and mass load splits under the various scenarios.
- Consider batch operation of certain components.
- Consider feed tanks and equalization tanks.

3.5.1   *Anaerobic Digestion*

Anaerobic digesters seek stabilization of sludge through anaerobic processes. When modeling anaerobic digesters, a separate fundamental model is used (often, IWA's Anaerobic Digestion Model No. 1 [ADM1]). The integration point between anaerobic digestion models and activated sludge models (ASMs) should be evaluated as part of the overall model. Like the ASM model, completely mixed tanks are typically assumed. Modelers should evaluate the actual mixing and effective digester volume. Geometry or incomplete mixing can result in inert debris accumulation (e.g., grit and struvite) in the corners of the tank, which, in turn, reduces the effective digester volume.

3.5.2   *Aerobic Digestion*

Aerobic digestion seeks stabilization of the sludge through aerobic processes. When modeling an aerobic digestion process, it is important to assess potential unaerated operation of the digester when decanting from the digester back to the head of the facility.

During the unaerated time, the contents will shift to either anoxic or anaerobic processes depending on the presence of nitrate and nitrite.

3.5.3   *Sludge Pretreatment*

In solids processes, there typically are holding tanks that allow hydraulic buffering to downstream processes. These tanks should be included in the model, both to determine dynamic interactions when solids processing is not constant and for biological effects within the holding tanks. Combining streams such as RAS and primary sludge in an unaerated environment could cause release of phosphorus from a biological phosphorus removal facility. This additional phosphorus load will be sent back to the activated sludge process.

3.6   Sidestream Treatment and Bioaugmentation

Return streams from the solids process can present a large mass loading on the liquid processing stream. Although models are typically set up for continuous operation, sidestream treatment should consider use of different holding tank steps that are used for equalization, combining and mixing streams, or for hold time. Dynamic modeling is required for these facilities to determine the effect on noncontinuous dewatering practices.

Sidestreams also present an opportunity for sidestream treatment to either reduce the loads returning to the liquid stream process or to grow biomass that will augment the main liquid stream biomass. There are multiple sidestream processes that have been developed. The modeler should consider the environment under which the sidestream treatment occurs. In bioaugmentation, additional biomass could be generated under higher temperatures than the main liquid stream process. The kinetics that result from adding this "seed" biomass should be considered when modeling. The seed biomass population that grows at higher temperatures will not be fully acclimated to the temperature of the main liquid treatment train. A large shift in temperature can reduce the biological activity of the seed biomass by a greater amount than suggested by conventional Arrhenius coefficients. This is a structural uncertainty in most commercially available models.

3.7   Tertiary Treatment

Tertiary treatment processes consist of advanced treatment processes that can include coagulation, flocculation, clarification, and filtration units. As part of a whole-facility model, these unit processes are often approximated strictly as a solids separation device.

Backwashing and counter-current continuous backwash systems may not be accurately represented in standard solids separation modules. The designer must determine which module or combination of modules best describe the treatment step being modeled.

3.8 Incorporating Regulatory Requirements

Different regulatory agencies have developed design criteria for wastewater treatment. It is the both the utility's and modeler's responsibility to meet with regulatory agencies prior to beginning the design process so the agency's specific requirements can be included and documented in required reports. Meeting these criteria is often required as part of a regulatory agency review of the project. In addition to identifying redundancy requirements, process design criteria can be dictated. These items should be tracked during process modeling to determine if the proposed process configuration meets state requirements. Often, when deviations occur, documenting the differences with an explanation will justify that the deviation does not put the treatment process at risk, but, rather, is the result of detailed design vs the "rules of thumb" used in typical state criteria.

As an example, *Recommended Standards for Wastewater Facilities* (Great Lakes–Upper Mississippi River Board, 2004) provides requirements for

- Total suspended solids and BOD contributions for new growth
- Surface overflow rate for primary and secondary clarifiers
- Solids loading rate for secondary clarifiers
- Volatile solids loading rate for anaerobic digesters for volatile solids loading rate and temperature
- Air requirements for mixing in aerobic digesters as airflow rate per volume of digester
- Maximum solids concentration for aerobic digestion
- Minimum solids production based on population equivalent
- Organic loading rate for activated sludge systems
- Oxygen transfer requirements based on BOD and TKN
- Air volume requirements based on BOD mass loading

In preparation for a modeling exercise, local regulations should be reviewed, and any parameter that is affected by process design should be tracked.

4.0 CONTROL STRATEGY SETUP

There are many control systems in WRRFs. Simple control systems that are typically included in many models include flow-controlled RAS and waste-activated sludge rates and set-dissolved oxygen concentrations in aeration basins. With the increasing sophistication of today's models, accompanied by the improving reliability and robustness of online analyzers and probes, many modelers are now looking for opportunities to implement real-time control opportunities. The majority of these control opportunities will be purposed to meet stringent effluent criteria or energy reduction.

4.1 Controller Functions Available in Current Model Packages

Available commercial models have varying degrees of controller functions built into the program. These functions are updated as new versions of the software are released. The controller functions can be as simple as varying wasting rates to maintain an SRT within an activated sludge system to modeling the aeration valve response to dissolved oxygen measurements in an aeration basin. When using any controller, it is important to understand the limitations of the controller as it relates to physical system response. For example, in aeration control based on dissolved oxygen, the response time of measured dissolved oxygen in the mixed liquor vs a change in aeration valve position needs to be considered. Further depth includes such considerations as valve travel time and positioned accuracy, which often fall outside of the ability of commercial simulators.

4.2 Locating Analytical Instruments

The critical element of a control strategy is measurement of a particular parameter. This measurement can be used for feed-back or feed-forward control of a control device in the treatment process. The location of the instrument will affect the accuracy and reliability of the control strategy. Simulations can be used to help determine the instrument location that provides the best control feedback to the process, but care must be taken to consider the physical conditions around the probe in addition to the simulation results.

4.3 Airflow Control Rates

In the activated sludge process, aeration of mixed liquor is typically the highest energy demand. Controlling air is also critical to the operation of the biological process. Excess aeration or inadequate oxygen transfer can affect the effluent quality and the operational costs of a facility. Process simulation can be used to provide valuable information about the aeration system.

4.3.1   Dissolved Oxygen Control

The most common method to control aeration is based on dissolved oxygen measurement in the basin. A dissolved oxygen concentration setpoint is used to vary the flowrate to achieve desired dissolved oxygen within the basin. As the facility gets more complex, the simple assumption that dissolved oxygen will be maintained at one point will challenged. Factors such as minimum mixing air requirements, minimum airflow per diffuser, inadequate number of dissolved oxygen measurements or air control valve, and blower turndown can all affect the actual performance of the aeration system. In instances where the effect may be large, models can be customized to reflect these variations in control.

4.3.2   Ammonia Control

A second method of aeration control is to vary the oxygen transfer to maintain a desired effluent ammonia concentration. This is typically achieved using a control algorithm that varies the dissolved oxygen setpoint based on the effluent ammonia concentration. This becomes an additional layer of the control loop for the aeration system. Using this control will reduce the overall aeration required to produce the quality of effluent desired. When using this control, the modeler should consider the effect of low dissolved oxygen concentration on mixing or, potentially, phosphorus release in facilities implementing biological phosphorus removal.

4.4   Supplemental Carbon Feed

External carbon feed can be used for a variety of reasons in several places in the treatment process. A readily biodegradable carbon source is required for biological phosphorus removal and/or denitrification. Various carbon sources such as waste sugars, methanol, manufactured carbon sources, glycerin, acetic acid, or molasses can be used for this purpose. The different carbon sources have different efficiencies in the treatment process, but can be applied either within the activated sludge process or to a subsequent fixed-film denitrification process. The carbon source selected needs to be modeled according to its effect on biomass population. Certain carbon sources such as methanol require a specialized bacteria for denitrification that have different growth parameters than the general biomass population. These bacteria and the carbon source need to be modeled separately from the general biomass and typical COD.

Other carbon sources can be used by a large cross section of bacteria. The response time for biomass growth for a denitrifying population grown using methanol will be different than one that has received sodium acetate. When an external feed of carbon is considered in process modeling, sensitivity to overdosing chemical should be considered.

4.5 Sidestream Return Control

Recycle streams generated from solids processing can produce a significant mass loading to the liquid treatment process. At facilities where thickening or dewatering activities are not continuous, the effect from intermittent loading needs to be considered, particularly if no equalization of the return streams is provided. At facilities where solids processing is continuous, the recycle load may be continuous; however, options to meter the recycle streams back to the main process during low loading periods should be considered. Process modeling can help determine when the best time to return these flows should be and suggest methods of doing so. Some facilities use online instrumentation that considers influent ammonia concentrations in the overall sidestream control strategy.

4.6 Chemical Dosing

Chemicals are dosed in a variety of locations within the treatment process. Typical chemical additions are metal dosing for phosphorus precipitation, iron salt addition for odor control, lime addition for alkalinity supplementation, or external carbon feed for BNR. In each of these instances, initial chemical feed quantities and patterns can be determined through use of the model. The model should reflect actual dosing strategy and should consider the potential for underdosing or overdosing chemicals.

4.7 Feed-Forward Control

Because models provide a representation of what occurs in an operating system, they can be used to simulate a potential control system. The accuracy of performance under the particular control system is reliant on the accuracy of tuning the control system. Examples of typical feed-forward control scenarios are dosing of alum for phosphorus removal in a tertiary treatment step based on the secondary process effluent concentration and dosing of carbon feed to a denitrification filter for nitrate removal.

5.0   USING MODELING RESULTS

It is important to keep in mind that models are not reality. Even the best calibrated model has a level of uncertainty associated with many aspects. In general, there are three categories of uncertainty:

- Quantifiable uncertainty—parameters that are known to be uncertain and the degree of uncertainty can be reduced. An example of this is uncertainty related to model structure. The model structure may have inherent assumptions that are only approximations of reality, such as the assumption that the autotrophic half-saturation constant for oxygen is a constant. These could potentially be made more accurate through further study and model development, provided the model allows for variable values.

- Irreducible uncertainty—uncertainty that cannot be reduced by any degree of study. An example of this is uncertainty related to future flows and loads. No amount of study would be able to quantify future flows and loads beyond a certain point.

- Total ignorance—parameters that the modeler does not know are uncertain. These items could be either reducible or not reducible. Any parameter might fall in this category, and the only method of reducing this is through learning and experience.

All these sources of uncertainty are present within models and contribute to each model's variation from reality. Project team members (stakeholders) need to be aware of this potential during projects so that appropriate measures can be taken to address uncertainty in the modeling results.

However, this does not mean that simulation results cannot be used. As discussed, simulations give valuable insight to how WRRFs function, although the exact accuracy of the output must be viewed in light of the uncertainties involved.

5.1   Interpreting Modeling Results

Reliability of modeling results is often directly dependent on the degree of effort expended by the modeler in reducing uncertainty. Once the model/simulator is selected, calibration and validation are the most common methods of reducing uncertainty of modeling results. The amount of calibration and validation done on a model should be selected according to use of the model and budget available.

In general, time-variant data add additional uncertainty to simulation results. As raw wastewater is inherently variable, this uncertainty is present in all modeling results that examine short-term effects. Predictions of future performance also are more uncertain because they are extended further into the future. It is up to the modeler to determine how best to interpret modeling results. Irreducible uncertainty ensures that the modeling results will never exactly match reality.

5.2 Communicating Modeling Results/Limitations

Management of a project team member's expectations in relation to the reliability and accuracy of modeling results is a key aspect of any modeling project. It is not uncommon for nonmodelers to have unrealistic expectations concerning level of effort and modeling accuracy; these expectations must be addressed early in the project and continue throughout the effort.

Presentation of modeling results can take many different routes. In general, the degree of accuracy that is presented in modeling results should reflect the level of calibration. For example, a facility option comparison with little calibration might only report the percentage difference between the options. If a highly calibrated model was developed looking only a short time into the future, it may be possible to report actual results with more significant digits in some of the results. Model results and limitations can be presented as follows:

- Mass balance data presentation—detailed mass balance data can be summarized and presented by a number of methods. In general, composite variables, such as BOD, COD, TSS, TKN, and total phosphorus should be presented in data summaries. Detailed mass balance data typically are only suitable for presentation in appendices.

- Model/simulator setup data—similar to mass balance data, only significant parameters should be presented in the main body of a project report or presentation. Detailed model parameters should be included in an appendix so that future users can reproduce the modeling results if necessary. Commercial simulators typically have reporting features that can produce such detailed information automatically.

- Model/simulator assumptions—a critical component of documenting a model includes assumptions that were used in the model/simulator. It is important to model results for the user to document the degree of conservatism used in determining model parameters.

- Identification of physical modifications necessary—the modeler should include discussion of the need to discuss other modifications that would go along with the modeling effort.

- Graphic display of dynamic models results is an effective presentation method. Care should be taken to not present too much information in a single graph. Comparing different time series data sets together is an effective method of displaying cause and effect. Figure 9.13 shows the effect of limited aeration air quantity under maximum day demand on aeration basin dissolved oxygen, airflow rate per diffuser, and the effect on final effluent quality.

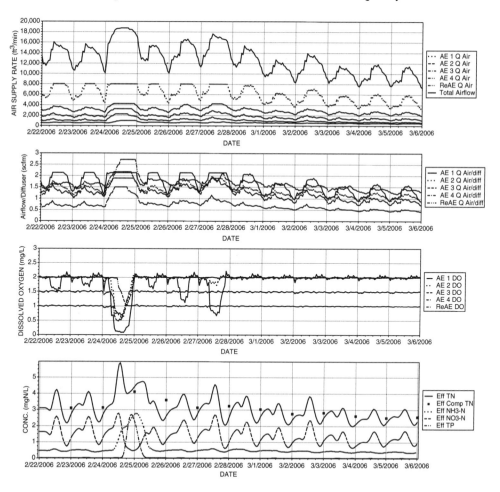

FIGURE 9.13   Graphic display of time series results for cause-and-effect comparisons.

5.3 Quality Assurance/Quality Control

Modeling of wastewater treatment processes is a complex task with many variables. As a result, quality assurance/quality control (QA/QC) is a critical part of any modeling project. Although QA/QC can take many different forms, all of these include participation by a person other than the modeler. Chapter 8 provided guidance on QA/QC for model setup, which is equally applicable to the following actual model:

- A "reality check" of modeling results by an external reviewer. This reality check can be through heuristic rules (general rules), by senior reviewers, or through simple spreadsheet models.
- A review of the model files and simulation results.

Regardless of the type of review selected, the following items should be reviewed as part of the QA/QC process:

- Modeling project structure
- Flows and loads analysis
- Interpretation of additional data collected through extra sampling and analysis
- Model default parameters
- Documentation of results
- Conclusions drawn from modeling results

5.3.1 Direct Model Quality Assurance/Quality Control

As the modeler documents decisions made throughout the modeling process, a reviewer can backtrack the results to review assumptions and standards along the way. This method requires that the reviewer be an equally qualified modeler. The benefit of this method is direct identification of any changes required in the modeling effort. However, limitations of the model can be missed during review.

5.3.2 Independent Checks on Sizing and Capacity

There are many different design and modeling tools available to a modeler. Using a second method may identify issues in assumptions used in the modeling effort. These checks could include different simulators or comparisons with typical design values. Discrepancies that are found during this effort should be evaluated against the original goal of the project. The chosen modeling method may better represent the conditions being evaluated.

6.0 REPORTING AND DOCUMENTATION

At the end of a modeling task, a report must be generated that captures the purpose and results of the simulations. The report can provide both historical documentation of assumptions and results from modeling to help decision makers understand the results of the various scenarios that were simulated, and background information for future users of the model.

6.1 Report Preparation

The report should contain information from each of the steps described in this manual. This may include information as presented in previously issued memorandums or reports. At a minimum, the report should contain the items listed in Section 5.2, where data are initially presented. The report as a whole should follow and document the steps of the project and include the following information:

- Objectives of the modeling effort
- Description of the existing facility
- Data collection
- Data analysis
- Setup and description of model
- Calibration and validation results
- Description of simulation scenarios
- Presentation of results
- Uncertainty analysis

During each of these steps, input and output files used in the model should be provided. Assumptions for each unit process should be tabulated and any deviation from default parameters should be documented. The goal of this is to accurately describe the decision making process that went into the project.

6.2 Documentation for Future Use

The report generated in the preceding section focuses mainly on results of the modeling effort. Appendices will also contain information describing modeling parameters (kinetic, stoichiometric, etc.) that were used for each simulation. After the model has been used for design or optimization, it may be used again in the future for forecasting capabilities, to change operating or influent characteristics to match changes in the facility, or to evaluate options developed later.

It is important to document all of the operating parameters of the system including default parameters that were not changed. As modeling tools progress, the capabilities of the model will improve and some parameters may be changed based on further research.

7.0   MODEL LIFE CYCLES

As more wastewater facilities are using models for operation, it is important that consultants coordinate their design efforts with the operators using the model. In addition, because different consultants will likely be involved in planning and design efforts over the years, it is important that modeling information is transferred to all parties involved. This will prevent information from being lost and consultants having to "reinvent the wheel" with every new project. In addition, it provides an opportunity to refine model accuracy as new data are collected. Special sampling campaigns should be repeated for different seasons and loading conditions to validate model accuracy over a model's life cycle. At the end of modeling efforts, consultants should provide model reports that document all inputs, including kinetic parameters and influent characteristics. Whether or not facility staff use the model, they retain ownership of all reports and files and should pass this information on to the next project.

Another factor to consider in a model's life cycle is the issue of new versions of software. Before a new model is used for design or operation, a scenario from the old model should be run in the new model to compare predictions. Changes in model fundamentals or default parameters may cause results to be different, and the model may need to be recalibrated. Latimer et al. (2008) describe a specific challenge they encountered when upgrading a model with single-step nitrification to two-step nitrification. Along with software developers, they identified that the model was "looping" nitrite back to nitrate in a simultaneous nitrification–denitrification process, which resulted in higher effluent nitrate predictions.

7.1   Maintaining and Refining the Model over Time

Once a modeling effort is complete, conclusions are used to make process or capital investment decisions. In some instances, the model may be reused to evaluate implemented changes to the process, troubleshoot performance issues, or re-evaluate the process based on changes to loading conditions.

Assumptions that were made in the initial modeling effort should be re-evaluated if project budget allows. Modelers should be aware of safety factors that were used in the initial evaluation. In practice, the performance may be better than anticipated based on conservative parameters initially used. If project budget allows, the model should be updated to reflect current performance, which allows it to be used as a tool for future activities. This may include some additional sampling and analysis of the operating WRRF to update the model.

7.2 Issues with Ownership

Modeling projects are often conducted for a facility by an independent consulting firm. At the end of the evaluation, ownership of modeling files can come into question if not otherwise stated in the contract. Unless specifically documented in the contract, if an owner has paid for the work, then he or she owns it. Modeling files developed as part of the effort become property of the client once the work is complete. In some circumstances, the company may not want to release a tool that they have developed in-house. Even if it states in the contract that these files are not to be transferred to the utility, it is the responsibility of the modeler to provide a comprehensive set of all calculations that support the conclusions reached as part of the modeling effort.

Models are often reused by other entities after a modeling project is completed. Models may be reused by another modeler within the same consulting company, by modelers from another consulting company, or by a member of a utility using a model developed by a consultant or in-house at the utility. Regardless of the circumstance, it is the responsibility of the modeler using a previously developed model to take charge of the model in much the same way a professional engineer stamps drawings based on work from another engineer. When using modeling work performed by others, it is important to understand the goals of the project under which it was developed and to identify assumptions that may not apply to the current modeling effort. As a modeler who may need to transition his or her work to another modeling effort, the reporting and documentation sections of this chapter are important to facilitate this exchange of information.

7.3 Documentation for Future Use

The documentation process provides information to recreate models for future projects. When preparing documentation, parameters and assumptions should be clearly presented to allow future users (or the current modeler) to understand what decisions were made. Use of a previously developed model does not serve as a guide to future users and it is the responsibility of future users to verify their own assumptions in implementing the model. The model was developed to be used for conditions defined in the original project. Deviations from these conditions, changes in industry understanding, or goals of the project are not necessarily transferable to any future modeling project.

7.4 Facility Data Management

The accuracy of a model varies greatly with the quality of data used to calibrate and create the model. As models are revisited during construction and operation of a new facility, model accuracy will increase as physical effects are further understood and influent waste characteristics can be better documented.

8.0 COMMON PITFALLS

There are many pitfalls that should be avoided when producing and using a process model. Shaw et al. (2007) described some of the most common pitfalls and suggested ways to avoid them. These pitfalls are summarized in the following sections.

8.1 No Safety Factors

Most engineering equations and typical practices used to design and operate wastewater facilities include safety factors. Process models, on the other hand, have no safety factors. Users must apply their own safety factors and good judgment in determining how close to the wire to push the facility based on modeling predictions. Actual facility operation can result in discrepancies in sludge age, temperature, and influent loading when compared to assumptions made during modeling.

8.2 Garbage in Equals Garbage Out

The accuracy of any model is highly dependent on the quality of the data used to develop the model. The accuracy of the model then directly affects confidence in the results it produces. This, in turn, affects the safety factor that should be applied to its predictions (see Section 8.1). Special sampling for influent characteristics and direct kinetic measurements are important steps to ensure that input data are good. Data checking also is an important step in this process.

8.3 Adjusting the Appropriate Parameters

Expert judgment must be used to adjust the proper parameters in the model. Often, it is possible to fit a model to facility data by adjusting any one of several parameters.

In modeling terms, the process models used in wastewater are considerably "over-parameterized," meaning that there are too many parameters than can be adjusted to obtain the same result, making it almost impossible to obtain a unique solution using purely numerical methods. It is possible for a modeler to adjust one parameter to get a model to fit the data. However, this may not properly represent mechanisms occurring in the facility, which renders the model useless for exploring conditions other than those used during calibration. In principle, the model can be forced to show anything the user wants. In practice, knowledge and modeling expertise must be used to discern correct adjustments to make to the model. In many instances, influent characteristics must be reevaluated and corrected or the mixing characteristics of the reactor need to be investigated if a model does not fit the facility data for a domestic WRRF. Calibration of the model to reasonably match the quality of the input source, the facility solids balance, and identification of the source of any discrepancies are critical initial steps that must be completed before other parameters are adjusted. Kinetics or sludge yields should only be adjusted from established default parameters as a last resort. Additional calibration information for critical parameters can be obtained by conducting facility-specific tests for parameters, such as nitrifier growth rates.

8.4 What Simulators Cannot Tell You

Simulators have traditionally been based on activated sludge models whose focus is on solids generation, effluent quality, and biomass growth. In operation, physical geometry and biology add additional complicating factors such as foaming, bulking, poor flow-splitting, and short-circuiting. In the clarifiers, a one-dimensional model is often used to describe the settling characteristics of the activated sludge, but cannot predict physical clarifier limitations. In these instances, computational fluid dynamic modeling can be implemented to help describe the clarifiers; however, this can be a costly task that does not directly integrate the rest of the facility model. Use of uncertainty analysis can help identify possible effects to population dynamics and overall process performance with changes in return streams.

Although simulators will not predict conditions such as bulking, the modeler should use data that indicate that the conditions may exist. Excessively long or short sludge ages will favor some filamentous organisms, as would inadequate aeration at the front of an aerobic zone.

8.5   Considering Minimum Conditions

In addition to considering the overall treatment capacity of the system, the modeler should consider how the facility will perform when first started up and during seasonal and diurnal minimum conditions. Under these conditions, the aeration system and pumping system are often oversized for loading. For example, the effect of too much basin volume could result in thin mixed liquor that does not settle well or long sludge ages that result in nitrification and aeration costs not necessary to meet permit requirements. The designer should focus on the range of scenarios that can be seen at the WRRF to determine treatment capabilities throughout the year.

8.6   False Precision in Control Capability

Whether or not the model is used to assess effects from controls, the engineer must consider actual operation of the facility including how precise dissolved oxygen must be controlled in each zone, flowrate control for influent splitting and recycle flow, and control of mixed liquor age based on limited solids analysis.

8.7   Defining Worst-Case Conditions as Highest Loading

The worst-case scenario is considered as the highest loading to the system under the most adverse environmental conditions. The maximum month or maximum week used for design to meet effluent quality may not be the same maximum month or maximum week used for design of the aeration system. These maximum loading conditions may also be the most advantageous for influent carbon with respect to nutrient removal. Worst-case conditions should examine potential combinations of loading characteristics such as high nitrogen and low influent BOD and extremes in temperature. Table 9.1 suggests a variety of operating conditions to be considered. Optimizing the design for worst case could result in an inefficient facility under average and startup loading conditions.

8.8   Focus on Configurations that Conform to Modeling Capabilities

Certain process configurations are inherently easier to model. Focusing on these treatment options because your predictive capabilities are highest could eliminate a more suitable process configuration. Examples of this are as follows:

- Focus on suspended growth-activated sludge process instead of fixed-film processes such as MBBR, IFAS, BAF, or trickling filters
- Use of chemical phosphorus removal because of uncertainties in implementation of biological phosphorus removal

9.0 REFERENCES

Benedetti, L.; Claeys, F.; Nopens, I.; Vanrolleghem, P. A. (2011) Assessing the Convergence of LHS Monte Carlo Simulations of Wastewater Treatment Models. *Water Sci. Technol.*, **63** (10), 2219–2224.

Eckhardt, R. (1987) Stan Ulam, John Von Neumann, and the Monte Carlo Method. *Los Alamos Science,* **15** (Special Issue), 131–137.

Great Lakes–Upper Mississippi River Board (2004) *Recommended Standards for Wastewater Facilities*; Health Research Inc., Health Education Services Division: Albany, New York.

Hall, A. (1873) On an Experimental Determination of PI. *Messenger of Mathematics*, **2**, 113–114.

Latimer, R. J.; Pitt, P.; Dold, P.; Takács, I.; Lynch, T. J. (2008) Kinetic Parameters for Modeling Two-Step Nitrification and Denitrification—Case Study. *Proceedings of the 1st International Water Association/Water Environment Federation Wastewater Treatment Modeling Seminar*; Mont-Sainte-Anne, Québec, Canada, June 1–3; International Water Association: London; p 19.

Melcer, H.; Dold, P. L.; Jones, R. M.; Bye, C. M.; Tákacs, I.; Stensel, H. D.; Wilson, A. W.; Sun, P.; Bury, S. (2003) *Methods for Wastewater Characterization in Activated Sludge Modeling*; Report 99-WWF-3; Water Environment Research Foundation: Alexandria, Virginia.

Metropolis, N.; Ulam, S. (1949) The Monte Carlo Method. *J. Amer. Stat. Assoc.*, **44**, 335–341.

Metropolis, N.; Rosenbluth, A.; Rosenbluth, M.; Teller, A.; Teller, E. (1953) Equation of State Calculations by Fast Computing Machines. *J. Chem, Phys.*, **21** (6), 1087.

Morokoff, W. J.; Caflisch, R. E. (1995) Quasi-Monte Carlo Integration. *J. Comput. Phys.*, **122** (2), 218–230.

Shaw, A.; Phillips, H. M.; Sabherwal, B.; deBarbadillo, C. (2007) Succeeding at Simulation. *Water Environ. Technol.*, **19** (4), 54–58.

Torvi, H.; Hertzberg, T. (1998) Methods of Evaluating Uncertainties in Dynamic Simulation—A Comparison of Performance. *Computers Chem. Eng.*, **22** (Supplement), S985–S988.

10.0 SUGGESTED READINGS

Bard, Y. A. (1974) *Nonlinear Parameter Estimation*; Academic Press: New York.

Bixio, D.; Parmentier, G.; Rousseau, D.; Verdonck, F.; Vanrolleghem, P. A.; Thoeye, C. (2003) Integrating Risk Analysis in the Design/Simulation of Activated Sludge Systems. *Proceedings of the 76th Annual Water Environment Federation Technical Exhibition and Conference*; Los Angeles, California, Oct 11–15; Water Environment Federation: Alexandria, Virginia.

Brun, R.; Kühni, M.; Siegrist, H.; Gujer, W.; Reichert, P. (2002) Practical Identifiability of ASM2d Parameters—Systematic Selection and Tuning of Parameter Subsets. *Water Res.*, **36**, 4113–4127.

De Pauw, D.; Vanrolleghem, P. A. (2003) Practical Aspects of Sensitivity Analysis for Dynamic Models. *Proceedings of IMACS 4th MATHMOD Conference*; Vienna, Austria, Feb 5–7.

Omlin, M.; Brun, R.; Reichert, P. (2001) Biogeochemical Model of Lake Zürich: Sensitivity, Identifiability and Uncertainty Analysis. *Ecol. Modeling*, **141** (1-3), 105–123.

Petersen, B. (2000) *Calibration, Identifiability and Optimal Experimental Design of Activated Sludge Models*. Ph.D. Thesis, Gent University, Gent, Belgium.

Ruano, M. V.; Ribes, J.; De Pauw, D.; Sin., G. (2007) Parameter Subset Selection for the Dynamic Calibration of Activated Sludge Models (ASMs): Experience Versus Systems Analysis. *Water Sci. Technol.*, **56** (8), 105–115.

Vanrolleghem, P. A.; Insel, G.; Petersen, B.; Sin, G.; De Pauw, D.; Nopens, I.; Dovermann, H.; Weijers, S.; Gernaey, K. (2003) A Comprehensive Model Calibration Procedure for Activated Sludge Models. *Proceedings of Water Environment Federation's 76th Annual Conference and Exhibition*, Los Angeles, California, Oct 11–15.

Wagner, S. (2007) Global Sensitivity Analysis of Predictor Models in Software Engineering. *Proceedings of the Third International Workshop on Predictor Models in Software Engineering*; Minneapolis, Minnesota, May 20–26.

Weijers, S.; Vanrolleghem, P. A. (1997) A Procedure for Selecting Best Identifiable Parameters in Calibrating Activated Sludge Model No. 1 to Full-Scale Plant Data. *Water Sci. Technol.*, **36** (5), 69–79.

Appendix A

A Systematic Approach for Model Verification— Application on Seven Published Activated Sludge Models

H. Hauduc,[a,b] L. Rieger,[b,c] I. Takács,[d] A. Héduit,[a] P. A. Vanrolleghem,[b] and S. Gillot[a*]

[a]Cemagref, UR HBAN, Parc de Tourvoie, BP 44, F-92163 Antony Cedex, France.
[b]model*EAU*, Université Laval, Département de génie civil, Pavillon Adrien-Pouliot, 1065 av. de la Médecine, Québec (QC), G1V 0A6, Canada.
[c]EnviroSim Associates Ltd., 7 Innovation Drive, Suite 205, Flamborough (ON), L9H 7H9, Canada.
[d]EnviroSim Europe, 15 Impasse Fauré, 33000 Bordeaux, France.

ABSTRACT

The quality of simulation results can be significantly affected by errors in the published model (typing, inconsistencies, gaps, or conceptual errors) and/or in the underlying numerical model description. Seven of the most commonly used activated sludge models have been investigated to point out typing errors, inconsistencies, and gaps in the model publications. These are (1) Activated Sludge Model No. 1

(*Continued on page 338*)

---

*Corresponding author (E-mail: Sylvie.gillot@cemagref.fr)

TABLE **A.1**   Activated Sludge Model No. 1

Original publication	Henze, M.; Grady, C. P. L., Jr.; Gujer, W.; Marais, G. v. R.; Matsuo, T. (2000) *Activated Sludge Model No. 1;*

		Soluble undegradable organics	Soluble biodegradable organics	Particulate undegradable organics from the influent	Particulate biodegradable organics	Ordinary heterotrophic organisms	Autotrophic nitrifying organisms (NH_4 + to NO_3^-)	Particulate undegradable endogenous products
	Original parameter symbol	S_I	S_S	X_I	X_S	$X_{B,H}$	$X_{B,A}$	X_P
	Units	g COD·m⁻³	g COD·m⁻³	g COD·m⁻³	g COD·m⁻³	g COD·m⁻³	g COD·m⁻³	g COD·m⁻³
	Parameter	S_U	S_B	$X_{U,Inf}$	XC_B	X_{OHO}	X_{ANO}	$X_{U,E}$
1	Aerobic growth of heterotrophs		$-1/Y_{OHO}$			1		
2	Anoxic growth of heterotrophs		$-1/Y_{OHO}$			1		
3	Aerobic growth of autotrophs						1	
4	Decay of heterotrophs				$1 - f_{XU\_Bio,lys}$	-1		$f_{XU\_Bio,lys}$
5	Decay of autotrophs				$1 - f_{XU\_Bio,lys}$		-1	$f_{XU\_Bio,lys}$
6	Ammonification of soluble organic nitrogen							
7	Hydrolysis of entrapped organics		1		-1			
8	Hydrolysis of entrapped organic nitrogen							

[1]$\mu_{OHO,Max}{}^*[S_B/(K_{SB,OHO} + S_B)]^*[S_{O_2}/(K_{O_2,OHO} + S_{O_2})]^*[S_{NHx}/(K_{NHx,OHO} + S_{NHx})]^*X_{OHO}$

[2]$\mu_{OHO,Max}{}^*[S_B/(K_{SB,OHO} + S_B)]^*[K_{O_2,OHO}/(K_{O_2,OHO} + S_{O_2})]^*[S_{NOx}/(K_{NOx,OHO} + S_{NOx})]^*[S_{NHx}/(K_{NHx,OHO} + S_{NHx})]^*n_{\mu OHO,Ax}{}^*X_{OHO}$

[3]$\mu_{ANO,Max}{}^*[S_{NHx}/(K_{NHx,ANO} + S_{NHx})]^*[S_{O_2}/(K_{O_2,ANO} + S_{O_2})]^*X_{ANO}$

[4]$b_{OHO}{}^*X_{OHO}$

Scientific and Technical Report No. 9; IWA Publishing: London, U.K.

Dissolved oxygen	Nitrate and nitrite (NO$_3$ + NO$_2$) (considered to be NO$_3$ only for stoichiometry)	Ammonia (NH$_4$ + NH$_3$)	Soluble biodegradable organic N	Particulate biodegradable organic N	Alkalinity (HCO$_3^-$)	Dissolved Nitrogen	Process rate
S_O	S_{NO}	S_{NH}	S_{ND}	X_{ND}	S_{ALK}		
$-$ g COD·m^{-3}	g N·m^{-3}	g N·m^{-3}	g N·m^{-3}	g N·m^{-3}	mol HCO$_3^-$·m^{-3}	g N·m^{-3}	
S_{O_2}	S_{NOx}	S_{NHx}	$S_{B,N}$	$XC_{B,N}$	S_{Alk}	S_{N_2}	
$-(1-Y_{OHO})/Y_{OHO}$		$-i_{N\_XBio}$			$-i_{N\_XBio}{}^*i_{Charge\_SNHx}$		(1)
	$-(1-Y_{OHO})/$ $(i_{NO_3,N_2}{}^*Y_{OHO})$	$-i_{N\_XBio}$			$-(1-Y_{OHO})/(i_{NO_3,N_2}{}^*Y_{OHO})^*$ $i_{Charge\_SNOx}-i_{N\_XBio}{}^*i_{Charge\_SNHx}$	$(1-Y_{OHO})/$ $(i_{NO_3,N_2}{}^*Y_{OHO})$	(2)
$-(-i_{COD\_NO_3}-Y_{ANO})/$ Y_{ANO}	$1/Y_{ANO}$	$-i_{N\_XBio}-$ $1/Y_{ANO}$			$-(i_{N\_XBio}+(1/Y_{ANO})^*i_{Charge\_SNHx}+$ $(1/Y_{ANO})^*i_{Charge\_SNOx}$		(3)
				$i_{N\_XBio}-f_{XU\_Bio,lys}{}^*i_{N\_XUE}$			(4)
				$i_{N\_XBio}-f_{XU\_Bio,lys}{}^*i_{N\_XUE}$			(5)
		1	-1		i_{Charge_SNHx}		(6)
							(7)
		1	-1				(8)

[5] $b_{ANO}{}^*X_{ANO}$

[6] $q_{am}{}^*S_{B,N}{}^*X_{OHO}$

[7] $q_{XCB\_SB,hyd}{}^*[(XC_B/X_{OHO})/(K_{XCB,hyd}+XC_B/X_{OHO})]^*([S_{O_2}/(K_{O_2,OHO}+S_{O_2})]+n_{qhyd,Ax}{}^*[K_{O_2,OHO}/(K_{O_2,OHO}+S_{O_2})]^*[S_{NOx}/(K_{NOx,OHO}+S_{NOx})])^*X_{OHO}$

[8] $q_{XCB,hyd}{}^*(XC_{B,N}/XC_B)^*[(XC_B/X_{OHO})/(K_{XCB,hyd}+XC_B/X_{OHO})]^*([S_{O_2}/(K_{O_2,OHO}+S_{O_2})]+n_{qhyd,Ax}{}^*[K_{O_2,OHO}/(K_{O_2,OHO}+S_{O_2})]^*[S_{NOx}/(K_{NOx,OHO}+S_{NOx})])^*X_{OHO}$

TABLE **A.1**   Activated Sludge Model No. 1 (*continued*)

Stoichiometry	Parameter notation		Units	Value* T = 20°C
	Original	Standardized		
Yield for X_{OHO} growth	Y_H	Y_{OHO}	g $X_{OHO} \cdot$ g XC_B^{-1}	0.67
Fraction of X_U generated in biomass decay	f_P	$f_{XU\_Bio,lys}$	g $X_U \cdot$ g X_{Bio}^{-1}	0.08
Yield of X_{ANO} growth per S_{NO_3}	Y_A	Y_{ANO}	g $X_{AUT} \cdot$ g $S_{NO_3}^{-1}$	0.24
N content of biomass (X_{OHO}, X_{PAO}, X_{ANO})	i_{XB}	i_{N_XBio}	g N $\cdot$ g X_{Bio}^{-1}	0.086
N content of products from biomass	i_{XP}	i_{N_XUE}	g N $\cdot$ g X_{UE}^{-1}	0.06
Conversion factor for NO_3 reduction to N_2	i_{NO_3,N_2}	i_{NO_3,N_2}	g COD $\cdot$ g N^{-1}	2.857
Conversion factor for NO_3 in COD	$i_{COD\_NO_3}$	$i_{COD\_NO_3}$	g COD $\cdot$ g N^{-1}	−4.571
Conversion factor for N_2 in COD	$i_{COD\_N_2}$	$i_{COD\_N_2}$	g COD $\cdot$ g N^{-1}	−1.714
Conversion factor for NHx in charge	$i_{Charge\_S_{NHx}}$	$i_{Charge\_S_{NHx}}$	Charge $\cdot$ g N^{-1}	0.071
Conversion factor for NO_3 in charge	$i_{Charge\_S_{NOx}}$	$i_{Charge\_S_{NOx}}$	Charge $\cdot$ g N^{-1}	−0.0714

*Henze, M.; Gujer, W.; Mino, T.; Matsuo, T.; Wentzel, M. C.; Marais, G. v. R.; van Loosdrecht, M. C. M. (2000) *Activated Sludge Model No. 2d*; Scientific and Technical Report No. 9; IWA Publishing: London, U.K.

TABLE A.1 Activated Sludge Model No. 1 (*continued*)

Kinetics	Parameter notation		Units	Value* T = 20°C
	Original	Standardized		
Maximum specific hydrolysis rate	k_h	$q_{XCB\_SB,hyd}$	g $XC_B \cdot$ g $X_{OHO}^{-1} \cdot d^{-1}$	3
Saturation coefficient for X_B/X_{OHO}	K_X	$K_{XCB,hyd}$	g $XC_B \cdot$ g X_{OHO}^{-1}	0.03
Correction factor for hydrolysis under anoxic conditions	η_h	$n_{qhyd,Ax}$		0.4
Maximum growth rate of X_{OHO}	μ_H	$\mu_{OHO,Max}$	d^{-1}	6
Reduction factor for anoxic growth of X_{OHO}	η_g	$n_{\mu OHO,Ax}$		0.8
Half-saturation coefficient for S_B	K_s	$K_{SB,OHO}$	g $S_B \cdot m^{-3}$	20
Decay rate for X_{OHO}	b_H	b_{OHO}	d^{-1}	0.62
Half-saturation coefficient for S_{O_2}	$K_{O,H}$	$K_{O2,OHO}$	g $S_{O_2} \cdot m^{-3}$	0.2
Half-saturation coefficient for S_{NOx}	K_{NO}	$K_{NOx,OHO}$	g $S_{NOx} \cdot m^{-3}$	0.5
Half-saturation coefficient for NH_4^*	$K_{NH,H}$	$K_{NHx,OHO}$	g $S_{NHx} \cdot m^{-3}$	0.05
Maximum growth rate of X_{ANO}	μ_A	$\mu_{ANO,Max}$	d^{-1}	0.8
Decay rate for X_{ANO}	b_A	b_{ANO}	d^{-1}	0.15
Rate constant for ammonification	k_a	q_{am}	$m^3 \cdot$ g $XC_{B,N}^{-1} \cdot d^{-1}$	0.08
Half-saturation coefficient for S_{O_2}	$K_{O,A}$	$K_{O2,ANO}$	g $S_{O_2} \cdot m^{-3}$	0.4
Half-saturation coefficient for S_{NHx}	K_{NH}	$K_{NHx,ANO}$	g $S_{NHx} \cdot m^{-3}$	1

*Same $K_{NH,H}$ value as ASM2d has been chosen.

TABLE A.2   Activated Sludge Model No. 2d

Original publication   Henze, M.; Gujer, W.; Mino, T.; Matsuo, T.; Wentzel, M. C.; Marais, G. v. R.; van Loosdrecht, M. C. M. (2000) *Activated Sludge*

		Dissolved oxygen	Fermentable organic matter	Fermentation product (volatile fatty acids)	Ammonium and ammonia nitrogen ($NH_4 + NH_3$)	Nitrate and nitrite ($NO_3 + NO_2$) (considered to be NO_3 only for stoichiometry)	Soluble inorganic phosphorus	Soluble undegradable organics	Alkalinity (HCO_3^-)
Original parameter symbol		S_{O_2}	S_F	S_A	S_{NH_4}	S_{NO_3}	S_{PO_4}	S_I	S_{ALK}
Units		$-$ g COD·m^{-3}	g COD·m^{-3}	g COD·m^{-3}	g N·m^{-3}	g N·m^{-3}	g P·m^{-3}	g COD·m^{-3}	mol HCO$_3^-$·m^{-3}
		S_{O_2}	S_F	S_{VFA}	S_{NHx}	S_{NOx}	S_{PO_4}	S_U	S_{Alk}
1	Aerobic hydrolysis		$1 - f_{SU\_XCB,hyd}$		$-((1 - f_{SU\_XCB,hyd})^* i_{N\_SF} + f_{SU\_XCB,hyd}^* i_{N\_SU} - i_{N\_XCB})$		$-((1 - f_{SU\_XCB,hyd})^* i_{P\_SF} + f_{SU\_XCB,hyd}^* i_{P\_SU} - i_{P\_XCB})$	$f_{SU\_XCB,hyd}$	$i_{Charge\_S_{NHx}}^* v_{1\_S_{NHx}} + i_{Charge\_S_{PO_4}}^* v_{1\_S_{PO_4}}$
2	Anoxic hydrolysis		$1 - f_{SU\_XCB,hyd}$		$-((1 - f_{SU\_XCB,hyd})^* i_{N\_SF} + f_{SU\_XCB,hyd}^* i_{N\_SU} - i_{N\_XCB})$		$-((1 - f_{SU\_XCB,hyd})^* i_{P\_SF} + f_{SU\_XCB,hyd}^* i_{P\_SU} - i_{P\_XCB})$	$f_{SU\_XCB,hyd}$	$i_{Charge\_S_{NHx}}^* v_{2\_S_{NHx}} + i_{Charge\_S_{PO_4}}^* v_{2\_S_{PO_4}}$
3	Anaerobic hydrolysis		$1 - f_{SU\_XCB,hyd}$		$-((1 - f_{SU\_XCB,hyd})^* i_{N\_SF} + f_{SU\_XCB,hyd}^* i_{N\_SU} - i_{N\_XCB})$		$-((1 - f_{SU\_XCB,hyd})^* i_{P\_SF} + f_{SU\_XCB,hyd}^* i_{P\_SU} - i_{P\_XCB})$	$f_{SU\_XCB,hyd}$	$i_{Charge\_S_{NHx}}^* v_{3\_S_{NHx}} + i_{Charge\_S_{PO_4}}^* v_{3\_S_{PO_4}}$
4	Aerobic growth on S_F	$-(1 - Y_{OHO})/Y_{OHO}$	$-1/Y_{OHO}$		$-(-1/Y_{OHO}^* i_{N\_SF} + i_{N\_XBio})$		$-(-1/Y_{OHO}^* i_{P\_SF} + i_{P\_XBio})$		$i_{Charge\_S_{NHx}}^* v_{4\_S_{NHx}} + i_{Charge\_S_{PO_4}}^* v_{4\_S_{PO_4}}$
5	Aerobic growth on S_A	$-(1 - Y_{OHO})/Y_{OHO}$		$-1/Y_{OHO}$	$-i_{N\_XBio}$		$-i_{P\_XBio}$		$i_{Charge\_S_{NHx}}^* v_{5\_S_{NHx}} + i_{Charge\_S_{PO_4}}^* v_{5\_S_{PO_4}} + i_{Charge\_S_{VFA}}^* v_{5\_S_{VFA}}$
6	Anoxic growth on S_F		$-1/Y_{OHO}$		$-(-1/Y_{OHO}^* i_{N\_SF} + i_{N\_XBio})$	$-(1 - Y_{OHO})/(i_{NO_3,N_2}^* Y_{OHO})$	$-(-1/Y_{OHO}^* i_{P\_SF} + i_{P\_XBio})$		$i_{Charge\_S_{NHx}}^* v_{6\_S_{NHx}} + i_{Charge\_S_{PO_4}}^* v_{6\_S_{PO_4}} + i_{Charge\_S_{NOx}}^* v_{6\_S_{NOx}}$
7	Anoxic growth on S_A			$-1/Y_{OHO}$	$-i_{N\_XBio}$	$-(1 - Y_{OHO})/(i_{NO_3,N_2}^* Y_{OHO})$	$-i_{P\_XBio}$		$i_{Charge\_S_{NHx}}^* v_{7\_S_{NHx}} + i_{Charge\_S_{PO_4}}^* v_{7\_S_{PO_4}} + i_{Charge\_S_{NOx}}^* v_{7\_S_{NOx}} + i_{Charge\_S_{VFA}}^* v_{7\_S_{VFA}}$
8	Fermentation		-1	1	i_{N_SF}		i_{P_SF}		$i_{Charge\_S_{NHx}}^* v_{8\_S_{NHx}} + i_{Charge\_S_{PO_4}}^* v_{8\_S_{PO_4}} + i_{Charge\_S_{VFA}}$
9	Lysis				$-(f_{XU\_Bio,lys}^* i_{N\_XU} + (1 - f_{XU\_Bio,lys})^* i_{N\_XCB} - i_{N\_XBio})$		$-(f_{XU\_Bio,lys}^* i_{P\_XU} + (1 - f_{XU\_Bio,lys})^* i_{P\_XCB} - i_{P\_XBio})$		$i_{Charge\_S_{NHx}}^* v_{9\_S_{NHx}} + i_{Charge\_S_{PO_4}}^* v_{9\_S_{PO_4}}$
10	Storage of X_{Stor}			-1			$Y_{PP\_Stor,PAO}$		$-i_{Charge\_S_{VFA}} + i_{Charge\_S_{PO_4}}^* Y_{PP\_Stor,PAO} - i_{Charge\_X_{PAO,PP}}^* Y_{PP\_Stor,PAO}$
11	Aerobic Storage of X_{PP}	$-(1/Y_{Stor\_PP})$					-1		$-i_{Charge\_S_{PO_4}} + i_{Charge\_X_{PAO,PP}}$
12	Anoxic Storage of X_{PP}					$-(1/Y_{Stor\_PP})^* (1/i_{NO_3,N_2})$	-1		$-i_{Charge\_S_{PO_4}} + i_{Charge\_S_{NOx}}^* v_{12\_S_{NOx}} + i_{Charge\_X_{PAO,PP}}$

Model No. 2d; Scientific and Technical Report No. 9; IWA Publishing: London, U.K.

Dissolved nitrogen gas	Particulate undegradable organics	Particulate biodegradable organics	Ordinary heterotrophic organisms	Phosphorus accumulating organisms	Stored polyphosphates in PAOs	Storage compound in PAOs	Autotrophic nitrifying organisms (NH₄ to NO₃⁻)	Total suspended solids	Metal hydroxide compounds	Metal phosphate compounds	Process rate
S_{N_2}	X_I	X_S	X_H	X_{PAO}	X_{PP}	X_{PHA}	X_{AUT}	X_{TSS}	X_{MeOH}	X_{MeP}	
g N·m⁻³	g COD·m⁻³	g COD·m⁻³	g COD·m⁻³	g COD·m⁻³	g P·m⁻³	g COD·m⁻³	g COD·m⁻³	g TSS·m⁻³	g TSS·m⁻³	g TSS·m⁻³	
S_{N_2}	X_U	XC_B	X_{OHO}	X_{PAO}	$X_{PAO,PP}$	$X_{PAO,Stor}$	X_{ANO}	X_{TSS}	X_{MeOH}	X_{MeP}	
		-1						$-i_{TSS\_XCB}$			(1)
		-1						$-i_{TSS\_XCB}$			(2)
		-1						$-i_{TSS\_XCB}$			(3)
			1					i_{TSS_XBio}			(4)
			1					i_{TSS_XBio}			(5)
$(1-Y_{OHO})/$ $(i_{NO3,N2}{}^*Y_{OHO})$			1					i_{TSS_XBio}			(6)
$(1-Y_{OHO})/$ $(i_{NO3,N2}{}^*Y_{OHO})$			1					i_{TSS_XBio}			(7)
											(8)
	$f_{XU\_Bio,lys}$	$1-f_{XU\_Bio,lys}$	-1					$f_{XU\_Bio,lys}{}^*i_{TSS\_XU} +$ $(1-f_{XU\_Bio,lys})^*$ $i_{TSS\_XCB} - i_{TSS\_XBio}$			(9)
					$-Y_{PP\_Stor,PAO}$	1		$-Y_{PP\_Stor,PAO}{}^*$ $i_{TSS\_X_{PAO,PP}} +$ $i_{TSS\_X_{PAO,Stor}}$			(10)
					1	$-(1/Y_{Stor\_PP})$		$i_{TSS\_X_{PAO,PP}} - (1/$ $Y_{Stor\_PP})^*i_{TSS\_X_{PAO,Stor}}$			(11)
$/Y_{Stor\_PP})^*$ $/i_{NO3,N2})$					1	$-(1/Y_{Stor\_PP})$		$i_{TSS\_X_{PAO,PP}} -$ $(1/Y_{Stor\_PP})^*$ $i_{TSS\_X_{PAO,Stor}}$			(12)

TABLE A.2 Activated Sludge Model No. 2d (*continued*)

		Dissolved oxygen	Fermentable organic matter	Fermentation product (volatile fatty acids)	Ammonium and ammonia nitrogen (NH$_4$ + NH$_3$)	Nitrate and nitrite (NO$_3$ + NO$_2$) (considered to be NO$_3$ only for stoichiometry)	Soluble inorganic phosphorus	Soluble undegradable organics	Alkalinity (HCO$_3^-$)
Original parameter symbol		S_{O_2}	S_F	S_A	S_{NH_4}	S_{NO_3}	S_{PO_4}	S_I	S_{ALK}
Units		− g COD·m⁻³	g COD·m⁻³	g COD·m⁻³	g N·m⁻³	g N·m⁻³	g P·m⁻³	g COD·m⁻³	mol HCO$_3^-$·m⁻³
		S_{O_2}	S_F	S_{VFA}	S_{NHx}	S_{NOx}	S_{PO_4}	S_U	S_{Alk}
13	Aerobic growth of X_{PAO}	$-(1 - Y_{PAO})/Y_{PAO}$			$-i_{N\_XBio}$		$-i_{P\_XBio}$		$i_{Charge\_S_{NHx}}{}^*v_{13\_S_{NHx}} - i_{P\_XBio}{}^*i_{Charge\_S_{PO_4}}$
14	Anoxic growth of X_{PAO}				$-i_{N\_XBio}$	$-(1 - Y_{PAO})/Y_{PAO}{}^*(1/i_{NO3,N2})$	$-i_{P\_XBio}$		$i_{Charge\_S_{NHx}}{}^*v_{14\_S_{NHx}} + i_{Charge\_S_{NOx}}{}^*v_{14\_S_{NOx}} - i_{P\_XBio}{}^*i_{Charge\_S_{PO_4}}$
15	Lysis of X_{PAO}				$-(f_{XU\_Bio,lys}{}^*i_{N\_XU} + (1 - f_{XU\_Bio,lys}){}^*i_{N\_XCB} - i_{N\_XBio})$		$-(f_{XU\_Bio,lys}{}^*i_{P\_XU} + (1 - f_{XU\_Bio,lys}){}^*i_{P\_XCB} - i_{P\_XBio})$		$i_{Charge\_S_{NHx}}{}^*v_{15\_S_{NHx}} + i_{Charge\_S_{PO_4}}{}^*v_{15\_S_{PO_4}}$
16	Lysis of X_{PP}						1		$i_{Charge\_S_{PO_4}} - i_{Charge\_X_{PAO,PP}}$
17	Lysis of X_{Stor}			1					$i_{Charge\_S_{VFA}}$
18	Aerobic growth of X_{AUT}	$-(-i_{COD\_NO_3} - Y_{ANO})/Y_{ANO}$			$-i_{N\_XBio} - 1/Y_{ANO}$	$1/Y_{ANO}$	$-i_{P\_XBio}$		$(-i_{N\_XBio} - 1/Y_{ANO}){}^*i_{Charge\_S_{NHx}} + 1/Y_{ANO}{}^*i_{Charge\_S_{NOx}} - i_{P\_XBio}{}^*i_{Charge\_S_{PO_4}}$
19	Lysis				$-(f_{XU\_Bio,lys}{}^*i_{N\_XU} + (1 - f_{XU\_Bio,lys}){}^*i_{N\_XCB} - i_{N\_XBio})$		$-(f_{XU\_Bio,lys}{}^*i_{P\_XU} + (1 - f_{XU\_Bio,lys}){}^*i_{P\_XCB} - i_{P\_XBio})$		$i_{Charge\_S_{NHx}}{}^*v_{19\_S_{NHx}} + i_{Charge\_S_{PO_4}}{}^*v_{19\_S_{PO_4}}$
20	Precipitation						−1		$-i_{Charge\_S_{PO_4}}$
21	Redissolution						1		$i_{Charge\_S_{PO_4}}$

[1] $q_{XCB\_SB,hyd}{}^*[S_{O_2}/(K_{O_2,hyd} + S_{O_2})]{}^*[(XC_B/X_{OHO})/(K_{XCB,hyd} + (XC_B/X_{OHO}))]{}^*X_{OHO}$

[2] $q_{XCB\_SB,hyd}{}^*n_{qhyd,Ax}{}^*[K_{O_2,hyd}/(K_{O_2,hyd} + S_{O_2})]{}^*[S_{NOx}/(K_{NOx,hyd} + S_{NOx})]{}^*[(XC_B/X_{OHO})/(K_{XCB,hyd} + (XC_B/X_{OHO}))]{}^*X_{OHO}$

[3] $q_{XCB\_SB,hyd}{}^*n_{qhyd,An}{}^*[K_{O_2,hyd}/(K_{O_2,hyd} + S_{O_2})]{}^*[K_{NOx,hyd}/(K_{NOx,hyd} + S_{NOx})]{}^*[(XC_B/X_{OHO})/(K_{XCB,hyd} + (XC_B/X_{OHO}))]{}^*X_{OHO}$

[4] $\mu_{OHO,Max}{}^*[S_{O_2}/(K_{O2OHO} + S_{O_2})]{}^*[S_F/(K_{SF,OHO} + S_F)]{}^*[S_F/(S_F + S_{VFA})]{}^*[S_{NHx}/(K_{NHx,OHO} + S_{NHx})]{}^*[S_{PO_4}/(K_{PO4,OHO} + S_{PO_4})]{}^*[S_{Alk}/(K_{Alk,OHO} + S_{Alk})]{}^*X_{OHO}$

[5] $\mu_{OHO,Max}{}^*[S_{O_2}/(K_{O2OHO} + S_{O_2})]{}^*[S_{VFA}/(K_{VFA,OHO} + S_{VFA})]{}^*[S_{VFA}/(S_F + S_{VFA})]{}^*[S_{NHx}/(K_{NHx,OHO} + S_{NHx})]{}^*[S_{PO_4}/(K_{PO4,OHO} + S_{PO_4})]{}^*[S_{Alk}/(K_{Alk,OHO} + S_{Alk})]{}^*X_{OHO}$

[6] $\mu_{OHO,Max}{}^*n_{\mu OHO,Ax}{}^*[K_{O2OHO}/(K_{O2OHO} + S_{O_2})]{}^*[S_{NOx}/(K_{NOx,OHO} + S_{NOx})]{}^*[S_F/(K_{SF,OHO} + S_F)]{}^*[S_F/(S_F + S_{VFA})]{}^*[S_{NHx}/(K_{NHx,OHO} + S_{NHx})]{}^*[S_{PO_4}/(K_{PO4,OHO} + S_{PO_4})]{}^*[S_{Alk}/(K_{Alk,OHO} + S_{Alk})]{}^*X_{OHO}$

[7] $\mu_{OHO,Max}{}^*n_{\mu OHO,Ax}{}^*[K_{O2OHO}/(K_{O2OHO} + S_{O_2})]{}^*[S_{NOx}/(K_{NOx,OHO} + S_{NOx})]{}^*[S_{VFA}/(K_{VFA,OHO} + S_{VFA})]{}^*[S_{VFA}/(S_F + S_{VFA})]{}^*[S_{NHx}/(K_{NHx,OHO} + S_{NHx})]{}^*[S_{PO_4}/(K_{PO4,OHO} + S_{PO_4})]{}^*[S_{Alk}/(K_{Alk,OHO} + S_{Alk})]{}^*X_{OHO}$

[8] $q_{SF\_VFA,Max}{}^*[K_{O2OHO}/(K_{O2OHO} + S_{O_2})]{}^*[K_{NOx,OHO}/(K_{NOx,OHO} + S_{NOx})]{}^*[S_F/(K_{fe} + S_F)]{}^*[S_{Alk}/(K_{Alk,OHO} + S_{Alk})]{}^*X_{OHO}$

[9] $b_{OHO}{}^*X_{OHO}$

[10] $q_{PAO,VFA\_Stor}{}^*[S_{VFA}/(K_{VFA,PAO} + S_{VFA})]{}^*[S_{Alk}/(K_{Alk,PAO} + S_{Alk})]{}^*[(X_{PAO,PP}/X_{PAO})/(K_{S,PP\_PAO} + (X_{PAO,PP}/X_{PAO}))]{}^*X_{PAO}$

[11] $q_{PAO,PO_4\_PP}{}^*[S_{O_2}/(K_{O2PAO} + S_{O_2})]{}^*[S_{PO_4}/(K_{PO4,PAO,upt} + S_{PO_4})]{}^*[S_{Alk}/(K_{Alk,PAO} + S_{Alk})]{}^*[(X_{PAO,Stor}/X_{PAO})/(K_{fStor\_PAO} + (X_{PAO,Stor}/X_{PAO}))]{}^*[(f_{PP\_PAO,Max} - X_{PAO,PP}/X_{PAO})/(K_{I,fPP\_PAO} + f_{PP\_PAO,Max} - (X_{PAO,PP}/X_{PAO}))]{}^*X_{PAO}$

Dissolved nitrogen gas	Particulate undegradable organics	Particulate biodegradable organics	Ordinary heterotrophic organisms	Phosphorus accumulating organisms	Stored polyphosphates in PAOs	Storage compound in PAOs	Autotrophic nitrifying organisms (NH₄ to NO₃⁻)	Total suspended solids	Metal hydroxide compounds	Metal phosphate compounds	Process rate
S_{N2}	X_I	X_S	X_H	X_{PAO}	X_{PP}	X_{PHA}	X_{AUT}	X_{TSS}	X_{MeOH}	X_{MeP}	
g N·m⁻³	g COD·m⁻³	g COD·m⁻³	g COD·m⁻³	g COD·m⁻³	g P·m⁻³	g COD·m⁻³	g COD·m⁻³	g TSS·m⁻³	g TSS·m⁻³	g TSS·m⁻³	
S_{N2}	X_U	XC_B	X_{OHO}	X_{PAO}	$X_{PAO,PP}$	$X_{PAO,Stor}$	X_{ANO}	X_{TSS}	X_{MeOH}	X_{MeP}	
				1		$-1/Y_{PAO}$		$i_{TSS\_XBio} - 1/Y_{PAO}*i_{TSS\_X_{PAO,Stor}}$			(13)
$(1-Y_{PAO})/Y_{PAO}*(1/i_{NO3,N2})$				1		$-1/Y_{PAO}$		$i_{TSS\_XBio} - 1/Y_{PAO}*i_{TSS\_X_{PAO,Stor}}$			(14)
	$f_{XU\_Bio,lys}$	$1-f_{XU\_Bio,lys}$		-1				$f_{XU\_Bio,lys}*i_{TSS\_XU} + (1-f_{XU\_Bio,lys})*i_{TSS\_XCB} - i_{TSS\_XBio}$			(15)
					-1			$-i_{TSS\_X_{PAO,PP}}$			(16)
						-1		$-i_{TSS\_X_{PAO,Stor}}$			(17)
							1	i_{TSS_XBio}			(18)
	$f_{XU\_Bio,lys}$	$1-f_{XU\_Bio,lys}$					-1	$f_{XU\_Bio,lys}*i_{TSS\_XU} + (1-f_{XU\_Bio,lys})*i_{TSS\_XCB} - i_{TSS\_XBio}$			(19)
								$f_{MeOH\_PO4,MW} + f_{MeP\_PO4,MW}$	$f_{MeOH\_PO4,MW}$	$f_{MeP\_PO4,MW}$	(20)
								$-(f_{MeOH\_PO4,MW} + f_{MeP\_PO4,MW})$	$f_{MeOH\_PO4,MW}$	$-f_{MeP\_PO4,MW}$	(21)

$_{PAO,PO4\_PP}*\eta_{\mu PAO}*[S_{NOx}/(K_{NOx,PAO}+S_{NOx})]*[K_{O2,PAO}/(K_{O2,PAO}+S_{O2})]*[S_{PO4}/(K_{PO4,PAO,upt}+S_{PO4})]*[S_{Alk}/(K_{Alk,PAO}+S_{Alk})]*[(X_{PAO,Stor}/X_{PAO})/(K_{fStor\_PAO}+(X_{PAO,Stor}/X_{PAO}))]*[(f_{PP\_PAO,Max}-$
$_{AO,PP}/X_{PAO})/(K_{I,fPP\_PAO}+f_{PP\_PAO,Max}-(X_{PAO,PP}/X_{PAO}))]*X_{PAO}$

$_{PAO,Max}*[S_{O2}/(K_{O2,PAO}+S_{O2})]*[S_{NHx}/(K_{NHx,PAO}+S_{NHx})]*[S_{PO4}/(K_{PO4,PAO,nut}+S_{PO4})]*[S_{Alk}/(K_{Alk,PAO}+S_{Alk})]*[(X_{PAO,Stor}/X_{PAO})/(K_{fStor\_PAO}+(X_{PAO,Stor}/X_{PAO}))]*X_{PAO}$

$_{PAO,Max}*\eta_{\mu PAO}*[K_{O2,PAO}/(K_{O2,PAO}+S_{O2})]*[S_{NOx}/(K_{NOx,PAO}+S_{NOx})]*[S_{NHx}/(K_{NHx,PAO}+S_{NHx})]*[S_{PO4}/(K_{PO4,PAO,nut}+S_{PO4})]*[S_{Alk}/(K_{Alk,PAO}+S_{Alk})]*[(X_{PAO,Stor}/X_{PAO})/(K_{fStor\_PAO}+(X_{PAO,Stor}/$
$_{AO}))]*X_{PAO}$

$_{PAO}*X_{PAO}*[S_{Alk}/(K_{Alk,PAO}+S_{Alk})]$

$_{PP\_PO4}*[S_{Alk}/(K_{Alk,PAO}+S_{Alk})]*[X_{PAO,PP}/X_{PAO}]*X_{PAO}$

$_{Stor\_VFA}*[S_{Alk}/(K_{Alk,PAO}+S_{Alk})]*[X_{PAO,Stor}/X_{PAO}]*X_{PAO}$

$_{ANO,Max}*[S_{O2}/(K_{O2,ANO}+S_{O2})]*[S_{NHx}/(K_{NHx,ANO}+S_{NHx})]*[S_{PO4}/(K_{PO4,ANO}+S_{PO4})]*[S_{Alk}/(K_{Alk,ANO}+S_{Alk})]*X_{ANO}$

$_{ANO}*X_{ANO}$

$_{2,pre}*S_{PO4}*X_{MeOH}$

$_{2,red}*X_{MeP}*[S_{Alk}/(K_{Alk,pre}+S_{Alk})]$

TABLE A.2 Activated Sludge Model No. 2d (*continued*)

Stoichiometry	Parameter notation		Units	Value* T = 20°C
	Original	Standardized		
Fraction of inert COD generated in hydrolysis	f_{SI}	$f_{SU\_XCB,hyd}$	g $S_U \cdot$ g XC_B^{-1}	0
Yield for X_{OHO} growth	Y_H	Y_{OHO}	g $X_{OHO} \cdot$ g XC_B^{-1}	0.625
Fraction of X_U generated in biomass decay	f_{XI}	$f_{XU\_Bio,lys}$	g $X_U \cdot$ g X_{OHO}^{-1}	0.1
Yield for X_{PAO} growth per $X_{PAO,Stor}$	Y_{PAO}	Y_{PAO}	g $X_{PAO} \cdot$ g X_{Stor}^{-1}	0.625
Yield for $X_{PAO,PP}$ storage (S_{PO_4} uptake) per $X_{PAO,Stor}$ utilized	Y_{PHA}	$1/Y_{Stor\_PP}$	g $X_{PP} \cdot$ g X_{Stor}^{-1}	0.2
Yield for $X_{PAO,PP}$ requirement (S_{PO_4} release) per $X_{PAO,Stor}$ stored (S_{VFA} utilized)	Y_{PO_4}	$Y_{PP\_Stor,PAO}$	g $X_{PP} \cdot$ g X_{Stor}^{-1} or g $X_{PO_4} \cdot$ g S_{VFA}^{-1}	0.4
Yield of X_{ANO} growth per S_{NO_3}	Y_A	Y_{ANO}	g $X_{AUT} \cdot$ g S_{NOx}^{-1}	0.24
X_{MeOH} requirement per S_{PO_4} utilized	$f_{MeOH\_PO_4,MW}$	$f_{MeOH\_PO_4,MW}$	g $X_{MeOH} \cdot$ g $S_{PO_4}^{-1}$	−3.45
X_{MeP} formation per S_{PO_4} utilized	$f_{MeP\_PO_4,MW}$	$f_{MeP\_PO_4,MW}$	g $X_{MeP} \cdot$ g $S_{PO_4}^{-1}$	4.86
Conversion factor for NO_3 reduction to N_2	$i_{NO3,N2}$	$i_{NO3,N2}$	g COD $\cdot$ g N^{-1}	2.857
Conversion factor for NO_3 in COD	i_{COD_NO3}	i_{COD_NO3}	g COD $\cdot$ g N^{-1}	−4.571
Conversion factor for N_2 in COD	i_{COD_N2}	i_{COD_N2}	g COD $\cdot$ g N^{-1}	−1.714
Conversion factor for S_{VFA} (CH_3COO^-) in charge	$i_{Charge\_S_{VFA}}$	$i_{Charge\_S_{VFA}}$	Charge $\cdot$ g COD^{-1}	−0.016
Conversion factor for NH_x in charge	$i_{Charge\_S_{NHx}}$	$i_{Charge\_S_{NHx}}$	Charge $\cdot$ g N^{-1}	0.071
Conversion factor for NO_3 in charge	$i_{Charge\_S_{NOx}}$	$i_{Charge\_S_{NOx}}$	Charge $\cdot$ g N^{-1}	−0.071
Conversion factor for PO_4 in charge	$i_{Charge\_S_{PO_4}}$	$i_{Charge\_S_{PO_4}}$	Charge $\cdot$ g P^{-1}	−0.048

TABLE A.2 Activated Sludge Model No. 2d (*continued*)

Stoichiometry	Parameter notation		Units	Value* T = 20°C
	Original	Standardized		
Conversion factor for $X_{PAO,PP}$ $(K_{0.33}Mg_{0.33}PO_3)_n$ in charge	$i_{Charge\_X_{PAO,PP}}$	$i_{Charge\_X_{PAO,PP}}$	Charge·g P^{-1}	−0.032
N content of S_F	$i_{N,SF}$	i_{N_SF}	g N·g S_F^{-1}	0.03
N content of S_U	$i_{N,SI}$	i_{N_SU}	g N·g S_U^{-1}	0.01
N content of X_U	$i_{N,XI}$	i_{N_XU}	g N·g X_U^{-1}	0.02
N content of X_B	$i_{N,XS}$	i_{N_XCB}	g N·g XC_B^{-1}	0.04
N content of biomass $(X_{OHO}, X_{PAO}, X_{ANO})$	$i_{N,BM}$	i_{N_XBio}	g N·g X_{Bio}^{-1}	0.07
P content of S_F	$i_{P,SF}$	i_{P_SF}	g P·g S_F^{-1}	0.01
P content of S_U	$i_{P,SI}$	i_{P_SU}	g P·g S_U^{-1}	0
P content of X_U	$i_{P,XI}$	i_{P_XU}	g P·g X_U^{-1}	0.01
P content of X_B	$i_{P,XS}$	i_{P_XCB}	g P·g X_B^{-1}	0.01
P content of biomass $(X_{OHO}, X_{PAO}, X_{ANO})$	$i_{P,BM}$	i_{P_XBio}	g P·g X_{Bio}^{-1}	0.02
Conversion factor for X_{MeP} $(FePO_4)$ in P	i_{P_MeP}	i_{P_XMeP}	g P·g X_{MeP}^{-1}	0.206
Conversion factor X_U in TSS	$i_{TSS,XI}$	i_{TSS_XU}	g TSS·g X_U^{-1}	0.75
Conversion factor X_B in TSS	$i_{TSS,XS}$	i_{TSS_XCB}	g TSS·g XC_B^{-1}	0.75
Conversion factor $X_{PAO,Stor}$ in TSS	$i_{TSS,XPHA}$	$i_{TSS\_X_{PAO,Stor}}$	g TSS·g X_{Stor}^{-1}	0.6
Conversion factor biomass in TSS	$i_{TSS,BM}$	i_{TSS_XBio}	g TSS·g X_{Bio}^{-1}	0.9
Conversion factor $X_{PAO,PP}$ in TSS	$i_{TSS,XPP}$	$i_{TSS\_X_{PAO,PP}}$	g TSS·g X_{PP}^{-1}	3.23

TABLE A.2 Activated Sludge Model No. 2d (*continued*)

Kinetics	Parameter notation		Units	Value* $T = 20°C$
	Original	Standardized		
Maximum specific hydrolysis rate	K_h	$q_{XC_B\_SB,hyd}$	g $XC_B \cdot$ g $X_{OHO}^{-1} \cdot d^{-1}$	3
Half-saturation coefficient for XC_B/X_{OHO}	K_X	$K_{XCB,hyd}$	g $XC_B \cdot$ g X_{OHO}^{-1}	0.1
Correction factor for hydrolysis under anoxic conditions	$\eta_{NO3,HYD}$	$n_{qhyd,Ax}$	–	0.6
Correction factor for hydrolysis under anaerobic conditions	η_{fe}	$n_{qhyd,An}$	–	0.4
Half-saturation/inhibition coefficient for S_{O2}	$K_{O2,HYD}$	$K_{O2,hyd}$	g $S_{O2} \cdot m^{-3}$	0.2
Half-saturation/inhibition coefficient for S_{NOx}	$K_{NO3,HYD}$	$K_{NOx,hyd}$	g N $\cdot m^{-3}$	0.5
Maximum growth rate of X_{OHO}	μ_H	$\mu_{OHO,Max}$	d^{-1}	6
Reduction factor for anoxic growth of X_{OHO}	$\eta_{NO3,H}$	$n_{\mu OHO,Ax}$	–	0.8
Rate constant for fermentation/ Maximum specific fermentation growth rate	q_{fe}	$q_{SF\_VFA,Max}$	g $S_F \cdot$ g $X_{OHO}^{-1} \cdot d^{-1}$	3
Half saturation parameter for S_F	K_F	$K_{SF,OHO}$	g $S_F \cdot m^{-3}$	4
Half saturation parameter for S_{VFA}	$K_{A,H}$	$K_{VFA,OHO}$	g $S_{VFA} \cdot m^{-3}$	4
Decay rate for X_{OHO}	b_H	b_{OHO}	d^{-1}	0.4
Half-saturation coefficient for fermentation of S_F	K_{fe}	K_{fe}	g $S_F \cdot m^{-3}$	4
Half-saturation coefficient for S_{O2}	$K_{O2,H}$	$K_{O2,OHO}$	g $S_{O2} \cdot m^{-3}$	0.2
Half-saturation coefficient for S_{NOx}	$K_{NO3,H}$	$K_{NOx,OHO}$	g $S_{NOx} \cdot m^{-3}$	0.5
Half-saturation coefficient for S_{NHx}	$K_{NH4,H}$	$K_{NHx,OHO}$	g $S_{NHx} \cdot m^{-3}$	0.05
Half-saturation coefficient for S_{PO4}	$K_{P,H}$	$K_{PO4,OHO}$	g $S_{PO4} \cdot m^{-3}$	0.01
Half-saturation coefficient for S_{Alk}	$K_{ALK,H}$	$K_{Alk,OHO}$	mol $HCO_3^- \cdot m^{-3}$	0.1
Rate constant for S_{VFA} uptake rate ($X_{PAO,Stor}$ storage)	q_{PHA}	q_{PAO,VFA_Stor}	g $X_{Stor} \cdot$ g $X_{PAO}^{-1} \cdot d^{-1}$	3
Rate constant for storage of $X_{PAO,PP}$	q_{PP}	$q_{PAO,PO4\_PP}$	g $X_{PP} \cdot$ g $X_{PAO}^{-1} \cdot d^{-1}$	1.5
Maximum ratio of $X_{PAO,PP}/X_{PAO}$	K_{PP}	K_{S,fPP_PAO}	g $X_{PP} \cdot$ g X_{PAO}^{-1}	0.01
Half-saturation coefficient for $X_{PAO,PP}/X_{PAO}$	K_{Max}	$f_{PP\_PAO,Max}$	g $X_{PP} \cdot$ g X_{PAO}^{-1}	0.34

| Kinetics | Parameter notation | | Units | Value* |
	Original	Standardized		T = 20°C
Half-inhibition coefficient for $X_{PAO,PP}/X_{PAO}$	K_{iPP}	K_{I,fPP_PAO}	g $X_{PP} \cdot$ g X_{PAO}^{-1}	0.02
Maximum growth rate of X_{PAO}	μ_{PAO}	$\mu_{PAO,Max}$	d^{-1}	1
Reduction factor for anoxic growth of X_{PAO}	$\eta_{NO3,PAO}$	$n_{\mu PAO}$	–	0.6
Saturation constant for $X_{PAO,Stor}/X_{PAO}$	K_{PHA}	K_{fStor_PAO}	g $X_{Stor} \cdot$ g X_{PAO}^{-1}	0.01
Endogenous respiration rate of X_{PAO}	b_{PAO}	b_{PAO}	d^{-1}	0.2
Rate constant for lysis of $X_{PAO,PP}$	b_{PP}	$b_{PP\_PO_4}$	d^{-1}	0.2
Rate constant for respiration of $X_{PAO,Stor}$	b_{PHA}	b_{Stor_VFA}	d^{-1}	0.2
Half-saturation coefficient for S_{VFA}	$K_{A,PAO}$	$K_{VFA,PAO}$	g $S_{VFA} \cdot m^{-3}$	4
Half-saturation coefficient for S_{O_2}	$K_{O2,PAO}$	$K_{O2,PAO}$	g $S_{O_2} \cdot m^{-3}$	0.2
Half-saturation coefficient for S_{NOx}	$K_{NO3,PAO}$	$K_{NOx,PAO}$	g $S_{NOx} \cdot m^{-3}$	0.5
Half-saturation coefficient for S_{NHx}	$K_{NH4,PAO}$	$K_{NHx,PAO}$	g $S_{NHx} \cdot m^{-3}$	0.05
Half-saturation coefficient for S_{PO_4} uptake ($X_{PAO,PP}$ storage)	K_{PS}	$K_{PO4,PAO,upt}$	g $S_{PO_4} \cdot m^{-3}$	0.2
Half-saturation coefficient for S_{PO_4} as nutrient (X_{PAO} growth)	$K_{P,PAO}$	$K_{PO4,PAO,nut}$	g $S_{PO_4} \cdot m^{-3}$	0.01
Half-saturation coefficient for S_{Alk}	$K_{ALK,PAO}$	$K_{Alk,PAO}$	mol $HCO_3^- \cdot m^{-3}$	0.1
Maximum growth rate of X_{ANO}	μ_{AUT}	$\mu_{ANO,Max}$	d^{-1}	1
Decay rate for X_{ANO}	b_{AUT}	b_{ANO}	d^{-1}	0.15
Half-saturation coefficient for S_{O_2}	$K_{O2,AUT}$	$K_{O2,ANO}$	g $S_{O_2} \cdot m^{-3}$	0.5
Half-saturation coefficient for S_{NHx}	$K_{NH4,AUT}$	$K_{NHx,ANO}$	g $S_{NHx} \cdot m^{-3}$	1
Half-saturation coefficient for S_{PO_4}	$K_{P,AUT}$	$K_{PO4,ANO}$	g $S_{PO_4} \cdot m^{-3}$	0.01
Half-saturation coefficient for S_{Alk}	$K_{ALK,AUT}$	$K_{Alk,ANO}$	mol $HCO_3^- \cdot m^{-3}$	0.5
Rate constant for P precipitation	k_{PRE}	$q_{P,pre}$	$m^3 \cdot$ g $Fe(OH)_3^{-1} \cdot d^{-1}$	1
Rate constant for redissolution	k_{RED}	$q_{P,red}$	d^{-1}	0.6
Half-saturation coefficient for alkalinity	$K_{ALK,PRE}$	$K_{Alk,pre}$	mol $HCO_3^- \cdot m^{-3}$	0.5

TABLE A.3   New General Model

Original publication   Barker, P. S.; Dold, P. L. (1997) General Model for Biological Nutrient Removal Activated Sludge Systems: Model

		Ordinary heterotrophic organisms	Autotrophic nitrifying organisms (NH_4^+ to NO_3^-)	Phosphorus accumulating organisms	Particulate undegradable endogenous products	Enmeshed slowly biodegradable substrate	Fermentable organic matter	Fermentation product (volatile fatty acids)	Stored poly-β-hydroxyalkanoate in PAOs	Particulate undegradable organics from the influent	Soluble undegradable organics
	Original parameter symbol	Z_H	Z_A	Z_P	Z_E	S_{ENM}	S_{BSC}	S_{BSA}	S_{PHB}	S_{UP}	S_{US}
	Units	g COD/m³	g COD/m³	g COD/m³	g COD/m³	g COD/m³	g COD/m³	g COD/m³	g COD/m³	g COD/m³	g COD/m³
		X_{OHO}	X_{ANO}	X_{PAO}	$X_{U,E}$	XC_B	S_F	S_{VFA}	$X_{PAO,Stor}$	$X_{U,Inf}$	S_U
1	Aerobic growth of X_{OHO} on S_F with S_{NHx}	1					$-1/Y_{OHO,Ox}$				
2	Anoxic growth of X_{OHO} on S_F with S_{NHx}	1					$-1/Y_{OHO,Ax}$				
3	Aerobic growth of X_{OHO} on S_F with S_{NOx}	1					$-1/Y_{OHO,Ox}$				
4	Anoxic growth of X_{OHO} on S_F with S_{NOx}	1					$-1/Y_{OHO,Ax} + i_{COD\_NO3}^* i_{N\_OHO}$				
5	Aerobic growth of X_{OHO} on S_{VFA} with S_{NHx}	1						$-1/Y_{OHO,Ox}$			
6	Anoxic growth of X_{OHO} on S_{VFA} with S_{NHx}	1						$-1/Y_{OHO,Ax}$			
7	Aerobic growth of X_{OHO} on S_{VFA} with S_{NOx}	1						$-1/Y_{OHO,Ox}$			
8	Anoxic growth of X_{OHO} on S_{VFA} with S_{NOx}	1						$-1/Y_{OHO,Ax} + i_{COD\_NO3}^* i_{N\_OHO}$			
9	Decay of X_{OHO}	-1			$f_{XU\_OHO,lys}$	$1 - f_{XU\_OHO,lys}$					
10	Aerobic hydrolysis of XC_B					-1	1				

Presentation. *Water Environ. Res.*, 69, 969–984.

Releasable stored polyphosphates in PAOs	Non-releasable stored polyphosphates in PAOs	Inorganic soluble phosphorus	Particulate biodegradable organic N	Soluble biodegradable organic N	Nitrate and nitrite (NO_3 + NO_2) (considered to be NO_3 only for stoichiometry)	Ammonia (NH_4^+ + NH_3)	Soluble inert organic N	Dissolved oxygen	Soluble nitrogen	Soluble hydrogen	Process rate
$P_{PP\text{-}LO}$	$P_{PP\text{-}HI}$	P_{O_4}	N_{BP}	N_{BS}	N_{O3}	N_{H3}	N_{US}	S_O	–	–	
g P/m³	g P/m³	g P/m³	g N/m³	g N/m³	g N/m³	g N/m³	g N/m³	g COD/m³	g N/m³	g H/m³	
$X_{PAO,PP,Lo}$	$X_{PAO,PP,Hi}$	S_{PO_4}	$XC_{B,N}$	$S_{B,N}$	S_{NOx}	S_{NHx}	$S_{U,N}$	S_{O_2}	S_{N_2}	S_{H_2}	
		$-i_{P\_OHO}$				$-i_{N\_OHO}$		$-(1-Y_{OHO,Ox})/Y_{OHO,Ox}$			(1)
		$-i_{P\_OHO}$			$-(1-Y_{OHO,Ax})/(i_{NO3,N2}{}^*Y_{OHO,Ax})$	$-i_{N\_OHO}$			$(1-Y_{OHO,Ax})/(i_{NO3,N2}{}^*Y_{OHO,Ax})$		(2)
		$-i_{P\_OHO}$			$-i_{N\_OHO}$			$-(1-Y_{OHO,Ox})/Y_{OHO,Ox}-i_{COD\_NO3}{}^*i_{N\_OHO}$			(3)
		$-i_{P\_OHO}$			$-i_{N\_OHO}-(1-Y_{OHO,Ax})/(i_{NO3,N2}{}^*Y_{OHO,Ax})$				$(1-Y_{OHO,Ax})/(i_{NO3,N2}{}^*Y_{OHO,Ax})$		(4)
		$-i_{P\_OHO}$				$-i_{N\_OHO}$		$-(1-Y_{OHO,Ox})/Y_{OHO,Ox}$			(5)
		$-i_{P\_OHO}$			$-(1-Y_{OHO,Ax})/(i_{NO3,N2}{}^*Y_{OHO,Ax})$	$-i_{N\_OHO}$			$(1-Y_{OHO,Ax})/(i_{NO3,N2}{}^*Y_{OHO,Ax})$		(6)
		$-i_{P\_OHO}$			$-i_{N\_OHO}$			$-(1-Y_{OHO,Ox})/Y_{OHO,Ox}-i_{COD\_NO3}{}^*i_{N\_OHO}$			(7)
		$-i_{P\_OHO}$			$-i_{N\_OHO}-(1-Y_{OHO,Ax})/(i_{NO3,N2}{}^*Y_{OHO,Ax})$				$(1-Y_{OHO,Ax})/(i_{NO3,N2}{}^*Y_{OHO,Ax})$		(8)
		$i_{P\_OHO}-f_{XU\_OHO,lys}{}^*i_{P\_XUE,OHO}$	$i_{N\_OHO}-f_{XU\_OHO,lys}{}^*i_{N\_XUE,OHO}$								(9)
											(10)

TABLE A.3 New General Model (*continued*)

		Ordinary heterotrophic organisms	Autotrophic nitrifying organisms (NH_4^+ to NO_3^-)	Phosphorus accumulating organisms	Particulate undegradable endogenous products	Enmeshed slowly biodegradable substrate	Fermentable organic matter	Fermentation product (volatile fatty acids)	Stored poly-β-hydroxyalkanoate in PAOs	Particulate undegradable organics from the influent	Soluble undegradable organics
Original parameter symbol		Z_H	Z_A	Z_P	Z_E	S_{ENM}	S_{BSC}	S_{BSA}	S_{PHB}	S_{UP}	S_{US}
Units		g COD/m³	g COD/m³	g COD/m³	g COD/m³	g COD/m³	g COD/m³	g COD/m³	g COD/m³	g COD/m³	g COD/m³
		X_{OHO}	X_{ANO}	X_{PAO}	$X_{U,E}$	XC_B	S_F	S_{VFA}	$X_{PAO,Stor}$	$X_{U,Inf}$	S_U
11	Anoxic hydrolysis of XC_B					−1	$Y_{hyd,Ax}$				
12	Anaerobic hydrolysis of XC_B					−1	$Y_{hyd,An}$				
13	Hydrolysis of organic N										
14	Ammonification										
15	Fermentation of S_F to S_{VFA}	$Y_{OHO,An}$					−1	$(1 - Y_{OHO,An}) \times Y_{fe}$			
16	Growth of X_{ANO}		1								
17	Decay of X_{ANO}		−1		$f_{XU\_ANO,lys}$	$1 - f_{XU\_ANO,lys}$					
18	Aerobic growth of X_{PAO} with S_{NHx}			1					$-1/Y_{PAO}$		
19	Aerobic growth of X_{PAO} with S_{NOx}			1					$-1/Y_{PAO}$		
20	Aerobic growth of X_{PAO} with S_{NHx}/S_{PO_4} limited			1					$-1/Y_{PAO}$		
21	Aerobic growth of X_{PAO} with S_{NOx}/S_{PO_4} limited			1					$-1/Y_{PAO}$		
22	Anoxic growth of X_{PAO} with S_{NHx}			1					$-1/Y_{PAO}$		

Releasable stored polyphosphates in PAOs	Non-releasable stored polyphosphates in PAOs	Inorganic soluble phosphorus	Particulate biodegradable organic N	Soluble biodegradable organic N	Nitrate and nitrite (NO3 + NO2) (considered to be NO3 only for stoichiometry)	Ammonia (NH4+ + NH3)	Soluble inert organic N	Dissolved oxygen	Soluble nitrogen	Soluble hydrogen	Process rate
P_{PP-LO}	P_{PP-HI}	P_{O_4}	N_{BP}	N_{BS}	N_{O3}	N_{H3}	N_{US}	S_O	–	–	
g P/m³	g P/m³	g P/m³	g N/m³	g N/m³	g N/m³	g N/m³	g N/m³	g COD/m³	g N/m³	g H/m³	
$X_{PAO,PP,Lo}$	$X_{PAO,PP,Hi}$	S_{PO_4}	$XC_{B,N}$	$S_{B,N}$	S_{NOx}	S_{NHx}	$S_{U,N}$	S_{O_2}	S_{N_2}	S_{H_2}	Process rate
										$(1 - Y_{hyd,Ax})/i_{COD\_H2}$	(11)
										$(1 - Y_{hyd,An})/i_{COD\_H2}$	(12)
			-1	1							(13)
				-1		1					(14)
		$-i_{P\_OHO}^* Y_{OHO,An}$				#NAME?				$(1 - (1 - Y_{OHO,An})^* Y_{fe} - Y_{OHO,An})/i_{COD\_H2}$	(15)
		$-i_{P\_ANO}$			$1/Y_{ANO}$	$-i_{N\_ANO} - 1/Y_{ANO}$		$-(-i_{COD\_NO3} - Y_{ANO})/Y_{ANO}$			(16)
		$i_{P\_ANO} - f_{XU\_ANO,lys} \times i_{P\_XUE,ANO}$	$i_{N\_ANO} - f_{XU\_ANO,lys} \times i_{N\_XUE,ANO}$								(17)
$f_{PP,Lo\_PP} \times Y_{Stor\_PP,Ox}/Y_{PAO}$	$(1 - f_{PP,Lo\_PP}) \times Y_{Stor\_PP,Ox}/Y_{PAO}$	$-Y_{Stor\_PP,Ox}/Y_{PAO} - i_{P\_PAO}$				$-i_{N\_PAO}$		$-(1 - Y_{PAO})/Y_{PAO}$			(18)
$f_{PP,Lo\_PP} \times Y_{Stor\_PP,Ox}/Y_{PAO}$	$(1 - f_{PP,Lo\_PP}) \times Y_{Stor\_PP,Ox}/Y_{PAO}$	$-Y_{Stor\_PP,Ox}/Y_{PAO} - i_{P\_PAO}$			$-i_{N\_PAO}$			$-(1 - Y_{PAO})/Y_{PAO} - i_{COD\_NO3}^* i_{N\_PAO}$			(19)
$-i_{P\_PAO}$						$-i_{N\_PAO}$		$-(1 - Y_{PAO})/Y_{PAO}$			(20)
$-i_{P\_PAO}$					$-i_{N\_PAO}$			$-(1 - Y_{PAO})/Y_{PAO} - i_{COD\_NO3}^* i_{N\_PAO}$			(21)
$_{PP,Lo\_PP} \times {}_{Stor\_PP,Ax}/{}_{PAO}$	$(1 - f_{PP,Lo\_PP}) \times Y_{Stor\_PP,Ax}/Y_{PAO}$	$-Y_{Stor\_PP,Ax}/Y_{PAO} - i_{P\_PAO}$			$-(1 - Y_{PAO})/(i_{NO3,N2} \times Y_{PAO})$	$-i_{N\_PAO}$			$(1 - Y_{PAO})/(i_{NO3,N2}^* Y_{PAO})$		(22)

TABLE A.3 New General Model (*continued*)

		Ordinary heterotrophic organisms	Autotrophic nitrifying organisms (NH_4^+ to NO_3^-)	Phosphorus accumulating organisms	Particulate undegradable endogenous products	Enmeshed slowly biodegradable substrate	Fermentable organic matter	Fermentation product (volatile fatty acids)	Stored poly-β-hydroxyalkanoate in PAOs	Particulate undegradable organics from the influent	Soluble undegradable organics
Original parameter symbol		Z_H	Z_A	Z_P	Z_E	S_{ENM}	S_{BSC}	S_{BSA}	S_{PHB}	S_{UP}	S_{US}
Units		g COD/m³	g COD/m³	g COD/m³	g COD/m³	g COD/m³	g COD/m³	g COD/m³	g COD/m³	g COD/m³	g COD/m³
		X_{OHO}	X_{ANO}	X_{PAO}	$X_{U,E}$	XC_B	S_F	S_{VFA}	$X_{PAO,Stor}$	$X_{U,Inf}$	S_U
23	Aerobic decay of X_{PAO}			-1	$f_{XU\_PAO,lys}$						$f_{SU\_PAO,lys}$
24	$X_{PAO,PP,Lo}$ lysis on aerobic decay										
25	$X_{PAO,PP,Hi}$ lysis on aerobic decay										
26	S_F lysis on aerobic decay							1	-1		
27	Anoxic decay of X_{PAO}			-1	$f_{XU\_PAO,lys}$						$f_{SU\_PAO,lys}$
28	$X_{PAO,PP,Lo}$ lysis on anoxic decay										
29	$X_{PAO,PP,Hi}$ lysis on anoxic decay										
30	S_F lysis on anoxic decay							1	-1		
31	Anaerobic decay of X_{PAO}			-1	$f_{XU\_PAO,lys}$						$1 - f_{XU\_PAO,lys}$
32	$X_{PAO,PP,Lo}$ lysis on anaerobic decay										

Releasable stored polyphosphates in PAOs	Non-releasable stored polyphosphates in PAOs	Inorganic soluble phosphorus	Particulate biodegradable organic N	Soluble biodegradable organic N	Nitrate and nitrite (NO₃ + NO₂) (considered to be NO₃ only for stoichiometry)	Ammonia (NH₄⁺ + NH₃)	Soluble inert organic N	Dissolved oxygen	Soluble nitrogen	Soluble hydrogen	Process rate
$P_{PP\text{-}LO}$	$P_{PP\text{-}HI}$	P_{O_4}	N_{BP}	N_{BS}	N_{O3}	N_{H3}	N_{US}	S_O	–	–	
g P/m³	g P/m³	g P/m³	g N/m³	g N/m³	g N/m³	g N/m³	g N/m³	g COD/m³	g N/m³	g H/m³	
$X_{PAO,PP,Lo}$	$X_{PAO,PP,Hi}$	S_{PO_4}	$XC_{B,N}$	$S_{B,N}$	S_{NOx}	S_{NHx}	$S_{U,N}$	S_{O_2}	S_{N_2}	S_{H_2}	
		$i_{P\_PAO} - f_{XU\_PAO,lys} \times i_{P\_XUE,PAO}$		$f_{SU\_PAO,lys} \times i_{N\_SU}$		$i_{N\_PAO} - f_{XU\_PAO,lys} \times i_{N\_XUE,PAO} - f_{SU\_PAO,lys} \times i_{N\_SU}$		$-(1 - f_{XU\_PAO,lys} - f_{SU\_PAO,lys})$			(23)
−1		1									(24)
	−1	1									(25)
											(26)
		$i_{P\_PAO} - f_{XU\_PAO,lys} \times i_{P\_XUE,PAO}$		$f_{SU\_PAO,lys} \times i_{N\_SU}$	$-(1 - f_{XU\_PAO,lys} - f_{SU\_PAO,lys})/i_{NO3,N2}$	$i_{N\_PAO} - f_{XU\_PAO,lys} \times i_{N\_XUE,PAO} - f_{SU\_PAO,lys} \times i_{N\_SU}$			$(1 - f_{XU\_PAO,lys} - f_{SU\_PAO,lys})/i_{NO3,N2}$		(27)
−1		1									(28)
	−1	1									(29)
											(30)
		$i_{P\_PAO} - f_{XU\_PAO,lys} \times i_{P\_XUE,PAO}$		$f_{SU\_PAO,lys} \times i_{N\_SU}$		$i_{N\_PAO} - f_{XU\_PAO,lys} \times i_{N\_XUE,PAO} - f_{SU\_PAO,lys} \times i_{N\_SU}$					(31)
−1		1									(32)

TABLE A.3   New General Model (*continued*)

		Ordinary heterotrophic organisms	Autotrophic nitrifying organisms (NH_4^+ to NO_3^-)	Phosphorus accumulating organisms	Particulate undegradable endogenous products	Enmeshed slowly biodegradable substrate	Fermentable organic matter	Fermentation product (volatile fatty acids)	Stored poly-β-hydroxyalkanoate in PAOs	Particulate undegradable organics from the influent	Soluble undegradable organics
Original parameter symbol		Z_H	Z_A	Z_P	Z_E	S_{ENM}	S_{BSC}	S_{BSA}	S_{PHB}	S_{UP}	S_{US}
Units		g COD/m³	g COD/m³	g COD/m³	g COD/m³	g COD/m³	g COD/m³	g COD/m³	g COD/m³	g COD/m³	g COD/m³
		X_{OHO}	X_{ANO}	X_{PAO}	$X_{U,E}$	XC_B	S_F	S_{VFA}	$X_{PAO,Stor}$	$X_{U,Inf}$	S_U
33	$X_{PAO,PP,Hi}$ lysis on anaerobic decay										
34	S_F lysis on anaerobic decay							1	−1		
35	Cleavage of $X_{PAO,PP,Lo}$ for anaerobic maintenance										
36	Sequestration of S_{VFA} by X_{PAO}							−1	$Y_{VFA\_Stor,PAO}$		

[1] $\mu_{OHO,Max} \times [S_{O_2}/(K_{O_2,OHO} + S_{O_2})] \times [S_F/(K_{SF,OHO} + S_F)] \times [S_F/(S_F + S_{VFA})] \times [S_{NHx}/(K_{NHx,OHO} + S_{NHx})] \times [S_{PO_4}/(K_{PO_4,Bio,nut} + S_{PO_4})] \times X_{OHO}$

[2] $\mu_{OHO,Max} \times \eta_{\mu OHO,Ax} \times K_{O_2,OHO}/(K_{O_2,OHO} + S_{O_2})] \times [S_F/(K_{SF,OHO} + S_F)] \times [S_F/(S_F + S_{VFA})] \times [S_{NOx}/(K_{NOx,OHO} + S_{NOx})] \times [S_{NHx}/(K_{NHx,OHO} + S_{NHx})] \times [S_{PO_4}/(K_{PO_4,Bionut} + S_{PO_4})] \times X_{OHO}$

[3] $\mu_{OHO,Max} \times [S_{O_2}/(K_{O_2,OHO} + S_{O_2})] \times [S_F/(K_{SF,OHO} + S_F)] \times [S_F/(S_F + S_{VFA})] \times [S_{NOx}/(K_{NOx,OHO} + S_{NOx})] \times [K_{NHx,OHO}/(K_{NHx,OHO} + S_{NHx})] \times [S_{PO_4}/(K_{PO_4,Bio,nut} + S_{PO_4})] \times X_{OHO}$

[4] $\mu_{OHO,Max} \times \eta_{\mu OHO,Ax} [K_{O_2,OHO}/(K_{O_2,OHO} + S_{O_2})] \times [S_F/(K_{SF,OHO} + S_F)] \times [S_F/(S_F + S_{VFA})] \times [S_{NOx}/(K_{NOx,OHO} + S_{NOx})] \times [K_{NHx,OHO}/(K_{NHx,OHO} + S_{NHx})] \times [S_{PO_4}/(K_{PO_4,Bio,nut} + S_{PO_4})] \times X_{OHO}$

[5] $\mu_{OHO,Max} \times [S_{O_2}/(K_{O_2,OHO} + S_{O_2})] \times [S_{VFA}/(K_{VFA,OHO} + S_{VFA})] \times [S_{VFA}/(S_F + S_{VFA})] \times [S_{NHx}/(K_{NHx,OHO} + S_{NHx})] \times [S_{PO_4}/(K_{PO_4,Bio,nut} + S_{PO_4})] \times X_{OHO}$

[6] $\mu_{OHO,Max} \times \eta_{\mu OHO,Ax} \times [K_{O_2,OHO}/(K_{O_2,OHO} + S_{O_2})] \times [S_{VFA}/(K_{VFA,OHO} + S_{VFA})] \times [S_{VFA}/(S_F + S_{VFA})] \times [S_{NOx}/(K_{NOx,OHO} + S_{NOx})] \times [S_{NHx}/(K_{NHx,OHO} + S_{NHx})] \times [S_{PO_4}/(K_{PO_4,Bio,nut} + S_{PO_4})] \times X_{OHO}$

[7] $\mu_{OHO,Max} \times [S_{O_2}/(K_{O_2,OHO} + S_{O_2})] \times [S_{VFA}/(K_{VFA,OHO} + S_{VFA})] \times [S_{VFA}/(S_F + S_{VFA})] \times [S_{NOx}/(K_{NOx,OHO} + S_{NOx})] \times [K_{NHx,OHO}/(K_{NHx,OHO} + S_{NHx})] \times [S_{PO_4}/(K_{PO_4,Bio,nut} + S_{PO_4})] \times X_{OHO}$

[8] $\mu_{OHO,Max} \times \eta_{\mu OHO,Ax} \times [K_{O_2,OHO}/(K_{O_2,OHO} + S_{O_2})] \times [S_{VFA}/(K_{VFA,OHO} + S_{VFA})] \times [S_{VFA}/(S_F + S_{VFA})] \times [S_{NOx}/(K_{NOx,OHO} + S_{NOx})] \times [K_{NHx,OHO}/(K_{NHx,OHO} + S_{NHx})] \times [S_{PO_4}/(K_{PO_4,Bio,nut} + S_{PO_4})] \times X_{OHO}$

[9] $m_{OHO} \times X_{OHO}$

[10] $q_{XCB\_SB,hyd} \times [S_{O_2}/(K_{O_2,OHO} + S_{O_2})] \times [(XC_B/X_{OHO})/(K_{XCB,hyd} + (XC_B/X_{OHO}))] \times X_{OHO}$

[11] $q_{XCB\_SB,hyd} \times \eta_{qhyd,Ax} \times [K_{O_2,OHO}/(K_{O_2,OHO} + S_{O_2})] \times [S_{NOx}/(K_{NOx,OHO} + S_{NOx})] \times [(XC_B/X_{OHO})/(K_{XCB,hyd} + (XC_B/X_{OHO}))] \times X_{OHO}$

[12] $q_{XCB\_SB,hyd} \times \eta_{qhyd,An} \times [K_{O_2,OHO}/(K_{O_2,OHO} + S_{O_2})] \times [K_{NOx,OHO}/(K_{NOx,OHO} + S_{NOx})] \times [(XC_B/X_{OHO})/(K_{XCB,hyd} + (XC_B/X_{OHO}))] \times X_{OHO}$

[13] $(r_{10} + r_{11} + r_{12}) \times XC_{B,N}/XC_B$

[14] $q_{am} * S_{B,N} * (X_{OHO} + X_{PAO})$

[15] $q_{SF\_VFA,Max} \times [S_F/(K_{SF,fe} + S_F)] \times [S_{NHx}/(K_{NHx,OHO} + S_{NHx})] \times [S_{PO_4}/(K_{PO_4,Bio,nut} + S_{PO_4})] \times [K_{O_2,OHO}/(K_{O_2,OHO} + S_{O_2})] \times [K_{NOx,OHO}/(K_{NOx,OHO} + S_{NOx})] \times X_{OHO}$

[16] $\mu_{ANO,Max} \times [S_{O_2}/(K_{O_2,ANO} + S_{O_2})] \times [S_{NHx}/(K_{NHx,ANO} + S_{NHx})] \times [S_{PO_4}/(K_{PO_4,Bio,nut} + S_{PO_4})] \times X_{ANO}$

[17] $b_{ANO} \times X_{ANO}$

[18] $\mu_{PAO,Max} \times [S_{O_2}/(K_{O_2,OHO} + S_{O_2})] \times [(X_{PAO,Stor}/X_{PAO})/(K_{fStor\_PAO} + (X_{PAO,Stor}/X_{PAO}))] \times [S_{NHx}/(K_{NHx,OHO} + S_{NHx})] \times [S_{PO_4}/(K_{PO_4,PAO,upt} + S_{PO_4})] \times X_{PAO}$

[19] $\mu_{PAO,Max} \times [S_{O_2}/(K_{O_2,OHO} + S_{O_2})] \times [(X_{PAO,Stor}/X_{PAO})/(K_{fStor\_PAO} + (X_{PAO,Stor}/X_{PAO}))] \times [S_{NOx}/(K_{NOx,OHO} + S_{NOx})] \times [K_{NHx,OHO}/(K_{NHx,OHO} + S_{NHx})] \times [S_{PO_4}/(K_{PO_4,PAO,upt} + S_{PO_4})] \times X_{PAO}$

Releasable stored polyphosphates in PAOs	Non-releasable stored polyphosphates in PAOs	Inorganic soluble phosphorus	Particulate biodegradable organic N	Soluble biodegradable organic N	Nitrate and nitrite (NO_3 + NO_2) (considered to be NO_3 only for stoichiometry)	Ammonia (NH_4^+ + NH_3)	Soluble inert organic N	Dissolved oxygen	Soluble nitrogen	Soluble hydrogen	Process rate
$P_{PP\text{-}LO}$	$P_{PP\text{-}HI}$	P_{O_4}	N_{BP}	N_{BS}	N_{O3}	N_{H3}	N_{US}	S_O	–	–	
g P/m³	g P/m³	g P/m³	g N/m³	g N/m³	g N/m³	g N/m³	g N/m³	g COD/m³	g N/m³	g H/m³	
$X_{PAO,PP,Lo}$	$X_{PAO,PP,Hi}$	S_{PO_4}	$XC_{B,N}$	$S_{B,N}$	S_{NOx}	S_{NHx}	$S_{U,N}$	S_{O_2}	S_{N_2}	S_{H_2}	
	−1	1									(33)
											(34)
−1		1									(35)
$-Y_{PP\_Stor,PAO}$		$Y_{PP\_Stor,PAO}$								$(1 - Y_{VFA\_Stor,PAO})/$ i_{COD_H2}	(36)

$\mu_{PAO,Max,Plim} \times [S_{O_2}/(K_{O_2,OHO} + S_{O_2})] \times [(X_{PAO,Stor}/X_{PAO})/(K_{fStor\_PAO,Plim} + (X_{PAO,Stor}/X_{PAO}))] \times [S_{NHx}/(K_{NHx,OHO} + S_{NHx})] \times [K_{PO_4,PAO,upt}/(K_{PO_4,PAO,upt} + S_{PO_4})] \times [X_{PAO,PP,Lo}/(K_{PP,PAO} + X_{PAO,PP,Lo})] \times X_{PAO}$

$\mu_{PAO,Max,Plim} \times [S_{O_2}/(K_{O_2,OHO} + S_{O_2})] \times [(X_{PAO,Stor}/X_{PAO})/(K_{fStor\_PAO,Plim} + (X_{PAO,Stor}/X_{PAO}))] \times [S_{NOx}/(K_{NOx,OHO} + S_{NOx})] \times [K_{NHx,OHO}/(K_{NHx,OHO} + S_{NHx})] \times [K_{PO_4,PAO,upt}/(K_{PO_4,PAO,upt} + S_{PO_4})] \times X_{PAO,PP,Lo}/(K_{PP,PAO} + X_{PAO,PP,Lo})] \times X_{PAO}$

$\mu_{PAO,Max} \times \eta_{\mu PAO} \times [K_{O_2,OHO}/(K_{O_2,OHO} + S_{O_2})] \times [(X_{PAO,Stor}/X_{PAO})/(K_{fStor\_PAO} + (X_{PAO,Stor}/X_{PAO}))] \times [S_{NOx}/(K_{NOx,OHO} + S_{NOx})] \times [S_{NHx}/(K_{NHx,OHO} + S_{NHx})] \times [S_{PO_4}/(K_{PO_4,PAO,upt} + S_{PO_4})] \times X_{PAO}$

$m_{PAO} \times [S_{O_2}/(K_{O_2,OHO} + S_{O_2})] \times X_{PAO}$

$m_{PAO} \times [S_{O_2}/(K_{O_2,OHO} + S_{O_2})] \times [X_{PAO,PP,Lo}/X_{PAO}] \times X_{PAO}$

$m_{PAO} \times [S_{O_2}/(K_{O_2,OHO} + S_{O_2})] \times [X_{PAO,PP,Hi}/X_{PAO}] \times X_{PAO}$

$m_{PAO} \times [S_{O_2}/(K_{O_2,OHO} + S_{O_2})] \times [X_{PAO,Stor}/X_{PAO}] \times X_{PAO}$

$m_{PAO} \times [K_{O_2,OHO}/(K_{O_2,OHO} + S_{O_2})] \times [S_{NOx}/(K_{NOx,OHO} + S_{NOx})] \times X_{PAO}$

$m_{PAO} \times [K_{O_2,OHO}/(K_{O_2,OHO} + S_{O_2})] \times [S_{NOx}/(K_{NOx,OHO} + S_{NOx})] \times [X_{PAO,PP,Lo}/X_{PAO}] \times X_{PAO}$

$m_{PAO} \times [K_{O_2,OHO}/(K_{O_2,OHO} + S_{O_2})] \times [S_{NOx}/(K_{NOx,OHO} + S_{NOx})] \times [X_{PAO,PP,Hi}/X_{PAO}] \times X_{PAO}$

$m_{PAO} \times [K_{O_2,OHO}/(K_{O_2,OHO} + S_{O_2})] \times [S_{NOx}/(K_{NOx,OHO} + S_{NOx})] \times [X_{PAO,Stor}/X_{PAO}] \times X_{PAO}$

$_{PAO} \times [K_{O_2,OHO}/(K_{O_2,OHO} + S_{O_2})] \times [K_{NOx,OHO}/(K_{NOx,OHO} + S_{NOx})] \times X_{PAO}$

$_{PAO} \times [K_{O_2,OHO}/(K_{O_2,OHO} + S_{O_2})] \times [K_{NOx,OHO}/(K_{NOx,OHO} + S_{NOx})] \times [X_{PAO,PP,Lo}/X_{PAO}] \times X_{PAO}$

$_{PAO} \times [K_{O_2,OHO}/(K_{O_2,OHO} + S_{O_2})] \times [K_{NOx,OHO}/(K_{NOx,OHO} + S_{NOx})] \times [X_{PAO,PP,Hi}/X_{PAO}] \times X_{PAO}$

$_{PAO} \times [K_{O_2,OHO}/(K_{O_2,OHO} + S_{O_2})] \times [K_{NOx,OHO}/(K_{NOx,OHO} + S_{NOx})] \times [X_{PAO,Stor}/X_{PAO}] \times X_{PAO}$

$_{P\_PO_4} \times [K_{O_2,OHO}/(K_{O_2,OHO} + S_{O_2})] \times [X_{PAO,PP,Lo}/(K_{PP,PAO} + X_{PAO,PP,Lo})] \times X_{PAO}$

$_{AO,VFA\_Stor} \times [S_{VFA}/(K_{VFA,PAO} + S_{VFA})] \times [X_{PAO,PP,Lo}/(K_{PP,PAO} + X_{PAO,PP,Lo})] \times X_{PAO}$

TABLE A.3   New General Model (*continued*)

Stoichiometry	Parameter notation		Units	Value* T = 20°C
	Original	Standardized		
Hydrolysis efficiency factor (anoxic)	E_{ANOX}	$Y_{hyd,Ax}$	g S_F/g XCB	0.9
Hydrolysis efficiency factor (anaerobic)	E_{ANA}	$Y_{hyd,An}$	g S_F/g XCB	0.6
Yield for X_{OHO} growth (aerobic)	$Y_{H,AER}$	$Y_{OHO,Ox}$	g XOHO/g XCB	0.666
Yield for X_{OHO} growth (anoxic)	$Y_{H,ANOX}$	$Y_{OHO,Ax}$	g XOHO/g XCB	0.666
Yield for X_{OHO} growth (anaerobic)	$Y_{H,ANA}$	$Y_{OHO,An}$	g XOHO/g XCB	0.1
Fraction of X_U generated in heterotrophic biomass decay	$f_{EP,H}$	$f_{XU\_OHO,lys}$	g XU/g XOHO	0.08
Yield for fermentation	Y_{AC}	Y_{fe}	g S_{VFA}/g SB	0.5
Yield for X_{PAO} growth per $X_{PAO,Stor}$	Y_P	Y_{PAO}	g XPAO/g XStor	0.639
Yield for $X_{PAO,PP}$ storage (S_{PO_4} uptake) per $X_{PAO,Stor}$ utilized (aerobic)	$f_{P,UPT1}$	$Y_{Stor\_PP,Ox}$	g XPP/g Xstor	0.95
Yield for $X_{PAO,PP}$ storage (S_{PO_4} uptake) per $X_{PAO,Stor}$ utilized (anoxic)	$f_{P,UPT2}$	$Y_{Stor\_PP,Ax}$	g XPP/g Xstor	0.55
Yield for $X_{PAO,Stor}$ storage per S_{VFA}	Y_{PHB}	$Y_{VFA\_Stor,PAO}$	g XStor/g S_{VFA}	0.889
Yield for $X_{PAO,PP}$ requirement (S_{PO_4} release) per $X_{PAO,Stor}$ stored (S_{VFA} utilized)	$f_{P,REL}$	$Y_{PP\_Stor,PAO}$	g XPP/g Xstor or g XPO$_4$/g S_{VFA}	0.52
Fraction of $X_{PAO,PP}$ that can be released	f_{PP}	f_{PP,Lo_PP}	g XPP,Lo/g XPP	0.94
Fraction of X_U generated in X_{PAO} decay	$f_{EP,P}$	$f_{XU\_PAO,lys}$	g XU/g XPAO	0.25
Fraction of S_U generated in X_{PAO} decay	$f_{ES,P}$	$f_{SU\_PAO,lys}$	g XU/g XPAO	0.2
Yield of X_{ANO} growth per S_{NO_3}	Y_A	Y_{ANO}	g XAUT/g S_{NOx}	0.15

TABLE A.3 New General Model (*continued*)

Stoichiometry	Parameter notation		Units	Value* T = 20°C
	Original	Standardized		
Fraction of X_U generated in X_{ANO} decay	$f_{EP,A}$	$f_{XU\_ANO,lys}$	g XU/g XANO	0.08
N content of S_U	$f_{N,SEP}$	i_{N_SU}	g N/g SU	0.07
N content of X_{OHO}	$f_{N,ZH}$	i_{N_OHO}	g N/g XOHO	0.07
N content of products from X_{OHO}	$f_{N,ZEH}$	$i_{N\_XUE,OHO}$	g N/g X_{UE}	0.07
N content of X_{PAO}	$f_{N,ZP}$	i_{N_PAO}	g N/g X_{PAO}	0.07
N content of products from X_{PAO}	$f_{N,ZEP}$	$i_{N\_XUE,PAO}$	g N/g X_{UE}	0.07
N content of X_{ANO}	$f_{N,ZA}$	i_{N_ANO}	g N/g X_{ANO}	0.07
N content of products from X_{ANO}	$f_{N,ZEA}$	$i_{N\_XUE,ANO}$	g N/g X_{UE}	0.07
P content of X_{OHO}	$f_{P,ZH}$	i_{P_OHO}	g P/g X_{OHO}	0.021
P content of products from X_{OHO}	$f_{P,ZEH}$	$i_{P\_XUE,OHO}$	g P/g X_{UE}	0.021
P content of X_{PAO}	$f_{P,ZP}$	i_{P_PAO}	g P/g X_{PAO}	0.021
P content of products from X_{PAO}	$f_{P,ZEP}$	$i_{P\_XUE,PAO}$	g P/g X_{UE}	0.021
P content of X_{ANO}	$f_{P,ZA}$	i_{P_ANO}	g P/g X_{ANO}	0.021
P content of products from X_{ANO}	$f_{P,ZEA}$	$i_{P\_XUE,ANO}$	g P/g X_{UE}	0.021
COD/VSS (OHO)	$f_{CV,H}$	i_{VSS_OHO}	g X_{OHO}/g VSS	1.48
COD/VSS (ANO)	$f_{CV,A}$	i_{VSS_ANO}	g X_{ANO}/g VSS	1.42
COD/VSS (PAO)	$f_{CV,P}$	i_{VSS_PAO}	g X_{PAO}/g VSS	1.42
Conversion factor for NO_3 reduction to N_2	$i_{NO3,N2}$	$i_{NO3,N2}$	g COD/g N	2.857
Conversion factor for NO_3 in COD	i_{COD_NO3}	i_{COD_NO3}	g COD/g N	−4.571
Conversion factor for N_2 in COD	i_{COD_N2}	i_{COD_N2}	g COD/g N	−1.714285714
Conversion factor for H_2 in COD	i_{COD_H2}	i_{COD_H2}	g COD/g H	8

TABLE A.3   New General Model (*continued*)

Kinetics	Parameter notation Original	Standardized	Units	Value* T = 20°C
Maximum specific hydrolysis rate	K_H	$q_{XCB\_SB,hyd}$	g XCB/g XOHO·d	2.81
Half-saturation coefficient for XC_B/X_{OHO}	K_X	$K_{XCB,hyd}$	g XCB/g XOHO	0.15
Correction factor for hydrolysis under anoxic conditions	$\eta_{S,ANOX}$	$n_{qhyd,Ax}$	–	1
Correction factor for hydrolysis under anaerobic conditions	$\eta_{S,ANA}$	$n_{qhyd,An}$	–	0.5
Half-saturation coefficient for S_{PO4} as nutrient	$K_{LP,GRO}$	$K_{PO4,Bio,nut}$	g S_{PO4}/m³	0.005
Maximum growth rate of X_{OHO}	μ_H	$\mu_{OHO,Max}$	d⁻¹	3.2
Reduction factor for anoxic growth of X_{OHO}	η_{gro}	$n_{\mu OHO,Ax}$	–	0.37
Rate constant for fermentation/ maximum specific fermentation growth rate	K_C	$q_{SF\_VFA,Max}$	g S_F/g XOHO·d	4
Half-saturation coefficient for S_F	$K_{S,H}$	$K_{SF,OHO}$	g S_F/m³	5
Half-saturation coefficient for S_{VFA}	$K_{S,H}$	$K_{VFA,OHO}$	g S_{VFA}/m³	5
Endogenous respiration rate of X_{OHO}	b_H	m_{OHO}	d⁻¹	0.62
Half-saturation coefficient for fermentation of S_F	$K_{S,ANA}$	$K_{SF,fe}$	g S_F/m³	2
Half-saturation coefficient for S_{O2}	$K_{O,HET}$	$K_{O2,OHO}$	g S_{O2}/m³	0.002
Half-saturation coefficient for S_{NOx}	K_{NO}	$K_{NOx,OHO}$	g S_{NOx}/m³	0.1
Half-saturation coefficient for S_{NHx}	K_{NA}	$K_{NHx,OHO}$	g S_{NHx}/m³	0.005
Rate constant for S_{VFA} uptake rate ($X_{PAO,Stor}$ storage)	K_P	q_{PAO,VFA_Stor}	g X_{Stor}/g XPAO·d	2

TABLE A.3   New General Model (*continued*)

Kinetics	Parameter notation		Units	Value* T = 20°C
	Original	Standardized		
Maximum growth rate of X_{PAO}	μ_{P1}	$\mu_{PAO,Max}$	d^{-1}	0.95
Maximum growth rate of X_{PAO} (when P is limiting)	μ_{P2}	$\mu_{PAO,Max,Plim}$	d^{-1}	0.42
Reduction factor for anoxic growth of X_{PAO}	η_P	$n_{\mu PAO}$	–	0.4
Half-saturation coefficient for $X_{PAO,Stor}/X_{PAO}$	K_{SP1}	K_{fStor_PAO}	g X_{Stor}/g X_{PAO}	0.1
Half-saturation coefficient for $X_{PAO,Stor}/X_{PAO}$ (P limit)	K_{SP2}	$K_{fStor\_PAO,Plim}$	g X_{Stor}/g X_{PAO}	0.05
Endogenous respiration rate of X_{PAO}	b_P	m_{PAO}	d^{-1}	0.04
Rate constant for lysis of $X_{PAO,PP}$	b_{PP}	b_{PP_PO4}	d^{-1}	0.03
Half-saturation coefficient for S_{VFA}	K_{SSEQ}	$K_{VFA,PAO}$	g S_{VFA}/m^3	2.5
Half-saturation coefficient for S_{PO4} uptake ($X_{PAO,PP}$ storage)	$K_{LP,UPT}$	$K_{PO4,PAO,upt}$	g S_{PO4}/m^3	0.25
Half-saturation coefficient for $X_{PAO,PP}$	K_{XP}	$K_{PP,PAO}$	g XPP/m^3	0.01
Maximum growth rate of X_{ANO}	μ_A	$\mu_{ANO,Max}$	d^{-1}	0.2–1.0
Decay rate for X_{ANO}	b_A	b_{ANO}	d^{-1}	0.04
Rate constant for ammonification	K_R	q_{am}	m^3/g XC_B,N·d	0.08
Half-saturation coefficient for S_{O2}	$K_{O,AUT}$	$K_{O2,ANO}$	g S_{O2}/m^3	0.5
Half-saturation coefficient for S_{NHx}	K_{NH}	$K_{NHx,ANO}$	g S_{NHx}/m^3	1

(Henze et al., 1987; republished in Henze et al. [2000a]); (2) Activated Sludge Model No. 2d (ASM2d) (Henze et al., 1999; republished in Henze et al. [2000b]; (3) Activated Sludge Model No. 3 (ASM3) (Gujer et al., 1999; corrected version published in Gujer et al. [2000]); (4) ASM3+Bio-P (Rieger et al., 2001); (5) ASM2d+TUD (Meijer, 2004); (6) New General Model (Barker and Dold, 1997); and (7) UCTPHO+ (Hu et al., 2007).

A systematic approach to verify models by tracking typing errors and inconsistencies in model development and software implementation is proposed. Then, stoichiometry and kinetic rate expressions are checked for each model and the errors found are reported in detail. An attached spreadsheet (see additional material) provides corrected matrices with the calculations of all stoichiometric coefficients for the discussed biokinetic models and gives an example of proper continuity checks.

REFERENCE

Hauduc, H.; Rieger, L.; Takács, I.; Héduit, A.; Vanrolleghem, P. A.; Gillot, S. (2010) A Systematic Approach for Model Verification—Application on Seven Published Activated Sludge Models. *Water Sci. Technol.*, **61** (4), 825–839.

Appendix B

Pseudo-Analytical Biofilm Model

The pseudo-analytical biofilm model solution is expressed using three dimensionless master variables: $\tilde{S}_{min,i}$, $\tilde{K}$, and $\tilde{S}_{LF,i}$ (Rittmann and McCarty, 2001). These dimensionless variables consolidate eight variables and provide insight to kinetic properties of the biofilm system. The dimensionless variables are expressed mathematically as Eq. B.1, B.2, and B.3, as follows:

$$\tilde{S}_{min,i} = \frac{S_{min,i}}{K_S} = \frac{b}{Y \cdot q_{max,i} - b} \tag{B.1}$$

where

$\tilde{S}_{min,i}$ = growth potential (dimensionless)

$q_{max,i}$ = maximum specific conversion rate of soluble substrate i ($g_{COD}/g_{CODx} \cdot d$)

When $\tilde{S}_{min,i}$ approaches the value one (1), there is a high growth potential; however, as the value is greater than one (1), low growth potential exists and there may be difficulty maintaining stable, steady-state biomass. Equation B.2 uses the parameter $\tilde{K}$ to compare external mass transport to the maximum dissolved component i utilization rate, as follows:

$$\tilde{K} = \frac{D}{L_L} \left[\frac{K_s}{q_{max,i} \cdot X_F \cdot D_F} \right]^{\frac{1}{2}} \tag{B.2}$$

A small value of $\tilde{K}$ (i.e., less than 1) suggests that resistance to external mass transfer is relatively slow and exerts significant control over the flux value. The dimensionless substrate concentration $\tilde{S}_{B,i}$, when normalized to K_s and having a value larger than 1 ($\tilde{S}_{B,i} \gg 1$), indicates that the utilization reaction is saturated (at least in the outer portions of the biofilm). The dimensionless substrate concentration in the influent stream $\left(\tilde{S}_{in,i} = \dfrac{S_{in,i}}{K_S} \right)$, bulk phase $\left(\tilde{S}_{B,i} = \dfrac{S_{B,i}}{K_S} \right)$, and at the biofilm–liquid interface is defined by Eq. B.3, as follows:

$$\tilde{S}_{LF,i} = \frac{S_{LF,i}}{K_S} \tag{B.3}$$

Equation B3 can be defined mathematically by Eqs. B.4, B.5, and B.6, as follows:

$$\tilde{S}_{in,i} = \frac{S_{in,i}}{K_s} \tag{B.4}$$

$$\tilde{S}_{B,i} = \frac{S_{B,i}}{K_s} \tag{B.5}$$

$$\tilde{S}_{LF,i} = \frac{S_{LF,i}}{K_s} \tag{B.6}$$

The basic pseudo-analytical biofilm model describes a steady-state biofilm with one microbial species and one rate-limiting substrate. Processes considered in the steady-state pseudo-analytical model include

- Soluble substrate utilization according to a single Monod rate expression
- One-dimensional mass transfer by diffusion perpendicular to the carrier (or growth medium)
- Soluble substrate concentration gradient external to the biofilm surface modeled as a mass-transfer resistance through a mass-transfer boundary layer
- Steady-state mass balance on active biomass inside the biofilm

Additional algebraic equations that comprise the pseudo-analytical solution of the steady-state biofilm reactor are defined by the following equations. Equation B.7 described a dimensionless mass-transfer resistance external to the biofilm surface, as follows:

$$\tilde{R} = \frac{D_{w,i}}{L_L} \frac{A_F}{Q} \tag{B.7}$$

where $\tilde{R}$ is the dimensionless mass-transfer resistance external to the biofilm surface.

Equation B.8 represents a total biofilm biomass loss rate (b), which is the sum of biomass lost inside the biofilm including inactivation (b_{ina}), respiration (b_{res}), and detachment (b_{det}), as follows:

$$b = b_{ina} + b_{res} + b_{det} \tag{B.8}$$

Equation B.9 mathematically describes the dimensionless steady-state flux of a soluble substrate i $\left(\tilde{J}_{LF,i}\right)$, as follows:

$$\tilde{J}_{LF,i} = \frac{J_{F,i}}{(K_s \cdot q_{max,i} \cdot X_F \cdot D_F)^{\frac{1}{2}}} \tag{B.9}$$

The pseudo-analytical solution to the one-dimensional biofilm problem was attained by fitting thousands of numerical solutions to create the following algebraic equation (i.e., the pseudo-analytical solution for a steady-state biofilm):

$$\tilde{J}_{LF,i} = f \cdot \tilde{J}_{LF,i}^{deep} \tag{B.10}$$

where

$\tilde{J}_{LF,i}^{deep}$ = soluble substrate i flux into a deep, partially penetrated biofilm (Eq. B.11)

$$= \tilde{J}_{LF,i}^{deep} = (2 \cdot [\tilde{S}_{LF,i} - \ln(1+\tilde{S}_{LF,i})])^{\frac{1}{2}} \tag{B.11}$$

f = flux reduction ratio, nondeep steady-state biofilm ($0 \le f \le 1$) (Eq. B.12)

$$= \tanh\left[\alpha \cdot \left(\frac{\tilde{S}_{LF,i}}{\tilde{S}_{min,i}} - 1 \right)^{\beta} \right] \text{ (according to Sáez and Rittmann [1992])} \tag{B.12}$$

$$\tanh = (e^x - e^{-x})/(e^x + e^{-x})$$

$$\alpha = \text{coefficient} = 1.5557 - 0.4117 \cdot \tanh[\log_{10} \tilde{S}_{min}] \tag{B.13}$$

$$\beta = \text{coefficient} = 0.5035 - 0.0257 \cdot \tanh[\log_{10} \tilde{S}_{min}] \tag{B.14}$$

Substituting Eqs. B.10, B.11, and B.12, the dimensionless flux-substrate relationship for the biofilm may be described as

$$\tilde{J}_{LF,C,i} = \tanh\left[\alpha \cdot \left(\frac{\tilde{C}_{LF,i}}{C_{min,i}} - 1\right)^{\beta}\right] \cdot (2 \cdot [\tilde{C}_{LF,i} - \ln(1 + \tilde{C}_{LF,i})])^{\frac{1}{2}} \qquad (B.15)$$

To ascertain S_{LF}, the submerged, completely mixed biofilm reactor mass balance must be coupled with Eq. B.7 by using the soluble substrate i flux across the mass-transfer boundary layer as a common boundary condition. The steady-state, dimensionless mass balance for a single, submerged, completely mixed biofilm reactor is then described by Eq. B.16, as follows:

$$0 = (\tilde{C}_{in,i} - \tilde{C}_{B,i}) - \tilde{R} \cdot (\tilde{C}_{B,i} - \tilde{C}_{LF,i}) \qquad (B.16)$$

Applying Fick's law for the mass-transfer boundary layer, the dimensionless flux into the biofilm can be expressed mathematically as

$$\tilde{J}_{LF,i} = \tilde{K} \cdot (\tilde{S}_{B,i} - \tilde{S}_{LF,i}) \qquad (B.17)$$

Substituting Eq. B.17 into Eq. B.16 connects the dimensionless flux $(\tilde{J}_{LF,i})$ to the dimensionless concentration of any soluble substrate i in the reactor influent stream $(\tilde{S}_{in,i})$ and yields the following equation:

$$\tilde{J}_{LF,C,i} = \frac{\tilde{K}}{1 + \tilde{R}}(\tilde{C}_{in,i} - \tilde{C}_{LF,i}) \qquad (B.18)$$

Once the dimensionless flux has been calculated (using Eq. B.17), the steady-state flux of any soluble substrate i can be calculated using Eq. B.19, as follows:

$$J_{LF,i} = \tilde{J}_{LF,i} \cdot (K_s \cdot q_{max,i} \cdot X_F \cdot D_{F,i})^{\frac{1}{2}} \qquad (B.19)$$

Finally, the steady-state biofilm thickness (L_F) can be calculated by applying Eq. B.20, as follows:

$$L_F = \frac{J_{LF,i} \cdot Y}{X_F \cdot b} \qquad (B.20)$$

Wanner et al. (2006) described methodology for adapting the pseudo-analytical biofilm model to include multiple species. Four scenarios were presented. The first scenario allowed for no interactions between heterotrophic and nitrifying bacteria; in essence, the bacteria behave completely independently. The second scenario presented the first level of interaction; that is, heterotrophic bacteria and autotrophic nitrifiers competed for ammonia-nitrogen in the bulk phase. The third scenario allowed for a different type of interaction, namely the competition for space. In this model, three different bacterial types exist in parallel layers. The layer nearest the liquid-biofilm interface contained heterotrophic bacteria. Behind the heterotrophs are the nitrifiers, and inert biomass lies behind the nitrifiers next to the substratum (Wanner et al., 2006). The fourth method, which is expressed mathematically, accounts for competitive space distribution. This method is presented because of its ease of implementation and application to the modified analytical biofilm model described in Appendix C. The total biofilm biomass density (X_F) is comprised of each biomass type k and is expressed as

$$X_F = \sum_K X_F = X_{F,H} + X_{F,N} + X_{F,I} + \ldots + X_{F,K} \tag{B.21}$$

where
X_F = total biofilm biomass concentration, g/m^3
$X_{F,H}$ = heterotrophic biofilm biomass concentration, g/m^3
$X_{F,N}$ = autotrophic nitrifier biofilm biomass concentration, g/m^3
$X_{F,I}$ = inert biofilm biomass concentration. g/m^3
$X_{F,k}$ = type k biofilm biomass concentration, g/m^3

REFERENCES

Rittmann, B. R.; McCarty, P. L. (2001) *Environmental Biotechnology: Principles and Application*; McGraw-Hill: Boston, Massachusetts.

Sáez, P. B.; Rittmann, B. E. (1992) Accurate Pseudo Analytical Solution for Steady State Biofilms. *Biotechnol. Bioeng.*, **39**, 790–793.

Wanner, O.; Eberl, H.; Morgenroth, E.; Noguera, D.; Picioreanu, C.; Rittman, B.; Loosdrecht, M. V. (2006) *Mathematical Modeling of Biofilms*; IWA Task Group on Biofilm Modeling, Scientific and Technical Report 18; IWA Publishing: London.

Appendix C

Analytical Biofilm Model

Biofilm models resulting from the analytical solution of Eq. 3.4 in Chapter 3 are based on a single rate-limiting soluble substrate (expressed as a Monod rate expression) in a homogeneous, one-dimensional biofilm under steady-state conditions. The kinetic expression upon which the solution is based is described mathematically as

$$r_{F,i} = q_{max,i} \cdot \underbrace{\left(\frac{S_{LF,i}}{K_i + S_{LF,i}} \right)}_{\substack{\text{Monod expression for} \\ \text{the rate–limiting substrate}}} \cdot X_F \qquad (C.1)$$

A technique that allows the analytical biofilm model user to incorporate complex kinetic expressions (i.e., multiplicative Monod expressions having more than one term) is presented. In addition, methodology is described that considers the competition for space between different bacterial types (inside the biofilm). The bacterial competition for space inside the biofilm does allow the model user to explicitly consider the effect that simultaneously occurring biochemical processes has on the flux of a soluble substrate i. While this approach accounts for the presence of multiple bacterial species and their effect on the flux of soluble substrate i, the effect that bacterial spatial distribution has on biological activity is unaccounted for in analytical biofilm models.

The maximum specific conversion rate, $q_{max,i}$, can be altered to collect all influencing expressions that are not the rate-limiting substrate. The product of several Monod rate expressions, other than that of the rate-limiting soluble substrate, cannot be directly included in the flux expression derived from the analytical biofilm. To overcome this limitation, biofilm growth may be limited by either the electron donor, electron acceptor, or a macronutrient by using the modified maximum specific conversion rate expression(s) $q_{max,mod,i}$, and dimensionless biofilm penetration correction factor(s) for each process j, β_j^* (Boltz, Johnson, Daigger, and Sandino, 2009).

The modified maximum specific conversion rate expression(s) is represented mathematically as

$$
q_{max,mod,i} = \overbrace{\nu_{j,i}}^{1} \overbrace{\mu_k}^{2} \cdot \overbrace{\eta_i}^{3} \cdot \overbrace{\prod_{i=1} \frac{S_{LF,i,ng}}{K_{i,ng}+S_{LF,i,ng}}}^{4} \cdot \overbrace{\prod_{\sigma=1} \frac{K_{i,\sigma,ng}}{K_{i,\sigma,ng}+S_{LF,\sigma,ng}}}^{5} \cdot \overbrace{\prod_{\tau=1} \frac{S_{LF,\tau,ng}}{S_{ENV,LF,\tau,ng}+S_{LF,\tau,ng}}}^{6}
$$

$$(C.2)$$

The following example of a modified rate expression is the anoxic growth of non-methanol-degrading facultative heterotrophic bacteria ($X_{F,H}$) caused by the transformation of fermentable substrates (S_F) on fermentation products (and has been presented as process $j = 6$ in Boltz, Johnson, Daigger, and Sandino [2009]):

$$
q_{max,mod,6} = \overbrace{\nu_{j,i}}^{1} \cdot \qquad\qquad \equiv -\frac{\overbrace{1}^{1}}{Y_H} \cdot
$$

$$
\overbrace{\mu_k}^{2} \cdot \qquad\qquad \equiv \overbrace{\mu_H}^{2} \cdot
$$

$$
\overbrace{\eta_i}^{3} \cdot \qquad\qquad \equiv \overbrace{\eta_{NO3H}}^{3} \cdot
$$

$$
\overbrace{\prod_{i=1} \frac{S_{LF,i,ng}}{K_{i,ng}+S_{LF,i,ng}}}^{4} \cdot \quad \equiv \overbrace{\frac{S_{LF,NO3}}{K_{NO3}+S_{LF,NO3}} \cdot \frac{S_{LF,NH4}}{K_{NH4}+S_{LF,NH4}} \cdot \frac{S_{LF,PO4}}{K_P+S_{LF,PO4}} \cdot \frac{S_{LF,ALK}}{K_{ALK}+S_{LF,ALK}}}^{4} \cdot
$$

$$
\overbrace{\prod_{\sigma=1} \frac{K_{i,\sigma,ng}}{K_{i,\sigma,ng}+S_{LF,\sigma,ng}}}^{5} \cdot \quad \equiv \overbrace{\frac{K_{O2}}{K_{O2}+S_{LF,O2}}}^{5} \cdot
$$

$$
\overbrace{\prod_{\tau=1} \frac{S_{LF,\tau,ng}}{S_{ENV,LF,\tau,ng}+S_{LF,\tau,ng}}}^{6} \quad \equiv \overbrace{\frac{S_{LF,F}}{S_{LF,F}+S_{LF,A}}}^{6}
$$

where

$\nu_{j,i}$ = stoichiometric coefficient for dissolved component i in process j (g_{COD}/g_X^1)

μ_k = maximum specific growth rate of biomass X_k (d^{-1})

$S_{LF,i,ng}$ = soluble substrate i concentration at biofilm surface, not rate-limiting, g/m^3

$K_{i,ng}$ = half-saturation coefficient for non-rate-limiting soluble substrate, g/m^3

η_i = correction factor for growth rates under anoxic conditions

σ = switching function(s) that reduces the rate in the presence of a material (e.g., dissolved oxygen)

τ = switching function(s) that reduces the rate in the presence of an inhibiting substrate

The following points are noteworthy:

- The single Monod expression described in Eq. C.1 represents the electron donor. Commonly, conditions found in water resource recovery facilities do not always result in the electron donor being the rate-limiting component; hence, the application of β_j^* for the assignment of an effective flux rate. This concept is described in greater detail later in this appendix.

- The Monod term corresponding to the substrate being modeled as the electron donor, e.g., $\dfrac{S_{LF,F}}{K_F + S_{LF,F}}$ for $S_{LF,F}$ (fermentable substrate), is evaluated in the analytical solutions to the one-dimensional biofilm and is thereby not included in $q_{max,mod,i}$.

The analytical solution to a one-dimensional biofilm can be solved only with zero- or first-order kinetics. The zero-order solution differs for completely or partially penetrated biofilms (or what are sometimes described as thin and thick biofilms, respectively). Strictly speaking, a biofilm is partially penetrated with respect to any soluble substrate i if it is gradient, $\dfrac{dC_{F,i}}{dz} = 0$, prior to $z = L_F$. A biofilm penetration depth δ_i can be computed by assuming steady state and applying zero-order kinetics. The biofilm penetration depth is described mathematically (after Harremoës [1978]) as

$$\delta_i = \sqrt{\frac{2 \cdot D_{F,i} \cdot S_{LF,i}}{q_{max,i} \cdot X_F}} \tag{C.3}$$

Here, $q_{max,i}$ should be substituted with $q_{max,mod,i}$ when complex rate expressions are being evaluated. Similarly, X_F should be replaced with $X_{F,k}$ when more than one bacterial species is competing for space inside the biofilm. To account for multiple substrate and biomass types in the biofilm, Rauch et al. (1999) presented a mathematical method for dealing with the fact that the biofilm penetration depth, δ_i, of any dissolved component i is influenced not only by the process being simulated, but also by any other process or processes that use the dissolved component i as an electron donor, electron acceptor, or macronutrient. Consequently, the actual biofilm penetration depth of dissolved component i might be less than that calculated using Eq. C.3. Boltz, Johnson, Daigger, and Sandino (2009) and Boltz, Johnson, Daigger, Sandino, and Elenter (2009) applied a variation of Eq. C.3 and the method of Rauch et al. (1999) to calculate the biofilm penetration depth, δ_i, of all potential electron donors, electron acceptors, and/or macronutrients because they are affected by all processes. This formula is described mathematically by Eq. C.4, as follows:

$$\delta_i = \sqrt{\frac{2 \cdot D_{F,i} \cdot S_{LF,i}}{\displaystyle\sum_{k=1} \sum_{i=1} q_{max,i|_i} \cdot X_{F,k}}} \qquad (C.4)$$

Equation C.5 (after Harremoës [1978]) defines the flux of any soluble substrate i across the surface of a partially penetrated, flat biofilm assuming zero-order reaction kinetics, $J^0_{F,i}$ ($g_{COD}\ m_F^{-2}\ d^{-1}$), as follows:

$$J^0_{F,i,pp} = \beta^*_j \cdot \sqrt{2 \cdot D_{F,i} \cdot q_{max,i} \cdot X_F} \cdot \sqrt{S_{LF,i}} \qquad (C.5)$$

where

$J^0_{F,i,pp}$ = partially penetrated dissolved component i flux, zero-order kinetics, g/m²·d

$\beta^*_j = \dfrac{1}{\gamma_{ED,EA|_i}} = \dfrac{\delta_{EA}}{\delta_{ED}}$, dissolved component is the penetration correction factor

$\gamma_{ED,EA} = \dfrac{\delta_{ED}}{\delta_{EA}} = \dfrac{v_{ED}}{v_{EA}} \cdot \dfrac{D_{ED}}{D_{EA}} \cdot \dfrac{S_{ED}}{S_{EA}}$ = electron donor to electron acceptor penetration depth

Typically, β is referred to as "the penetration ratio," and represents a dimensionless number defined as

$$\frac{\text{substrate penetration depth}}{\text{biofilm thickness}}, \frac{\delta_i}{L_F}$$

In this instance, however, β^*_j is defined as the "penetration correction factor" for every process j. Again, $q_{max,i}$ should be substituted with $q_{max,mod,i}$ when complex rate expressions are being evaluated, and X_F should be replaced with $X_{F,k}$ when more than one bacterial species is competing for space inside the biofilm. When the biofilm thickness is less than the penetration depth, Eq. C.5 is invalid and the following relationship (Harremoës, 1978; Rittmann and McCarty, 2001) can be used to describe zero-order flux:

$$J^0_{F,i,cp} = L_F \cdot q_{max,i} \cdot X_F \qquad (C.6)$$

where $J^0_{F,i,cp}$ is the completely penetrated soluble substrate i flux (zero-order kinetics, g/m²·d).

Consistent with the other solutions to the analytical biofilm model, $q_{max,i}$ should be substituted with $q_{max,mod,i}$ when complex rate expressions are being evaluated, and X_F should be replaced with $X_{F,k}$ when more than one bacterial species is competing for space inside the biofilm. Equation C.7 defines the flux of dissolved component i into a flat biofilm assuming first-order kinetics, $J_{F,i}^1$ (g_{COD} m_F^{-2} d^{-1}) (after Harremoës [1978]; Rittmann and McCarty, 2001).

$$J_{F,i}^1 = \frac{\beta_j^* \cdot q_{max,i} \cdot X_F \cdot L_F \cdot S_{B,i}}{K_i} \cdot \varepsilon \tag{C.7}$$

where

$$\varepsilon = \frac{\tanh\left(\dfrac{L_F}{L_{crit}}\right)}{\left(\dfrac{L_F}{L_{crit}}\right)} = \frac{\tanh\alpha}{\alpha} = \frac{J_{LF,i}^1 \text{ with diffusion}}{J_{LF,i}^1 \text{ without diffusion}} = \text{mass transport limitation}$$
effect

$$L_{crit} \approx \sqrt{\frac{D_{F,i} \cdot K_S}{q_{max,i} \cdot X_{F,k}}} = \text{characteristic length (m)}$$

$$\alpha = \left(\frac{L_F}{L_{crit}}\right) = \sqrt{\frac{q_{max,i} \cdot X_{F,k} \cdot L_F^2}{D_{F,i} \cdot K_S}} = \text{ratio indicating limitation by mass transfer}$$

Consistent with the other solutions to the analytical biofilm model, $q_{max,i}$ should be substituted with $q_{max,mod,i}$ when complex rate expressions are being evaluated, and X_F should be replaced with $X_{F,k}$ when more than one bacterial species is competing for space inside the biofilm. The efficiency factor, ε, is a ratio of flux with diffusional resistance explicitly considered a hypothetical flux without consideration given to diffusional resistance. The dimensionless first-order effectiveness factor, α, in which $L_{crit} \gg L_F$, indicates the biofilm is probably fully penetrated. On the other hand, $L_{crit} \ll L_F$ indicates the biofilm is probably mass-transfer-limited. No biofilm penetration depth can be calculated using this expression because the first-order flux expression does not reach zero.

Rauch et al. (1999) presented a simplified mixed-culture biofilm model based on dividing the biofilm into two compartments that include the solid matrix (which consists of several bacterial species, particulate substrates, and inert material) and liquid phase. Rauch et al. (1999) stated that by decoupling the liquid phase in which diffusion occurs and the solid matrix in which biochemical reaction occurs, a zero-order kinetic rate can be applied consistently. When considering the biofilm as a single compartment, as in the approach applied to the analytical model described in this chapter, a compromise between either the zero- or first-order kinetic assumptions is the assignment of a weighted average. A method of composite kinetics presented by Pérez et al. (2005), with corrigendum by Gapes et al. (2006), is described as

$$J_{F,i}^{avg} = \left(\frac{S_{LF,i}}{S_{LF,i} + K_i} \right) \cdot J_{F,i}^0 + \left(1 - \frac{S_{LF,i}}{S_{LF,i} + K_i} \right) \cdot J_{F,i}^1 \tag{C.8}$$

where $J_{F,i}^{avg}$ is the weighted average of the first- and zero-order kinetic fluxes ($g_{COD}\ m_F^{-2}\ d^{-1}$).

Pérez et al. (2005) demonstrated that the composite kinetic method can be successfully applied without segregating the biofilm component into multiple compartments. Another approach to select the flux of soluble substrate i by considering each of the different analytical flux equations is to select the minimum of the three analytical solutions (described by Eqs. C.5, C.6, and C.7) according to the following equation:

$$J_{LF,i}(S_{LF,i}) = \min\{J_{LF,i}^1(S_{LF,i}), J_{LF,i,cp}^0(S_{LF,i}), J_{LF,i,pp}^0(S_{LF,i})\} \tag{C.9}$$

Analytical solutions can also be used to describe mixed-culture, one-dimensional biofilm; a biofilm biomass density balance may be applied. The total biofilm biomass density (X_F) is comprised of each biomass type k and is expressed as

$$X_F = \sum_k X_F = X_{F,H} + X_{F,N} + X_{F,I} + \ldots + X_{F,k} \tag{C.10}$$

where
   X_F = total biofilm biomass concentration, g/m^3
   $X_{F,H}$ = heterotrophic biofilm biomass concentration, g/m^3
   $X_{F,N}$ = autotrophic nitrifier biofilm biomass concentration, g/m^3
   $X_{F,I}$ = inert biofilm biomass concentration, g/m^3
   $X_{F,k}$ = type k biofilm biomass concentration, g/m^3

REFERENCES

Boltz, J. P.; Johnson, B. R.; Daigger, G. T.; Sandino, J. (2009) Modeling Integrated Fixed Film Activated Sludge (IFAS) and Moving Bed Biofilm Reactor (MBBR) Systems I: Mathematical Treatment and Model Development. *Water Environ. Res.*, **81**, 555–575.

Boltz, J. P.; Johnson, B. R.; Daigger, G. T.; Sandino, J.; Elenter, D. (2009) Modeling Integrated Fixed Film Activated Sludge (IFAS) and Moving Bed Biofilm Reactor (MBBR) Systems II: Evaluation. *Water Environ. Res.*, **81**, 576–586.

Gapes, D.; Pérez, J.; Picioreanu, C.; van Loosdrecht, M. C. M. (2006) Corrigendum to "Modeling Biofilm and Floc Diffusion Processes Based on Analytical Solution of Reaction-Diffusion Equations". *Water Res.*, **40**, 2997–2998.

Harremoës, P. (1978) *Biofilm Kinetics in Water Pollution Microbiology*, Vol. 2; Michell, R., Ed.; Wiley & Sons: New York.

Pérez, J.; Picioreanu, C.; van Loosdrecht, M. C. M. (2005) Modeling Biofilm and Floc Diffusion Processes Based on Analytical Solution of Reaction-Diffusion Equations. *Water Res.*, **39**, 1311–1323.

Rauch, W.; Vanhooren, H.; Vanrolleghem, P. A. (1999) A Simplified Mixed-Culture Biofilm Model. *Water Res.*, **33** (9), 2148–2162.

Rittmann, B. R.; McCarty, P. L. (2001) *Environmental Biotechnology: Principles and Application*; McGraw-Hill: Boston, Massachusetts.

Appendix D

Numerical Biofilm Model

Mathematical treatment of biofilms using the numerical one-dimensional biofilm model is based on assignment of a system consisting of three phases. The three phases distinguished in this model include a particulate component in the solid phase (which form the biofilm solid matrix), the particulate component in the pore volume liquid phase, and soluble substrate(s) in the pore volume liquid phase. The particulate component in the solid phase is described mathematically as

$$X_{F,k}^{\text{attached}} = \rho_{S,i} \cdot \varepsilon_{S,i} \tag{D.1}$$

where

$X_{F,k}^{\text{attached}}$ = attached particulate component k concentration (in the solid matrix), g_X/m^3

$\rho_{S,i}$ = density of the attached particulate, or solid, component k, g_X/m^3

$\varepsilon_{S,i}$ = volume fraction of the attached particulate component k or solid phase, g_X/m^3

The biofilm phase identified as the pore volume has both a liquid volume and a particulate component k that is suspended in the pore volume and identified with concentration ($X_{F,k}^{\text{suspended}}$). The volume fraction of the biofilm solid matrix is $\Sigma \varepsilon_{S,i}$ and the porosity or biofilm pore volume fraction θ is defined mathematically by Eq. D.2, as follows:

$$\theta = 1 - \sum_{i=1}^{n_x} \frac{X_{F,k}^{\text{attached}}}{\rho_{S,i}} \tag{D.2}$$

Here, n_x is the number of particulate components considered in the model. The pore volume is distinguished by two phases, namely the phase of the suspended particulate components with concentrations ($X_{F,k}^{\text{suspended}}$) and the biofilm liquid phase. The liquid phase volume fraction is defined mathematically as

$$\varepsilon_{L,f} = \theta - \sum_{i=1}^{n_x} \frac{X_{F,k}^{\text{suspended}}}{\rho_{S,i}} \tag{D.3}$$

If the concentration gradients being modeled occur in only one direction (perpendicular to the substratum), the one-dimensional mass balance equations for attached particulate-, suspended particulate-, and dissolved components inside the biofilm can be modeled by Fick's law. Equation D.4 mathematically describes the simultaneous diffusion and biochemical reaction of the dissolved component $S_{F,i}$, as follows:

$$J_{F,i} = -D_{F,i} \frac{dS_{F,i}}{dz} \tag{D.4}$$

where $J_{F,i}$ is the flux of soluble substrate i inside the biofilm (g/m²·d). Local mass balances can be written that are differential equations that express the variation of concentration of a component in time at a point in space as a result of mass transport and biochemical transformation processes. The local mass balances are the mathematical form of equality, which, in Cartesian space (that is, with ortho-normal unit vectors), can be written as

$$\frac{\partial S}{\partial t} = \frac{\partial J_x}{\partial x} - \frac{\partial J_y}{\partial y} - \frac{\partial J_z}{\partial z} + r \tag{D.5}$$

where

t = the time, d

$x, y,$ and z = spatial coordinates, m

S = the concentration, g/m³

$J_x, J_y,$ and J_z = components of the mass flux J (g/m²·d) along the coordinates

r = the net production rate (g/m³) of the component

Equation D.5 is known as the *equation of continuity for a component*, either soluble or particulate. Substituting Eq. D.4 into Eq. D.5 yields a mass balance that describes development in time and the spatial profile inside the biofilm of the concentration $S_{F,i}$ (for soluble substrates) as Eq. D.6, as follows, which is presented for complete development of the numerical one-dimensional biofilm model:

$$\frac{\partial S_{F,i}}{\partial t} = D_{F,i} \frac{\partial^2 S_{F,i}}{\partial z^2} - r_{F,i} \tag{D.6}$$

which is applicable to the following boundary conditions (1 and 2):

$$BC1 : \frac{\partial S_{F,i}}{\partial z} = 0$$

$$BC2 : (S_{F,i} = S_{LF,i}) \text{ at } z = 0$$

The equations presented by Wanner and Reichert (1996) and Reichert and Wanner (1997) define the substrate concentration, $C_{F,i}$, which is defined as substrate mass per unit biofilm liquid phase and relates to $S_{F,i}$ by Eq. D.7, as follows:

$$S_{F,i} = \varepsilon_{l,F} \cdot C_{F,i} \tag{D.7}$$

For attached particulate components which form the solid biofilm matrix, particle transport is assumed to be the result of microbial growth and loss by decay inside the biofilm. Growing or shrinking cells led to a volume expansion or concentration of the biofilm solid matrix, respectively, and to a displacement of adjacent neighboring bacterial cells (Wanner 1989). This displacement can be interpreted as advective mass transport and is formally described as a specific mass flux $J_{M,i}$ (g/m²·d) by Eq. D.8, as follows:

$$J_{M,i} = u_F \cdot X_{M,i} \tag{D.8}$$

where, u_F is the distance by which cells are displaced per unit time (m/d). By definition, $J_{M,i}$ is the advective mass flux of particulate matter in the biofilm matrix. The displacement velocity, u_F, of a cell at a location z is equal to the added net specific mass production of all microbial species in the biofilm matrix between the substratum and this location defined by Eq. D.9, as follows:

$$u_F(z) = \frac{1}{1-\theta} \int_0^z \sum_{i=1}^{n_s} \frac{r_{M,i}}{\rho_{S,i}} \, dz \tag{D.9}$$

Based on Eq. D.8 and Eq. D.9, a mass balance analogous to Eq. D.6 can be derived and used to describe development in time and the spatial profile in the biofilm of the attached particulate component i, as follows:

$$\frac{\partial X_{M,i}}{\partial t} = \frac{\partial (u_F, X_{M,i})}{\partial z} + r_{M,i} \tag{D.10}$$

To solve Eq. D.10, a boundary condition must be established that allows for a no-flux condition at the substratum, that is, $J_{M,i} = 0$ when $z = 0$. Equation D.10 is used to calculate the relative abundance, spatial distribution, and development in time of microbial species and particulate components inside the biofilm. Development of biofilm thickness (L_F) in time is the result of the net production of biomass in the biofilm, as described by Eq. D.9, of the attachment at the biofilm surface of microbial cells and particles suspended in the bulk phase and of the detachment of microbial cells and particles from the biofilm surface to the bulk of the liquid. This movement of solids is modeled according to the following equation:

$$\frac{dL_F}{dt} = \underbrace{u_F(L_F)}_{\text{growth-decay}} - u_{det} + u_{att} \tag{D.11}$$

where

u_{det} = global detachment velocity, m/d

u_{att} = global attachment velocity, m/d

Global infers that all species detach at the same rate and that the detached biomass of a microbial species is proportional to its concentration at the biofilm surface. The velocity u_{det} yields a description of the decrease of biofilm thickness L_F per unit time as a result of detachment processes (e.g., erosion, sloughing, and predation). This phenomenon can be described mathematically by the following equation:

$$u_{det} = u_{det}(L_F, u_F(L_F), t, \tau_{LF} \ldots) \tag{D.12}$$

In addition to modeling detachment as a global phenomenon, individual detachment of various particulate components from the biofilm solid matrix is achievable (Wanner and Reichert, 1996). Microbial cells attach to a biofilm surface by *adsorption*, which is a generic term for many processes that contribute to particulates attaching to a biofilm surface. A significant process included in adsorption is bioflocculation (Boltz and La Motta, 2007). The process of microbial cells suspended in the bulk phase being adsorbed to the biofilm surface is modeled as an attachment velocity described mathematically by the following equation:

$$u_{att} = \frac{1}{1-\theta(L_F)} \sum_{i=1}^{n_x} \frac{k_{att,i} \cdot X_{L,k}}{\rho_{S,i}} \tag{D.13}$$

where

k_{att} = the attachment rate coefficient, m/d

$X_{L,k}$ = the concentration on the bulk-liquid side of the biofilm surface of the suspended particulate component k, g_x/m^3

The environment of the biofilm is modeled as a completely mixed bulk liquid. Conversion processes in the bulk liquid can be equally important as those that take place in the biofilm. For each dissolved and particulate component considered, an additional mass balance is required and is expressed mathematically as Eq. 3.13 and Eq. 3.14 (see Chapter 3) for soluble substrate and particulate components, respectively. The exchange of soluble substrates and particulate components between the biofilm and the bulk of the liquid is still modeled as a mass transfer boundary layer described in Eq. 3.9 (see Chapter 3).

REFERENCES

Boltz, J. P.; La Motta, E. J. (2007) The Kinetics of Particulate Organic Matter Removal as a Response to Bioflocculation in Aerobic Biofilm Reactors. *Water Environ. Res.*, **79** (7), 725–735.

Reichert, P.; Wanner, O. (1997) Movement of Solids in Biofilms: Significance of Liquid Phase Transport. *Water Sci. Technol.*, **36** (1), 321–328.

Wanner, O. (1989) Modeling Population Dynamics. In *Structure and Functions of Biofilms*; Characklis, W. G., Wilderer, P. A.; Wiley & Sons: New York; pp 91–110.

Wanner, O.; Reichert, P. (1996) Mathematical-Modelling of Mixed-Culture Biofilms. *Biotechnol. Bioeng.*, **49** (2), 172–184.

Appendix E

Project Modeling Scope of Services

PURPOSE

This scope of services describes tasks to be undertaken by _____ ("the consultant") to develop a site-specific process model for _____ ("the client").

TASK 100—MODEL DEVELOPMENT

Task 101—Kickoff Meeting

An initial meeting will be held with the client to determine their expectations and requirements for the process modeling effort. A goal of this meeting will be to determine what data and reports are available for this project. It may also be determined that special sampling is required to properly characterize facility streams or performance. At this meeting, the baseline model will be defined (configuration and loadings) and scenarios to be developed in Task 104 will be identified. Details of the model interface required for Task 105 will also be decided.

Task 102—Procure and Analyze Available Data and Reports

The client will provide all data identified in Task 101 to the consultant in electronic format. They will also make available any reports that they think will be helpful in describing the facility and its performance. Baseline model loadings, influent characteristics, and facility performance will be established from information made available to the consultant.

Task 103—Develop Baseline Model

A baseline model will be developed by the consultant. The model will be set up using physical (tank sizes and layout) and operating information supplied by the client and kinetic and stoichiometric parameters based on the consultant's defaults. The model will be calibrated using influent characteristics, loadings, and facility performance data established in Task 102.

Task 104—Develop Scenarios

Up to X different scenarios will be set up based on requirements established at the kickoff meeting. Scenarios to be developed will be established by the consultant and the client at the kickoff meeting. The scenarios may include seasonal loadings, storm events, process modifications, and diurnal profile. Additional scenarios may be developed at extra cost, outside of the scope of this proposal.

Task 105—Refine Model Interface

The number and types of inputs to and outputs from the model will be set up according to the client's requirements that were agreed on at the kickoff meeting.

Task 106—Write Summary Memorandum

A memorandum will be prepared describing the baseline model setup and scenarios, assumptions made in doing so, and guidance on how to use the model.

Deliverables

In addition to the summary memorandum described in Task 106, the consultant will provide the client with a baseline model layout file, any layout or data files required to run the scenarios, and any spreadsheets required to view any output data. All of these files will be provided in electronic format on a compact disk, via e-mail, or on a Web site accessible by the client.

Exclusions

The consultant will not provide a copy of the modeling software needed to run the model layouts.

Appendix F

Glossary

absorption   Assimilation of molecules or other substances into the physical structure of a liquid or solid without chemical reaction.

accuracy   Closeness of computations, estimates, or measurements to exact or true values; see *trueness* and *precision*.

acetogenesis   The metabolic process that converts volatile acids to acetate, the primary substrate of acetoclastic methanogenesis.

acidogenenisis   The conversion of soluble organic matter to simple organic acids under anaerobic conditions.

activated sludge   The biologically active solids in an activated sludge process.

ADM   Anaerobic digestion model; the model developed by the International Water Association for anaerobic digestion.

adsorption   The process of transferring molecules of gas, liquid, or a dissolved substance to the surface of a solid, where it is bound by weak chemical (hydrogen bonding) or physical forces.

advanced wastewater treatment   Treatment processes designed to remove pollutants that are not adequately removed by conventional secondary treatment processes.

aeration   Addition of air or oxygen to water or wastewater, typically by mechanical means, to increase dissolved oxygen levels and maintain aerobic conditions.

aerobe   An organism that requires free oxygen for respiration.

aerobic SRT   Sludge retention time within the oxygenated portion of a biological treatment system; see *solids residence time*.

aerobic   Condition characterized by the presence of free oxygen.

algorithm   A procedure for solving a problem.

365

alkalinity   The ability of water to neutralize an acid via the presence of carbonate, bicarbonate, and hydroxide ions. Alkalinity is often given in the units milliequivalent per liter (meq/L) or in the units mg $CaCO_3$/L.

alum   Aluminum sulfate, frequently used as a precipitant of phosphorus and coagulant in water and wastewater treatment. Chemical formula: $Al_2(SO_4)_3 \cdot 14H_2O$.

aluminum sulfate   See *alum*.

ammonification   Bacterial decomposition of organic nitrogen to ammonia.

anaerobic   Condition characterized by the absence of free oxygen and other electron receptors such as nitrate and sulfate. As a process, *anaerobic* implies the active presence of strictly anaerobic organisms. As an activated sludge process component, *anaerobic* implies the active presence of anaerobic and facultative organisms.

anaerobic digestion   Solids stabilization process operated specifically without oxygen in which much of the organic waste feed is converted to methane and carbon dioxide.

analytical models   Models that can be solved mathematically in closed form. For example, some model algorithms that are based on relatively simple differential equations can be solved analytically to provide a single solution.

ANNAMMOX   Anaerobic ammonia oxidizing autotrophs; a chemolithothrophic autotrophic species that oxidizes ammonia to nitrogen gas using nitrite as the electron acceptor. A small amount of nitrate might also be produced.

anoxic   Refers to the liquid condition where there is little to no oxygen present, but there are appreciable levels of nitrate and/or nitrite.

AOB   Ammonia-oxidizing bacteria, an older term for ammonia-oxidizing organism.

AOO   Ammonia-oxidizing organism.

ash   The nonvolatile inorganic solids that remain after incineration.

ASM   Activated sludge model; typically refers to the series of biological models developed by the International Water Association.

ASP   Activated sludge process.

attached growth   See *biofilm*.

autotroph   Plants and bacteria that can synthesize organic compounds from inorganic nutrients.

average annual daily flow   The average flowrate occurring over a 24-hour period based on annual flowrate data.

average daily flow   The total flow past a physical point over a period of time divided by the number of days in that period. Can be segregated into dry weather and wet weather averages.

average dry-weather flow (ADWF)   The average of the daily flows sustained during dry-weather periods with limited infiltration.

average flow   The arithmetic average of flows measured at a given point.

average wet-weather flow (AWWF)   The average of the daily flows sustained during wet-weather periods when infiltration is a factor.

averaging periods   The unit of time over which a measurement is taken and then subsequently averaged.

batch reactor   A reactor in which the contents are completely mixed and flow neither enters nor leaves the reactor vessel during the period of mixing.

bias   Systematic deviation between a measured (i.e., observed) or computed value and its "true" value. Bias is affected by faulty instrument calibration and other measurement errors, systematic errors during data collection, and sampling errors such as incomplete spatial randomization during design of sampling programs.

bioaugmentation   The addition of a particular viable biomass to encourage certain performance criteria.

biochemical oxygen demand (BOD$_5$)   A standard measure of wastewater strength that quantifies the oxygen consumed in a stated period of time (typically 5 days as signified by the subscript) at 20 °C; also BOD and CBOD.

biodegradable   Capable of undergoing biological decomposition.

biofilm   An agglomeration of microorganisms embedded in a matrix of biological origin called *extracellular polymeric substances* attached to a surface. Biofilms contain soluble compounds, particulate matter (organic and inorganic), and higher life forms such as stalked protozoa.

biofilm biomass density   (1) areal density is the dry weight of biofilm particulate components per unit volume or (2) a COD-based measurement per unit volume commensurate with oxygen-based concentration.

biofilm discretization   Mathematical modeling of biofilms requires a priori assumption of biofilm composition. One-dimensional biofilm models that account for heterogeneity are typically situated as discrete monolayers. Dividing the one-dimensional biofilm into monolayers is referred to as *discretization*.

biological nutrient removal   The reduction of nutrients in a wastewater stream using biological techniques.

biological phosphorus removal   The removal of soluble phosphorus by biological organisms. Consists of both uptake caused by growth and enhanced biological phosphorus removal.

biological process   Any process in which microorganisms metabolically convert complex materials into simpler, more stable substances.

biomass   The mass of microorganisms contained in a system.

Bio-P   See *enhanced biological phosphorus removal.*

bioreactor   A vessel in which a biological process occurs.

biosolids   Solids that have been removed from wastewater and stabilized (e.g., digested or composted) to meet criteria in U.S. Environmental Protection Agency's (U.S. EPA's) 40 CFR 503 regulations and therefore can be beneficially used.

birthday cake analysis   An artificial loading pattern that combines the annual average, maximum monthly average, and peak day loading, which resembles a birthday cake when plotted.

black box model   See *statistical model.*

blanket   Typically refers to the sludge layer on the bottom of a gravity solids separation device.

BOD   See *biochemical oxygen demand.*

boundary conditions   In mathematics, used for differential equation systems (restraints on values the state variables can assume, that is, negative concentrations are outside the boundary); sets of values for state variables and their rates along problem domain boundaries sufficient to determine the state of the system within the problem domain.

BPR   See *biological phosphorus removal.*

bulk-liquid volume   Biofilms are systematically described with compartments. In most instances, a bulk-liquid compartment lies over the biofilm. Water is incorporated to more than one of the compartments used to describe biofilm systems. Water treated as a compartment independent of the biofilm is referred to as the *bulk of the liquid.* Bulk-liquid volume is analogous to the water volume in a continuous flow stirred tank reactor (CFSTR).

calibration   The process of adjusting model parameters within physically defensible ranges until the resulting prediction gives the best practical fit to observed data. In some disciplines, calibration is also referred to as *parameter estimation*.

calibration, dynamic   The calibration of a model to specific influent and/or environmental conditions that vary dynamically with time.

calibration, steady state   The calibration of a model to a fixed set of influent, operating, and environmental conditions.

carbon oxidation   The conversion (mineralization) of organic compounds to carbon dioxide.

carbonaceous biochemical oxygen demand   The portion of biochemical oxygen demand that consumes oxygen via carbon oxidation; typically measured after a sample has been incubated for 5 days; also called *first-stage biochemical oxygen demand*.

CBOD   See *carbonaceous biochemical oxygen demand*.

CFD   See *computational fluid dynamics*.

CFSTR   See *continuous-flow stirred tank reactor*.

chemical equilibrium   A condition in which there is no net transfer of mass or energy between system components. It occurs in a reversible chemical reaction when the rate of the forward reaction equals the rate of the reverse reaction.

chemical oxygen demand (COD)   Measurement of the oxidation potential of a water or wastewater using a chemical oxidizing agent. For modeling purposes, this measure is used to describe the concentration of organic matter (mg/L) in a wastewater or sludge sample and is the unit of measure most typically used in modeling programs that simulate biochemical treatment systems.

code   Instructions, written in the syntax of a programming language, that provide the computer with a logical process. Code may also be referred to as a *computer program*. The term *code* describes the fact that computer languages use a different vocabulary and syntax than algorithms that may be written in standard language.

coefficient   A numerical quantity, interposed in a formula that expresses the relationship between two or more variables to include the effect of special conditions or to correct a theoretical relationship to one found by experiment or actual practice.

collinearity   The condition where variables are so highly correlated that it is impossible to come up with reliable estimates of their individual regression coefficients.

colloidal Refers to a particle that is not removed from the liquid phase through settling or filtration by a particular pore size filter. Often defined as those solids that will pass a 0.45-micron filter paper, but can be removed from solution by coagulation and subsequent solid/liquid separation.

combined variable See *composite variable*.

completely mixed stirred-tank reactor (CSTR) An ideal reactor in which concentrations are uniform throughout. The effluent concentration equals the reactor concentration.

complexity The opposite of simplicity. Complex systems tend to have a large number of variables, multiple parts, and mathematical equations of a higher order and are more difficult to solve. In relation to computer models, *complexity* typically refers to the level of difficulty in solving mathematically posed problems as measured by time, number of steps or arithmetic operations, or memory space required (time complexity, computational complexity, and space complexity, respectively).

composite sample Multiple samples (time, volume, or flow weighted) taken over a defined time interval combined into one sample for analytical processing.

composite variable A combination of state variables typically to form variables that can actually be measured in the facility (e.g., BOD_5, total COD, total Kjeldahl nitrogen, total phosphorus, total suspended solids, and volatile suspended solids). Also called *combined variable*.

computational fluid dynamics A branch of fluid mechanics that uses numerical methods and algorithms to solve and analyze problems that involve fluid flows. Can be expanded to include soluble and particulate components in the calculations.

conceptual model A description of reality in terms of verbal descriptions, equations, governing relationships, and "natural laws" that describe reality. This is the user's perception of the study area and corresponding simplifications and numerical accuracy limits that are assumed acceptable to achieve the purpose of the modeling.

confidence Instead of estimating the parameter by a single value, an interval likely to include the parameter is given. How likely the interval is to contain the parameter is determined by the confidence level. Increasing the desired confidence level will widen the confidence interval.

constant A quantity with a fixed value (e.g., the speed of light or gravitational force) representing known physical, biological, or ecological activities.

continuity equation   An equation describing the conservation of mass, energy, charge, and so on.

continuous-flow stirred tank reactor   See *completely mixed stirred-tank reactor.*

covariance matrix   A matrix whose element in the i, j position is the covariance between the ith and jth elements of a random vector (that is, of a vector of random variables). Covariance is a measure of how much two random variables change together. The dimensionless, normalized version of the covariance, the correlation coefficient, shows, by its magnitude, the strength of the linear relation.

CSTR   See *completely mixed stirred-tank reactor.*

data reconciliation   The application of one or several independent checks to verify the consistency of the data.

death-regeneration   A biological modeling approach where decay products can be subsequently used by biomass for growth.

denitrification   The biological conversion of nitrate to nitrogen gas.

denitritation   The biological conversion of nitrite to nitrogen gas.

design criteria   (1) Engineering guidelines specifying construction details and materials; (2) objectives, results, or limits that must be met by a facility, structure, or process in the performance of its intended functions.

design standards   Standards established for design of equipment and structures. The standards may or may not be mandatory.

detention time   The theoretical time required to displace the contents of a tank or unit at a given rate of discharge.

deterministic model   A model that provides a single solution for the state variables. Changes in model outputs are solely the result of changes in model components (i.e., model inputs).

dispersed suspended solids   The solids remaining after 30 minutes of settling secondary clarifier effluent.

dissolved solids   Solids in solution that cannot be removed by filtration; may or may not include colloidal solids depending on analytical method.

diurnal   A daily fluctuation in flow or composition.

diurnal pattern   A repeating daily fluctuation in flow or composition that is of similar pattern from one 24-hour period to another.

domain boundaries (spatial and temporal)   Spatial and temporal domains of a model are the limits of extent and resolution with respect to time and space for which the model has been developed and over which it should be evaluated.

domestic wastewater   Wastewater originating in sanitation devices (e.g., sinks and toilets) in residential dwellings, office buildings, and institutions; also called *sanitary wastewater.*

dry-weather flow   A measure of total flows in the sanitary collection system under dry-weather conditions. It is the sum of the volume of wastewater discharges and the volume of groundwater infiltrating the collection system.

DSS   See *dispersed suspended solids.*

dynamic model   A model in which time is an independent variable.

dynamic simulation   Time-varying solution of the ordinary differential equation system; typically a simulation in which inputs to the model and outputs from the model vary with time.

EBPR   See *enhanced biological phosphorus removal.*

electron acceptor   A chemical entity that accepts electrons transferred to it from another compound. In wastewater treatment, examples are oxygen, nitrate, and nitrite.

electron donor   A chemical entity that donates electrons to another compound. In wastewater treatment, examples are acetic acid and soluble biodegradable organics in biological metabolism.

empirical knowledge   Experience-based knowledge.

empirical model   A model for which the structure is derived from mathematical analyses of relationships among observed data rather than from knowledge on the physical, biological, or ecological processes.

endogenous residue   Decay products produced after cell lysis.

endogenous respiration   Bacterial growth phase during which microbes metabolize their own protoplasm without creating more protoplasm because not enough food is available.

enhanced biological phosphorus removal   The biological removal of phosphorus from a wastewater through cultivation and wasting of bacteria that retain phosphorus in excess of their minimum biologically required amount; a subset of biological phosphorus removal.

equalization   The process of dampening hydraulic or organic variations in a flow so nearly constant conditions can be achieved.

extrapolation   A process that uses assumptions about fundamental causes underlying observed phenomena to project beyond the range of available data; in general, extrapolation is not considered a reliable process for prediction.

facility model   Combination of submodels to simulate a facility or part of a facility. A typical activated sludge facility model consists of at least an input, a transport or hydraulic model, a biokinetic model, an aeration model, and a clarifier model.

facility-wide model   In Europe, the model used to describe all unit processes and their interconnections in a facility mass balance. In North America, often referred to as *whole-facility model*.

fermentation   See *acidogenesis*.

ffCOD   See *flocculated and filtered COD*.

final effluent   The effluent from the final unit of a treatment process at a water resource recovery facility.

fixed film   See *biofilm*.

floc   (1) Small, gelatinous masses formed in water by adding a coagulant or in wastewater via biological activity; (2) collections of smaller particles agglomerated into larger, more easily settleable particles via chemical, physical, or biological treatment.

flocculated and filtered COD   Flocculated and filtered chemical oxygen demand; ffCOD is a subset of total COD that refers to the truly soluble COD components. This is typically defined as the material present after coagulation and filtration using a 0.45-micron filter paper. It differs from the soluble COD (SCOD) in that the colloidal materials have been removed.

flocculated suspended solids   The suspended solids result from the supernatant of a flocculation and settling test.

flow rate   The volume or mass of a gas, liquid, or solid material that passes a point in a stated period of time.

flux   Flow or mass rate per unit area; for membranes, the volumetric filtration rate for a given area of membrane. A typical unit of flux is liters per square meter of membrane area per day or gallons per square foot per day.

forcing/driving variables   External or exogenous (outside the model framework) factors that influence state variables calculated within the model.

FSS   See *flocculated suspended solids.*

function   A mathematical relationship between variables.

GMP   See *good modeling practice.*

good modeling practice   The set of protocols encompassing a rigorous approach to wastewater treatment modeling; an International Water Association task group focused on activated sludge models.

grab sample   A single water or wastewater sample taken at a particular time and place.

Greenfield   Refers to construction on an undeveloped site.

Gujer matrix   Tabular representation of state variables and their interactions within the model. Formerly known as a Petersen matrix.

half-saturation coefficient   A parameter used in the Monod equation defining the shape of the resulting curves and corresponding to the substrate (or other component) concentration at which the value of the Monod saturation/inhibition function is 0.5.

half-saturation concentration   The concentration at which the process rate is half of its maximum rate.

Henry's law   The principle that, at a constant temperature, the concentration of a gas dissolved in a fluid (with which it does not combine chemically) is almost directly proportional to the partial pressure of the gas at the surface of the fluid.

heterotrophic bacteria   A type of bacteria that derives its cell carbon from organic carbon; most pathogenic bacteria are heterotrophic bacteria.

Heuristic rules   Experience-based rules.

hydraulic loading   Total volume of liquid applied per unit of time to a specific tank or treatment process.

hydraulic retention time   The length of time that a given hydraulic loading of wastewater or solids will be retained in a pipe, reactor, unit process, or facility.

hydrolysis   Enzyme-mediated reactions that convert complex organic compounds (i.e., particulate) into simple compounds or reduced-mass materials; often refers to solids particle breakdown in the first step of anaerobic digestion.

identifiability analysis   Evaluation of the uniqueness of the estimates of model parameters from measured data. Structural (theoretical, a priori) identifiability analysis assesses the uniqueness of parameter estimates from ideal data for a given experimental frame; practical (a posteriori) identifiability analysis assesses the accuracy with which parameters can be estimated with a given data set. In the latter instance, identifiability is not an objective property, but depends on the required accuracy.

IFAS   See *integrated fixed film activated sludge.*

inerts   Constituents that are assumed not to react in the model; inerts may be soluble or particulate and organic or inorganic.

influent characterization   See *influent stoichiometry.*

influent fractionation   See *influent stoichiometry.*

influent stoichiometry   The breakdown of influent constituents into state variables; also called *influent fractionation* and *influent characterization.*

inorganic matter   Substances of mineral origin (not containing organic carbon) that are not subject to decay.

inorganic suspended solids   Particulate inorganic matter.

input variable   For modeling purposes, variables may be divided into state variables (or internal variables), input variables, and output variables.

instantaneous peak   Highest record flowrate occurring for a period consistent with recording equipment. In many situations, the recorded peak flow may be considerably below the actual peak flow because of metering and recording equipment limitations.

insoluble   Incapable of being dissolved.

integrated fixed-film activated sludge   The activated sludge process supplemented with biofilm carriers.

integrated model   Model that includes several domains, for example, collection system, wastewater treatment, and receiving waterbody.

interface model   A model that describes the way in which output variables are passed from a model of one type to a model that uses different variables for its inputs.

irreducible uncertainty   Uncertainty that cannot be reduced by any degree of study; an example of this would be the uncertainty related to future flows and loads. No amount of study would be able to quantify future flows and loads beyond a certain point.

ISS   See *inorganic suspended solids.*

Jar test   A test procedure using laboratory glassware to evaluate coagulation, flocculation, and sedimentation in a series of parallel comparisons.

Kinetics   The reaction rates of biological and chemical processes and how they are determined.

kinematical   Related to *kinematics*, the branch of mechanics that deals with motion without reference to force or mass.

lumped model   A model that considers several reactors (or process trains) as one unit without explicitly accounting for the spatial variability of its characteristics. Parameters are considered to be valid for the system taken as a whole.

lysis   A cell rupture that results in loss of its contents.

mass balance   Balance of material flows, including input, output, production, or loss using the law of conservation of mass.

mass transfer   The movement of atoms or molecules by diffusion or convection from an area of high concentration to one of low concentration.

mass-transfer boundary layer   Particulate and soluble substrates are subject to concentration gradients as the materials diffuse into the biofilm from the bulk of the liquid. A concentration gradient external to the biofilm surface is described as resistance to mass transfer. The distance over which a substrate concentration in the bulk of the liquid is reduced is called the *mass-transfer boundary layer*.

material balance   See *mass balance*.

maximum day   The maximum single day value in the evaluation period.

maximum month   The maximum single month period in the evaluation period; often considered a 30-day period.

MBBR   See *moving bed biofilm reactor.*

MBR   See *membrane bioreactor.*

measurement error   Errors in observed data that are functions of human or instrumental errors.

mechanistic model   A model that has a structure that explicitly represents an understanding of physical, chemical, and/or biological processes; mechanistic models quantitatively describe the relationship between some phenomenon and underlying first principles of cause. Hence, in theory, they are robust and useful for inferring solutions outside of the domain in which initial data were collected and used to parameterize the mechanisms.

membrane bioreactor   An activated sludge process using membranes for solid–liquid separation compared to conventional gravity settling.

metabolic models   Models based on the metabolic processes of biological treatment using transformations of intermediate compounds.

methanogenisis   The metabolic conversion of organic acids or hydrogen and carbon dioxide to methane; the primary methanogenic populations associated with anaerobic digestion are acetoclastic methanogens and hydrogenotrophic methanogens.

methylotrophs   Specialized organisms capable of degrading methanol.

minimum day   Average of the minimum flows sustained for the period of a day in the record examined (typically from 2 a.m. to 6 a.m.).

minimum hour   The average of the minimum flows sustained for a period of 1 hour in the record examined (typically based on 10-minute increments).

minimum month   The average of the minimum daily flows sustained for the period of 1 month in the record examined.

mixed-culture   A biological culture consisting of a mixture of organism types.

mixed liquor   The mixture of wastewater and activated sludge in the bioreactor.

mixed liquor suspended solids   Suspended solids in a mixture of wastewater and activated sludge.

mixed liquor volatile suspended solids   The volatile fraction of the mixed liquor suspended solids as measured by flash incineration at 550 °C.

MLSS   See *mixed liquor suspended solids*.

MLVSS   See *mixed liquor volatile suspended solids*.

model   A representation of the behavior of an object or process, often in mathematical or statistical terms; models can be physical or conceptual.

model calibration   See *calibration*.

model error   The difference between observed and simulated variables; model error can be calculated in different ways as cumulative, absolute, quadratic, and so on.

model prediction accuracy   A measure of how closely a model matches observed data.

model prediction error   See *model prediction accuracy*.

model setup   Procedure to put together required submodels during the construction of a larger system model.

model testing   The process of comparing model predictions to independent data.

model validation   See *validation*.

modeler   A specialist who undertakes a technical modeling activity.

modular modeling approach   The coupling of different unit process models; output data are typically transferred between different components or submodels. Often, a modeling interface facilitates interactions between different types of models; see also *interface model*.

Monod kinetics   Kinetic expressions including one or more Monod terms.

Monod term   A mathematical function commonly used in models to describe the variation in kinetics of biological growth as a function of a model component (i.e., substrate, nutrient, or pH). It is identical in form to the Michaelis–Menten equation often referred to in industrial applications; also called *switching function*.

Monte Carlo simulation   A simulation technique used to approximate the probability of certain outcomes that involves running multiple simulations with randomized inputs.

moving bed biofilm reactor   A biological contact zone containing suspended biofilm carriers with no solids recycle.

multidimensional numerical model   The increasing amount of experimental evidence regarding biofilm heterogeneity has led to development of two- and three-dimensional biofilm models that capture the spatial and temporal heterogeneity of a biofilm's physical, chemical, and biological environment. Multidimensional biofilm models allow users to assess biofilm activity and interactions at the microscale. These models can incorporate flow or neglect flow, they are computationally intensive, and are presently used as research tools.

net yield   Net mass of solids produced in a biological process divided by mass of substrate removed, typically expressed in TSS, BOD, or COD units; is equal to the synthesis yield minus decay.

nitratation   The biological conversion of nitrite to nitrate by autotrophic bacteria.

nitrification   A biological process in which ammonia is converted first to nitrite and then to nitrate by autotrophic bacteria.

nitritation   The biological conversion of ammonia to nitrite by autotrophic bacteria.

nitrogen uptake rate   The uptake rate of nitrate/nitrite by biomass.

NOB   Nitrite-oxidizing bacteria, and older term for *nitrite-oxidizing organism.*

noise   Inherent variability data that the model does not characterize (see definition for *variability*).

NOO   Nitrite-oxidizing organism.

numerical solver   Mathematical routine included in a simulator to solve differential equations in a model.

NUR   See *nitrogen uptake rate.*

nutrient   Any substance that is assimilated by organisms to promote or facilitate their growth; examples are nitrogen and phosphorus, when considering their potential to result in excess biological growth in the environment.

objective function   A function to quantify deviations among model outputs and observations.

observations   Measurements at a water resource recovery facility.

one-dimensional numerical model   A mathematical biofilm model that accounts for the simultaneous diffusion and biochemical reaction of soluble substrate(s) in one direction along the z-coordinate, which is perpendicular to the biofilm growth medium or substratum.

online data   Data derived from continuous monitoring instrumentation.

ordinary differential equation (ODE)   An equation in which the derivative of a variable depends on the variable itself.

organic loading   The amount of organic matter fed to a treatment process over a time period.

OUR   See *oxygen-uptake rate.*

oxygen uptake rate   The oxygen used during biological oxidation, typically expressed as mg O_2/(L·h) in the activated sludge process.

PAO   See *polyphosphate-accumulating organism.*

parameters   Terms in a model that are fixed during a model run or simulation but can be changed in different runs as a method for conducting sensitivity analysis or to achieve calibration goals.

particulate   Typically considered to be a solid particle larger than 0.45 μm or large enough to be removed by filtration.

peak hour   The average of the peak flows sustained for a period of 1 hour in the record examined (typically based on 10-minute increments).

Petersen matrix   See *Gujer matrix.*

PFR   Plug flow reactor.

PHA   Poly-hydroxy-alkanoate; the organic storage product used by polyphosphate-accumulating organisms.

plug flow   Flow condition in which a fluid package passes through a tank without longitudinal mixing; packages are discharged in the same sequence in which they entered. Opposite of completely mixed, where the reactor concentration equals the effluent concentration. Ideal plug flow does not exist and the actual hydraulic behavior is somewhere in between ideal plug flow and completely mixed.

Polyphosphate-accumulating organism   The bacterial group responsible for enhanced biological phosphorus removal.

precision   For measurements, a term for random errors. Opposite of *trueness*, or lack of systematic errors. For model predictions, precision of simulation results is the degree to which several simulation results are similar to each other.

primary   Typically refers to the first solids/liquid separation step in liquid wastewater treatment.

probabilistic modeling   Modeling where randomness is present and state variables are not described by unique values, but rather by probability distributions.

probability density function   A statistical function that describes the relative likelihood for a random variable to take on a given value. The probability for the random variable to fall within a particular region is given by the integral of this variable's density over the region. The probability density function is nonnegative everywhere, and its integral over the entire space is equal to 1.

process model   A model describing the behavior of a certain unit process (e.g., activated sludge reactor).

pseudo-analytical model   Pseudo-analytical solutions to flux are comprised of a small set of algebraic equations that can be solved directly by hand or in a spreadsheet. The pseudo-analytical solutions were obtained by fitting appropriately chosen algebraic equations to thousands of numerical solutions of the equations comprising a steady-state model.

quality assurance (QA)   Planned, systematic production processes that provide confidence in a product's suitability for its intended purpose.

quality control (QC)   A system for verifying and maintaining a desired level of quality in a product or process by careful planning, use of proper equipment, continued inspection, and corrective action as required.

quantifiable uncertainty   Parameters that are known to be uncertain and the degree of uncertainty can be reduced. An example of this is uncertainty related to model structure. The model structure may have inherent assumptions that are only approximations of reality, such as the assumption that the autotrophic half-saturation constant for oxygen is a constant. These potentially could be made more accurate through further study and model development.

RBCOD   Refers to the readily biodegradable portion of chemical oxygen demand. See flocculated and filtered chemical oxygen demand. *Readily biodegradable* is perhaps a misnomer because this material is actually a combination of truly soluble biodegradable and non-biodegradable COD.

redox   Refers to the oxidation–reduction environment.

reliability   The confidence that (potential) users have in a model and in information derived from the model such that they are willing to use the model and derived information. Specifically, reliability is a function of the performance record of a model and its conformance to best available, proven science.

residence time   The period of time that a volume of liquid or solids remains in a tank or system.

respirometry   Measurement and interpretation of the oxygen uptake rate by biomass.

SAE   See *standard aeration efficiency.*

SBOD   Refers to the soluble fraction of biochemical oxygen demand.

SBR   See *sequencing batch reactor.*

scenario   A set of conditions representing a likely future happening.

SCOD   Refers to the soluble fraction of chemical oxygen demand.

scope creep   The slow growth of the scope of work over the course of the project, typically in an unplanned manner.

secondary   Most often refers to the biological treatment and subsequent solids separation step in wastewater treatment.

sedimentation   A gravity-based process for removing settleable suspended solids from water or wastewater; typically occurs in a quiescent basin or clarifier.

semi-empirical model   Empirical models that have had mechanistic aspects incorporated to their structure.

sensitivity   The degree to which model outputs are affected by changes in selected parameters.

sensitivity analysis   A method for evaluating how changes in a model's input parameters affect its output.

sequencing batch reactor   A biological treatment process with five phases: fill, react, settle, draw, and idle. It typically includes a system of multiple tanks so one can be filled while another is treating wastewater or being drained.

sidestream   Liquid streams generated during solids processing or odor control that typically are returned to the head of the facility for reprocessing in the wastewater treatment train.

simulation   A model run providing outputs based on model inputs.

simulator   Software used to run a model, typically with interactive inputs.

SLR   See *solids loading rate.*

sludge   Any residual produced during primary, secondary, or advanced wastewater treatment that has not undergone any process to reduce pathogens or vector attraction; also called *raw sludge.* The term *sludge* should be used with a specific process descriptor (e.g., primary sludge, waste activated sludge, or secondary sludge).

sludge age   See *solids retention time.*

sludge volume index   A test that quantifies the settleability of an activated sludge mixture; the volume (in milliliters) occupied by 1 g of settled sludge after settling for 30 minutes in a 1-L graduated cylinder.

solids loading rate   The mass of solids applied to a solids–liquid separation process divided by the area of that process.

solids retention time   The average period of time that solids remain in a system; also called *sludge age.*

solubility   The amount of a substance that can dissolve in a solution under a given set of conditions.

SOR   See *surface overflow rate.*

SOTE   See *standard oxygen-transfer efficiency.*

SOTR   See *standard oxygen-transfer rate.*

SPA   See *state-point analysis.*

SRT   See *sludge retention time.*

stakeholders   A person or group that has an investment, share, or interest in something (e.g., a business or industry).

standard aeration efficiency   The standard oxygen-transfer rate divided by the power input.

Standard Methods for the Examination of Water and Wastewater (Standard Methods)   A publication published jointly by the American Public Health Association, the American Water Works Association, and Water Environment Federation; it contains descriptions of analytical techniques commonly accepted for use in water and wastewater treatment.

standard oxygen-transfer efficiency   The standard oxygen-transfer rate (SOTR) divided by the mass flow of oxygen in the gas stream.

standard oxygen-transfer rate   The oxygen-transfer mass rate at standard conditions of 20 °C water temperature, zero dissolved oxygen concentration, and atmospheric pressure.

state-point analysis   A one-dimensional approach to modeling flocculent settling, such as secondary clarification.

state variable   Fundamental components (e.g., ammonia) in a model; variables may be divided into state (or internal), input, and output variables.

statistical model   An alternative to mechanistic models that uses statistical rules based on historical data to determine the likely behavior of a process rather than using fixed equations to do so. Also called a *black box model.*

steady-state model   A model that is not dynamic, that is, a model in which inputs are considered stationary in time.

stiff models   A model is called stiff when the range of the time constants within the model is large. This would occur when some of the systems react quickly and some are significantly slower.

stochastic model   A model that includes variability in model parameters and variables. Therefore, solutions obtained by the model are a function of deterministic input, model structure, and random variability.

stoichiometric coefficients   Factors used to convert mass units between different state variables; typically obtained from stoichiometric (mass balance) chemical equations or empirical observations that describe the transformations; not to be confused with influent stoichiometry.

stoichiometry   The relative quantities of the components taking part in biological and chemical processes.

stop criteria   Those parameters, that, when met, signal the end of the calibration and/or validation steps.

submodel   A model within a model; typically used to describe a particular facet of a larger or complex process.

substrate   (1) Wastewater or solids constituents used to promote biological growth and (2) any surface to which a coating is applied, sometimes used in conjunction with biofilms.

supermodel   Model describing the entire water resource recovery facility with one common set of state variables (as opposed to *interface model*); also called *integrated facility model.*

supplemental carbon   Typically a biodegradable carbon source, such as methanol or acetic acid, added to biological treatment to improve denitrification or biological phosphorus removal.

surface overflow rate   The flowrate overflowing a gravity settling process divided by the settling process area.

suspended growth process   A biological treatment process in which microbes and substrate are suspended in the wastewater.

suspended solids   Solids captured by filtration through a glass fiber filter or 0.45-μm filter membrane.

sustained flow (and load)   The value (flowrate or mass loading) sustained or exceeded for a given period of time (e.g., 1 hour, 1 day, or 1 month).

SVI   See *sludge volume index.*

switching function   Used to turn process rate equations on and off, depending on environmental conditions. Monod kinetic equations are typically used; however, other switching functions are also available (e.g., Haldane and Andrews).

system SRT   Solids retention time within a processing system. In an activated sludge process, system SRT will include the length of time that solids are held in the anaerobic, anoxic, and aerobic portions of aeration tanks and in secondary clarifiers.

tertiary   Typically refers to the treatment processes downstream of secondary treatment.

three-dimensional numerical model   See *multidimensional numerical model.*

time series   Temporal sequence of consecutive data.

time step (of a given model)   Unit interval of time used by a discrete model for time series simulations (frequency variable).

total ignorance   The state of uncertainty where there are parameters that the modeler does not know are uncertain; these items could be either reducible or not reducible. Any parameter might fall in this category, and the only method of reducing this is through learning and experience.

total solids   (1) The sum of dissolved and suspended solids in water or wastewater and (2) the matter remaining on a weighed dish after the fluid has been evaporated at 103 to 105 °C.

total suspended solids   The solids that can be removed from solution through filtration, typically with a 0.45-micron filter paper.

tracer test   Carried out by injecting one pulse (or several) of an inert tracer or increasing its concentration for a specified time at the inlet of the system under study; the tracer substance should not be degradable and not adsorb to the sludge. The time series of recovered tracer is measured at specified locations within the system and at its outlet.

trickling filter   An aerobic biofilm treatment process in which wastewater flows across a bed of highly permeable media. As the wastewater disperses, its organic matter is degraded by microorganisms in the slime on the media surface.

trueness   Term for systematic errors; the closeness of agreement between a measurement and an accepted reference value. The measure of trueness is typically expressed in terms of bias.

two-dimensional numerical model   See *multidimensional numerical model.*

UASB   See *upflow anaerobic sludge blanket.*

ultimate biochemical oxygen demand   The amount of oxygen required to completely satisfy carbonaceous and nitrogenous biochemical oxygen demand.

uncertainty   Lack of knowledge about models, parameters, constants, data, and beliefs. There are many sources of uncertainty, including the science underlying a model, uncertainty in model parameters and input data, observation error, and code uncertainty. Additional study and collecting more information allow errors that stem from uncertainty to be minimized/reduced (or eliminated). In contrast, variability is irreducible, but can be better characterized or represented with further study.

underflow   The flow stream removed from the bottom of a tank or basin. Often, this is concentrated solids, but, in dissolved air floatation, *underflow* refers to the liquid product.

unit process   A step in the treatment process that accomplishes a particular treatment goal.

upflow anaerobic sludge blanket   An anaerobic biological treatment process where wastewater flows upward through a blanket of anaerobic solids.

validation   Substantiation that a model has sufficient accuracy consistent with the intended application of the model, typically through the evaluation of a calibrated model on an independent set of relevant data.

variability   Spread in a variable or probability distribution.

variable   A quantity that varies in time; variables in the model may be divided into state (or internal), input, and output variables.

velocity gradient   A measure of the degree of mixing imparted to water or wastewater during flocculation. Also called *G value*.

volatile solids   Organic matter subject to decomposition; it is ignitable at 550 °C. Typically, this is used to represent the organic fraction of the sludge or other solids material.

volatile suspended solids   Organic, biodegradable matter suspended in water or wastewater; the percentage of this matter in a total suspended solids sample is determined by heating the sample to 600 °C.

waste activated sludge   Excess activated sludge that is discharged from an activated sludge treatment process.

whole-facility model   Model used to describe all unit processes and their interconnections in a facility mass balance; in Europe, often called *facility-wide* model.

yield   The amount of particulate produced per unit amount of soluble material.

Index